T0303872

IMMUNOGOLD-SILVER STAINING

Principles, Methods, and Applications

Edited by

M.A. Hayat

CRC Press

Boca Raton   New York   London   Tokyo

Library of Congress Cataloging-in-Publication Data

Hayat, M. A.
    Immunogold-silver staining : principles, methods, and applications / edited by
M. A. Hayat.
        p. cm.
    Includes bibliographical references and index.
    ISBN 0-8493-2449-1
    1. Immunogold labeling.   2. Silver staining (Microscopy).
    I. Title.
    QR187.I482H39   1995
    574.8'028—dc20   95-1831
                                                                                              CIP

No claim to original U.S. Government works
International Standard Book Number 0-8493-2449-1
Library of Congress Card Number 95-1831
1 2 3 4 5 6 7 8 9 0

Preface

This is the first book devoted exclusively to the subject of immunogold-silver staining (IGSS). This volume is authored by 47 distinguished scientists representing 12 countries. The primary objective of this book is to discuss principles, methods, and applications of IGSS. An effort was made to assemble researchers with strong interest in the development and applications (both applied and basic) of this important methodology. Emphasis is placed on the comprehensiveness of the methodology in that it is presented in a self-explanatory form, so that the reader can practice it without outside help. Applications of this technique were selected that represented diversified areas of biomedical sciences. The contents of the chapters are a testament to the acceptance of the IGSS method as an enormously useful tool in basic research as well as in the fields of diagnostic histology and cytology. The advantages of IGSS over immunogold labeling are explained. Future applications of IGSS are also discussed.

The first contribution is by Gorm Danscher and colleagues, who present an overview of the IGSS (autometallography) methodology, including a discussion on avoiding and/or removing nonspecific background staining. They also summarize future trends of this technique. Gerhard Hacker and associates present a brief review of silver staining, including the demonstration of nucleolar organizer regions. Detailed discussion comparing different silver salts used for silver-enhancement (amplification) of colloidal gold is given. These authors also include reliable step-by-step protocols for light and electron microscopical applications of IGSS for immunocytochemistry *and in situ* DNA hybridization.

Alex Kalyuzhny and colleagues recommend monitoring the optical density of the physical developer as a relevant predictor of its ability to induce silver-enhancement. This procedure provides a simple and accurate way to standardize silver intensification of colloidal gold. Tibor Krenács and László Krenács compare the resin media used for embedding the specimens for IGSS. They discuss the physicochemical characteristics of commonly used resins, and suggest optimal combinations of the resin and the IGSS method to obtain satisfactory results in postembedding labeling.

James Hainfeld and Frederic Furuya discuss silver-enhancement of small Nanogold (1.4 nm) and ultrasmall undecagold (0.8 nm) clusters. They also describe the physicochemical characteristics of these gold clusters and their applications to studies using light and electron microscopy, confocal microscopy, dot blots, and running gold-labeled proteins on gels. The advantages and limitations of this method are also presented. York-Dieter Stierhof and colleagues describe the application of 1 nm colloidal gold markers in the TEM, SEM, and STEM. Details of the instrumentation are included. They also present advantages and limitations of using 1- to 3-nm colloidal gold particles, and practical aspects of silver-enhancement.

Lone Bastholm and Folmer Elling discuss the application of immunocytochemistry on thin cryosections using silver-enhanced colloidal gold particles as markers and the practical advantages and limitations encountered during practicing this procedure. Step-by-step protocols for indirect IGSS employing 1- or 5-nm gold particles are also presented.

Using colloidal gold particles of different dimensions and various silver salts, Jiang Gu and colleagues present comparative and quantitative evaluations of these variables. They also include optimal preparatory conditions for obtaining reliable results.

Paul Monaghan presents the combination of rapid cooling methods, followed by freeze-substitution, with IGSS. Various cooling methods are summarized, and parameters of freeze-substitution and resin embedding are discussed in detail. Gary Login and Ann Dvorak discuss the theory and practice of microwave irradiation in conjunction with IGSS. They explain how microwave procedures facilitate specimen fixation, IGSS, antigen retrieval, and stimulation of primary and secondary antibody reactions. Charles Taban and Maria Cathieni present the preparation of the colloidal gold-protein-substance P complex and the procedure for binding studies of unfixed, cryostat sections of the brain. They use a microwave oven for stabilizing cryostat sections and obtaining rapid preincubation, incubation, and fixation.

Gold toning causes the replacement of metallic silver precipitated in tissue by metallic gold for light and electron microscopy. This modified approach transforms the particles of immunogold-silver to aggregates of fine grains. Ryohachi Arai and Ikuko Nagatsu examine the usefulness of this procedure. Richard Burry discusses the use of nanogold clusters for preembedding IGSS. He recommends the use of MES buffer (pH 6.0) and N-propyl-gallate as a reducing agent to control the reaction. This procedure is presented in a step-by-step fashion. Gian Carlo Manara and colleagues discuss the advantages and limitations of preembedding IGSS. They describe the application of this technique to the plasma membrane-associated antigens.

Jean-Charles Cailliez and Eduardo Dei-Cas discuss the usefulness of IGSS in localizing and characterizing fungal molecules of biological significance, especially in the medical field. The IGSS method has a considerable potential in virology. Simultaneous localization of two antigens in the same section can be accomplished. Antonio Marchetti and Generoso Bevilacqua describe the application of this technique, in conjunction with light microscopy and scanning and transmission electron microscopy, to the detection of viral antigens in infected cells.

Dennis Goode presents the methodology for the visualization of microtubules with IGSS and backscattered scanning electron microscopy. Six techniques for specimen preparation are given, followed by their evaluations. Advantages and limitations of backscattered electron imaging are also discussed. Kuixiong Gao and Alvin Gao describe the use of epipolarization microscopy in conjunction with IGSS. A microscope equipped with epi-illumination of polarized light provides more sensitive detection of immunogold-silver stained antigenic sites of a low abundance. This approach allows simultaneous visualization of the label and the tissue profile, and bridges the gap between light microscopy and electron microscopy.

Cornelia Hauser-Kronberger and colleagues discuss the usefulness of silver amplification of colloidal gold in non-microscopical field. They discuss the application of this methodology to the detection of proteins on blots and immunoadsorbent assay.

Hiroshi Shimizu and Takeji Nishikawa discuss the use of an image analyzer to impart various colors and to enhance the size of small colloidal gold particles in the immunoreaction product. Although silver staining as yet is not included in this procedure, the gold particles are easier to visualize with this approach because they are of different color from that of the other background structures.

The chapters of this book are written by the authors who pioneered the development of the methods, or who refined earlier techniques and have extensive experience in their application. Individual chapters are designed to provide the maximum practical information necessary to reproduce the techniques described. Typical results of the methods are presented in the form of photomicrographs and electron micrographs. The theoretical basis of the methods are briefly presented in the introduction of most chapters. References are provided to articles that review the applications.

It is hoped that this volume will be helpful to scientists in a variety of fields of cell and molecular biology, including diagnostic pathology, and that it will contribute to a wider application of the IGSS methodology. I am grateful to the authors for their enthusiasm and cooperation.

M. A. Hayat
May 1994

Contributors

Ryohachi Arai
Department of Anatomy
Fujita Health University School of
 Medicine
Toyoake, Japan

Erich Arrer
Central Laboratory
Salzburg General Hospital
Salzburg, Austria

Lucilla Badiali-De Giorgi
Institute of Clinical Electron Microscopy
University of Bologna
Bologna, Italy

Lone Bastholm
Institute of Pathological Anatomy
University of Copenhagen
Copenhagen, Denmark

Generoso Bevilacqua
Institute of Pathology
University of Pisa
Pisa, Italy

Richard W. Burry
Department of Cell Biology,
 Neurobiology, and Anatomy
College of Medicine
The Ohio State University
Columbus, Ohio

Jean-Charles Cailliez
Faculté Libre de Sciences
Institut Catholique de Lille
France

Nancyleigh Carson
Deborah Research Institute
Browns Mills, New Jersey

Maria M. Cathieni
Laboratory of Neurobiology
Psychiatric Institutions of the University
 of Geneva
Chêne-Bourg, Switzerland

Michael D'Andrea
Deborah Research Institute
Browns Mills, New Jersey

Gorm Danscher
Department of Neurobiology
Institute of Anatomy B
University of Aarhus
Aarhus C, Denmark

Eduardo Dei-Cas
Unité 42 INSERM
Domaine du CERTA
France

Otto Dietze
Institute of Pathological Anatomy
Salzburg General Hospital
Salzburg, Austria

Ann M. Dvorak
Departments of Pathology
Harvard Medical School
Beth Israel Hospital
Boston, Massachusetts

R. Elde
Department of Cell Biology and
 Neuroanatomy
University of Minnesota Medical School
Minneapolis, Minnesota

Folmer Elling
Institute of Pathological Anatomy
University of Copenhagen
Copenhagen, Denmark

Corrado Ferrari
Department of Pathology
University of Parma
Parma PR, Italy

Michele Forte
Deborah Research Institute
Browns Mills, New Jersey

Frederic Furuya
Brookhaven National Laboratory
Biology Department
Upton, New York

Alvin Wei Gao
Physics Department
Davidson College
Davidson, North Carolina

Kuixiong Gao
Department of Cell Biology,
    Neurobiology, and Anatomy
College of Medicine
University of Cincinnati
Cincinnati, Ohio

Dennis Goode
Department of Zoology
University of Maryland
College Park, Maryland

Lars Grimelius
Institute of Pathology
University Hospital
Uppsala, Sweden

Jiang Gu
Deborah Research Institute
Browns Mills, New Jersey

Gerhard W. Hacker
Institute of Pathological Anatomy
Immunohistochemistry and Biochemistry
    Unit
Salzburg General Hospital
Salzburg, Austria

M. A. Hayat
Department of Biology
Kean College of New Jersey
Union, New Jersey

James F. Hainfeld
Brookhaven National Laboratory
Biology Department
Upton, New York

Cornelia Hauser-Kronberger
Institute of Pathological Anatomy
Immunohistochemistry and Biochemistry
    Unit
Salzburg General Hospital
Salzburg, Austria

René Hermann
Labor für Elektronenmikrospie I
Institut für Zellbiologie
Eidgenössische Technische Hochschule
    Zürich
Zürich, Switzerland

Bruno M. Humbel
Department of Molecular Cell Biology
Utrecht University
Utrecht, The Netherlands

Alex E. Kalyuzhny
Department of Pharmacology
University of Minnesota Medical School
Minneapolis, Minnesota

László Krenács
Department of Pathology
Albert Szent-Györgyi Medical University
Szeged, Hungary

Tibor Krenács
Department of Pathology
Albert Szent-Györgyi Medical University
Szeged, Hungary

Gary R. Login
Departments of Pathology
Harvard School of Dental Medicine
Beth Israel Hospital
Boston, Massachusetts

H. H. Loh
Department of Pharmacology
University of Minnesota Medical School
Minneapolis, Minnesota

Gian Carlo Manara
Department of Dermatology
University of Parma
Parma PR, Italy

Antonio Marchetti
Institute of Pathology
University of Pisa
Pisa, Italy

Paul Monaghan
Institute of Cancer Research
Haddow Laboratories
Sutton, England

Wolfgang H. Muss
Institute of Pathological Anatomy
Salzburg General Hospital
Salzburg, Austria

Ikuko Nagatsu
Department of Anatomy
Fujita Health University
School of Medicine
Toyoake, Japan

Takeji Nishikawa
Department of Dermatology
Keio University School of Medicine
Tokyo, Japan

Gianandrea Pasquinelli
Institute of Clinical Electron Microscopy
University of Bologna
Bologna, Italy

Robyn Rufner
Deborah Research Institute
Browns Mills, New Jersey

Angelika Schiechl
Institute of Pathological Anatomy
Salzburg General Hospital
Salzburg, Austria

Heinz Schwarz
Max-Planck-Institut für
 Entwicklungsbiologie
Tübingen, Germany

Hiroshi Shimizu
Department of Dermatology
Keio University School of Medicine
Tokyo, Japan

York-Dieter Stierhof
Max-Planck-Institut für Biologie
Tübingen, Germany

Charles H. Taban
Laboratory of Neurobiology
Psychiatric Institutions of the University
 of Geneva
Chêne-Bourg, Switzerland

Contents

Ryohachi Arai

Erich Arrer

Lucilla Badiali-De Giorgi

Lone Bastholm

Generoso Bevilacqua

Richard W. Burry

Jean-Charles Cailliez

Nancyleigh Carson

Maria M. Cathieni

Michael D'Andrea

Gorm Danscher

Eduardo Dei-Cas

Otto Dietze

Ann M. Dvorak

R. Elde

Folmer Elling

Corrado Ferrari

Michele Forte

Frederic Furuya

Alvin Wei Gao

Kuixiong Gao

Dennis Goode

Lars Grimelius

Jiang Gu

Gerhard W. Hacker

M. A. Hayat

James F. Hainfeld

Cornelia Hauser-Kronberger

René Hermann

Bruno M. Humbel

Alex E. Kalyuzhny

László Krenács

Tibor Krenács

Gary R. Login

H. H. Loh

Gian Carlo Manara

Antonio Marchetti

Paul Monaghan

Wolfgang H. Muss

Ikuko Nagatsu

Takeji Nishikawa

Gianandrea Pasquinelli

Robyn Rufner

Angelika Schiechl

Heinz Schwarz

Hiroshi Shimizu

York-Dieter Stierhof

Charles H. Taban

Overview

M. A. Hayat
Immunogold-Silver Staining: Principles, Methods, and Applications.
CRC Press, Boca Raton, FL, 1995.

Immunogold labeling is the most important and widely used immunocytochemical method for in situ study of antigens because of its superior reliability, practicality, specificity, sensitivity, efficiency, accuracy, precision, and simplicity. Both intracellular and cell surface antigens can be detected with this technique. The method is useful for preembedding as well as postembedding labeling of antigens. A wide variety of antigens can be detected on frozen, paraffin, and resin sections (semithin and thin), and thin cryosections, employing monoclonal and polyclonal antibodies, along with colloidal gold-conjugated secondary antibodies, protein A, IgG, or avidin streptavidin.

Within the last decade immunogold staining has been established as a powerful tool to obtain detailed information about antigenicity from all kinds of specimens. Colloidal gold is being employed extensively for localizing antigens in plant and animal cells as well as in yeast, fungi, bacteria, viruses, viroids, and prions. Immunogold staining facilitates the study of the antigenicity of not only single cells but also of its fragments.

Colloidal gold particles ranging in diameter from 1 to 40 nm are commercially available. Alternatively, they can be produced in a laboratory. The preparation of colloidal gold particles of various sizes by using different methods is presented by Handley (1989) and Hayat (1989).

Various modes of operation of the electron microscope are useful for detecting antigens, depending on the objective of the study. Conventional, high resolution, and high voltage transmission electron microscopes can be used in conjunction with this methodology. By virtue of the large numbers of secondary and backscattered electrons generated by the dense metal, colloidal gold is an excellent marker for scanning electron microscopy. Scanning transmission electron microscopy can also be employed with immunogold staining. The methodology additionally can be used in conjunction with correlative light and electron microscopy.

Immunonegative staining in combination with colloidal gold markers is also useful for the immunological studies of small particles such as viruses and bacteria. The fracture-label or label-fracture method can be combined with colloidal gold for high resolution labeling of cell surfaces. Replica immunogold cytochemistry allows the visualization of the inner surface of isolated and labeled plasma membranes at high resolution in the transmission electron microscope.

Immunogold labeling is an ideal approach to achieve "quantitation" of antigenicity. However, an absolute quantitation presently is not possible, and is limited to semiquantitation (Hayat, 1992). Recently, using silver-enhanced 1-nm gold particles and electron microscopy, semiquantitative study was carried out with a computerized morphometric image analyzer (Gu et al., 1993). The use of gold particles as markers for antigenicity has become truly universal.

Immunogold-Silver Staining (IGSS)

The fundamental disadvantage of a light microscope is its limited resolving power. The size of gold particles is too small to be visualized with this microscope. Gold particles of a small size are also unresolvable in the conventional transmission electron microscope (TEM). The reasons for using gold particles of a small size are discussed later.

The limitation of gold particles of a small size in terms of their unresolvability can be circumvented by increasing their dimension by silver-enhancement (amplification). Immunogold and IGSS have replaced peroxidase-antiperoxidase and other immunocytochemical methods. The former are the techniques of choice because they provide positive immunostaining where other methods fail. Immunogold labeling and IGSS are able to demonstrate even a small amount of antigen present in the specimen. The sensitivity of immunolocalization with IGSS can be increased 200 times that of standard immunogold procedures. Immunogold-silver staining bridges the gap between light and electron microscopy using semithin and thin sections, respectively. The mechanism underlying the IGSS methodology is discussed by Danscher et al. (1993). The historical development of the IGSS (autometallography) has been reviewed by Scopsi (1989) and Danscher et al. (1993).

Although gold particles after silver-enhancement can be imaged in the light microscope using darkfield illumination, this system is not very useful for imaging cellular features. Epi-illumination with polarized light is another technique for detecting colloidal gold in the light microscope (De Waele et al., 1988). The staining appears as bright granules on a dark background. The efficiency of this detection is better than that obtained with bright field microscopy because electromagnetic radiation increases the amount of free silver in the solution which increases the background staining. Antigens in fine-needle aspirate biopsies (cytocentrifuge preparations) can also be localized with IGSS, especially when combined with epi-polarization microscopy (Hughes et al., 1988). Better detection efficiency can also be obtained with reflection contrast microscopy (Hoefsmit et al., 1986). Gold particles as small as 5 nm can be visualized after IGSS with this type of microscopy. When combined with a low brightness transmitted light, a light microscope can simultaneously image cellular structure and gold label. Although IGSS can be combined with fluorescent probes, this double labeling may present a problem of separating the two labels (Goodman et al., 1991; Stierhof et al., 1992).

After silver-enhancement of gold particles, paraffin or semithin resin sections can be counterstained with organic dyes such as hematoxylin and eosin (Hughes et al., 1991). Drug uptake and distribution in tissues can also be studied with IGSS in the light microscope (Henneberry and Aherne, 1992). Despite the use of various systems of light microscopy, low resolving power still remains a limitation.

The resolving power can be enormously increased by using the IGSS method in combination with the TEM. This method can also be employed for the secondary and backscattered electron imaging modes of the scanning electron microscope (SEM). Recently, atomic force microscopy was used for studying human lymphocyte surface antigens that were immunogold-silver stained (Neagu et al., 1994). In this study silver-enhancement of 1-nm gold particles facilitated high resolution detection of cell surface structures. A direct comparison between atomic force microscopy, bright field and fluorescence microscopy, and flow cytometry can be carried out.

Antigenicity can be detected with pre- or postembedding IGSS. The latter approach is preferred, however. Indirect IGSS methods are better than direct protein A-gold or streptavidin-gold-silver methods to achieve highest detection efficiency. Not only single but also double or triple immunolabeling are feasible with IGSS. A modification of the IGSS technique involves the use of combined secondary and tertiary gold-labeled antibodies (Giffin et al., 1993). This approach reportedly increases the labeling density by depositing greater amount of silver in less time which results in signal amplification with minimal background staining. However, the advantage of this procedure is yet to be confirmed. In addition, it has not been tested for electron microscopy.

Immunogold-silver staining can also be applied in conjunction with in situ hybridization for light and electron microscopy (Roth et al., 1992; Amikura et al., 1993). Intensification of IGSS can be carried out with gold toning for light and electron microscopy (Arai et al., 1992). Yellowish brown or black deposits after IGSS can be intensified to a dark black deposit by using gold chloride (0.05% for 10 min at 4°C) for light microscopy. This procedure allows postfixation with OsO_4 for electron microscopy, since the final reaction product is protected from the oxidizing effect of osmication.

Standard microwave irradiation can be used to expedite IGSS (van de Kant et al., 1993). Fixation, washing, en bloc staining with uranyl acetate, and partial dehydration (70% ethanol) for electron microscopy can also be accomplished with microwave irradiation at 4°C by placing specimens vials in crushed ice in the oven (Wagenaar et al., 1993). All of these steps can be completed within 65 min. No published report is available to indicate the use of this rapid procedure for IGSS.

Manual capillary action staining has been reported to increase the efficiency of antigen detection with IGSS for light microscopy. This staining system was developed by Brigati et al. (1988) to perform all stages of the IGSS procedure for achieving increased speed of immunostaining and decreased background labeling for light microscopy.

Light Microscopy

Geoghehan et al. (1978) were the first to use the red or pink color of colloidal gold sols for light microscopical immunogold staining using paraffin sections. In semithin resin sections red color of light scattered from gold particles as small as 14 nm was seen in cell organelles containing high concentrations of labeled antigens in the light microscope (Lucocq and Roth, 1984). Since the sensitivity of immunogold staining in light microscopy is inferior in comparison with other immunocytochemical techniques, the former did not gain general acceptance; the pinkish color of the gold deposit is difficult to visualize.

The real breakthrough for immunogold staining for light microscopy came with the introduction of silver-enhancement of colloidal gold particles (20 nm) bound to immuno-globulin in paraffin sections (5 µm) (Holgate et al., 1983). This approach significantly enhanced the sensitivity, efficiency, and accuracy of antigen detectability in the light microscope. Using IGSS, gold particles as small as 1 nm in diameter can be visualized in the light microscope. Thin sections subjected to IGSS can also be viewed with the light microscope, especially by using phase contrast or epi-polarization illumination (Stierhof et al., 1992).

Surface properties of gold particles differ depending on the preparatory conditions. Some evidence indicates that gold particles prepared with white phosphorus produce less nonspecific staining in paraffin sections than gold particles produced with tannic acid. This difference has been observed by using gold particles of 5 nm in diameter for labeling β-casein in the sections (5 to 8 µm) of paraffin-embedded mammary epithelium of pregnant mice (Breter and Erdmann, 1993). Very efficient labeling was achieved using a much lower concentration (OD_{520} = 0.004) of gold colloid prepared with white phosphorus than that obtained with other gold probes (OD_{520} = 0.04) used for IGSS. The adverse effect of tannic acid, however, can be overcome by treatment with hydrogen peroxide and separation of freshly prepared gold particles from tannic acid by gradient centrifugation (Slot and Geuze, 1985), or treatment with gelatin (Behnke et al., 1986).

Colloidal gold particles of a relatively large size have mostly uniform diameter, circular outline, and an homogeneous interior, whereas silver-enhanced gold particles may show an irregular outline, a variable diameter, and an heterogeneous interior. In other words, silver-enhancement of gold particles is not uniform. The size of "1 nm" colloidal gold particles sold commercially ranges in size from 1 to 3 nm. Unconjugated 1 nm gold particles are unstable, but they are stable when coupled to proteins.

Transmission Electron Microscopy

Colloidal gold particles as immunolabels have been used most commonly and effectively for transmission electron microscopy. Unenhanced gold particles of ~3 nm in diameter can be visualized in the conventional transmission electron microscope (TEM). Gold particles of 1 nm in diameter need silver-enhancement to be visualized in this microscope. Unenhanced gold particles of 1 nm in diameter can be imaged in the scanning transmission electron microscope (STEM) (Stierhof et al., 1992). Undecagold (0.8 nm in diameter) is difficult to visualize in the TEM, but can be imaged in the STEM. Nanogold (1.4 nm in diameter) is visible in the TEM with or without silver-enhancement.

Scanning Electron Microscopy

The immunogold staining method for scanning electron microscopy was introduced by Horisberger et al. (1975). This microscopy can provide valuable information which would otherwise be difficult to obtain with the TEM when a large surface area of the tissue or a number of cells needs to be studied. Colloidal gold can be used for both secondary and backscattered electron imaging. Gold-labeled cell surfaces can also be viewed by mixing the secondary and backscattered electron signals (de Harven and Soligo, 1989). Multiple IGSS for scanning electron microscopy can also be carried out.

Secondary Electron Imaging

Secondary electrons mainly provide a topographic image of the specimen surface. The high secondary emission of the gold particles enables visualization of cell surfaces with or without metal or carbon coating. Colloidal gold particles >15 nm in diameter can be viewed in the secondary electron imaging mode. The sensitivity of this method can be improved by using smaller gold particles responsive to silver-enhancement. When IGSS is applied, for example, to 5.7-nm gold particles, they are clearly observed in the secondary electron imaging mode (Scopsi et al., 1986). Recently, 6-nm gold particles were silver-enhanced and examined with secondary and backscattered electron imaging (Herter et al., 1993).

Backscattered Electron Imaging

The possibility of imaging gold markers in the backscattered electron imaging mode was first reported by Trejdosiewicz et al. (1981). This procedure is useful for the demonstration of cell surface antigens (Namork, 1991) as well as intracellular epitopes (Gross and De Boni, 1990). In the backscattered electron imaging mode the contrast is not generated by the specimen topography. It originates instead from the presence of elements of higher atomic number on or below the specimen surface. Gold, with an atomic number of 79, produces a high contrast when viewed in the backscattered electron imaging mode. Organic molecules consist of elements (C, H, O, N) of low atomic number, and thus generate very little backscattered electron imaging signal.

When colloidal gold-labeled cells are fixed only with aldehydes and coated with carbon, optimum backscattered electron imaging of gold markers is obtained (Walther et al., 1983; de Harven and Soligo, 1989). Imaging is somewhat diminished when OsO_4 fixation or heavy metal conductive coating is used. Quantitation of gold labeling is more accurate with the backscattered electron imaging than with the secondary electron imaging. All gold particles are clearly visualized in the former mode, whereas only a small percentage are identified in the latter images of the same cell.

The resolution obtained with the backscattered electron imaging is better than that provided by the light microscope but, because of the reduced efficiency of the backscattered electron detector, resolution is inferior to that given by the secondary electron imaging mode, and much poorer than that obtained with the TEM. Most backscattered electron detectors do not resolve gold particles of a smaller size (<5 nm). However, very small gold particles of ~1 nm in diameter can be resolved by using in-lens-field emission scanning electron microscope (Hermann et al., 1991).

The interpretation of secondary electron images of gold markers is somewhat ambiguous since the images are never the exclusive making of secondary electrons. Electron backscattering makes a significant contribution to the so-called secondary electron images by emitting low-energy electrons at points of exit of high energy backscattered electrons. Low energy electrons reach the secondary electron detector and participate in image formation. Mixed images thus obtained provide direct correlation between patterns of distribution of gold-labeled molecules and surface features of labeled cells.

Another limitation of secondary electron imaging is the difficulty of visualizing structures located under other components in cells and tissues. Backscattered electron imaging, on the other hand, provides images of metal-labeled components lying under other cellular or extracellular structures. Since backscattered electrons can retain most of their original energy, they can escape from much deeper layers in a specimen and can be detected separately from the lower energy secondary electrons emitted from orbits in the atoms of the specimen. In theory, any intracellular component with discrete localization sites can be studied using the backscattered electron imaging, provided a highly specific antibody is available.

Coating with chromium (Peters, 1986) allows simultaneous imaging of both cell surface morphology and labeling topography in the backscattered electron imaging mode (Hermann and Müller, 1991; Herter et al., 1993). This approach is useful for identifying specific cell types in heterogeneous tissues. It is possible to specifically label and study one type of structure with backscattered electron imaging and relate it to the position of cell components of other types in the secondary electron mode of the same specimen.

To minimize steric hindrance and increase the labeling efficiency, it is desirable to use as small gold particles as possible. Gold particles as small as 5 nm in diameter can be detected in the backscattered electron imaging mode by employing improved detectors (Walther and Müller, 1985). Such gold particles can be silver-enhanced and visualized with the backscattered electron mode (Goode and Maugel, 1987). Cells can be double-labeled, for example, with 5- and 20-nm gold particles and silver-enhanced before viewing them in the backscattered imaging mode (Namork and Heier, 1989; Namork, 1991).

Image analysis techniques can be used in conjunction with IGSS, to digitize, enhance, and process secondary and backscattered electron images of the same field of view (Stump et al., 1988). This approach allows density analysis and distribution of gold-labeled ligands on its target cell.

Size of Colloidal Gold Particles

The size of colloidal gold particles is too small to be visualized with the light microscope, and even the conventional electron microscope (TEM) is unable to image gold particles of a very small size. The question arises — why are gold particles of a small size preferred over those of a relatively large size? The primary reasons for choosing gold particles of a small size are that they provide better specificity, efficiency, accuracy, and precision of antigen detection. The degree of penetration and binding in general are an inverse function of the size of the gold particles; thus, the sensitivity of the second antibody is greater with small gold conjugates.

The number of gold particles staining an area of cell surface decreases as the size of the particles increases. In other words, an increase in particle size decreases labeling density. The decrease in labeling density can occur due to poor access of antigens and/or repulsion forces. It has been established that small gold particles encounter less steric hindrance, yielding a high labeling density. Certain binding sites are inaccessible to large gold particles owing to a narrow spacing of glycoproteins on the cell surface. Small gold particles allow the labeled antibody to move more freely, leading to increased labeling efficiency. Small gold particles, furthermore, require less binding force between antibodies and antigens to hold them on site.

The resolution of the distribution pattern of target proteins is lowered with increasing gold particle size. Therefore, small gold particles are preferred for analyzing the binding pattern of antibodies at high magnifications. The possibility that protein structure may be obscured by superimposition by gold particles is decreased in the presence of small gold particles. Small gold particles block relatively few antigenic sites. Large gold particles, on the other hand, may cover many cell surface binding sites in a cluster when viewed in the SEM. Large gold particles may also mask an excessive amount of the underlying surface structure which makes precise topographical relationships difficult in the SEM. Antigens in the preembedding as well as in the postembedding labeling are more accessible to gold particles of a small size, resulting in higher labeling densities. Decreasing the size of gold particles is not accompanied by an increase in the background staining.

Notwithstanding the above-mentioned advantages of the gold particles of a small size, the choice of the size of the particles is determined by many other factors which are discussed in detail by Kellenberger and Hayat (1991).

Considerations for Immunogold-Silver Staining

The IGSS technique is extremely sensitive since only two gold atoms can initiate the process of silver amplification of gold. The involvement of silver in the reaction makes this procedure highly sensitive to impurities from varied sources, such as anything that comes in contact with the specimen, including glassware. At least the following precautions must be taken to avoid background staining and other artifacts.

1. Optimal fixation is determined by the type of specimen and antigenicity under study. For light microscopy, as an average, 4% neutral formaldehyde in 0.1 M PIPES or HEPES buffer (pH 7.2 to 7.4) is recommended for fixation (see Griffiths, 1993).
2. For electron microscopy, a mixture of 4% formaldehyde and ~0.5% glutaraldehyde in the same buffer is recommended.
3. Fixation with OsO_4 should be avoided.
4. Depending upon the prefixation, tissue sections can be fixed for 2 min with 1 to 2% glutaraldehyde in PBS (pH 7.2) to prevent loss of label after incubation with second antibody.
5. For preserving labile, sensitive antigens, low temperature polymerization of the embedding medium is recommended.
6. If LR White resin sections fail to easily adhere to glass slides and a loss of sections becomes a problem especially when exposed to detergents (e.g., Tween 20), this resin should be replaced with glycol methacrylate. The sections of the latter resin adhere firmly to the slides when dried on a hotplate at 60 to 70°C for 30 to 60 min (Tacha et al., 1993).
7. All chemicals must be of analytical grade.

8. Only deionized glass-distilled water should be used.
9. All glassware must be meticulously clean. This can be accomplished by washing it with 10% Farmer's solution for at least 20 to 30 min. This solution consists of 9 parts of 10% sodium thiosulfate and 1 part of 10% potassium ferricyanide. If pure chemicals and clean glassware are used, the developing solution remains stable for at least 10 min. This period is long enough to enlarge gold particles as much as five thousand times, producing dark brown to black signal that can be monitored easily with the light microscope.
10. Metallic tools and containers should be avoided. Plastic or teflonized forceps are preferred over steel forceps.
11. Copper grids for collecting thin sections must be avoided since oxidation of this metal occurs during incubation with the gold complex and gives rise to contamination. This is true especially when overnight incubations and/or buffers containing sodium azide are used. Nickel or platinum grids are preferred.
12. Immunogold-silver staining for electron microscopy should be carried out by floating the grids on drops of reagents placed on top of dental wax or Parafilm. Total immersion of grids during incubation must be strictly avoided since this will induce considerable increase in nonspecific labeling.
13. Washing the grids with PBS is done carefully in order to remove as completely as possible the excess, unbound gold conjugate.
14. During the whole protocol, grids should not be allowed to dry since drying would result in a nonspecific aggregation of gold particles.
15. The buffer used prior to immunogold incubation should be adjusted to pH 8.2 to stabilize the gold. The pH of the buffer for other washes should be 7.6, containing 0.1% gelatin to avoid nonspecific labeling.
16. To minimize background staining, the primary antibody should be affinity-purified, showing specificity.
17. Optimal dilutions of all solutions must be determined. Both the antibodies and the protein A- or IgG-gold complexes need to be optimally diluted. Second antibody-gold dilutions of 1:20 to 1:40 in PBS or TBS are recommended as a starting concentration, but should be determined by titration.
18. During development, 0.2 M citrate buffer (pH 3.5) is recommended, which is prepared as follows:

> Citric acid monohydrate 25.5 g
> Trisodium citrate dihydrate 23.5 g
> Deionized glass-distilled water to make 100 ml

19. Silver lactate developer is a suitable choice, but is light sensitive. On the other hand, silver acetate developer is less light sensitive, and is recommended for development by checking, after a certain degree of gray staininghas already been achieved, at intervals with the light microscope.
20. A controlled, reasonably slow process of silver-enhancement is recommended for optimal staining. Optimal silver-enhancement duration can be determined by running pilot specimens together with each group of preparations and removing them at intervals to examine the progress of development. Specimens should be withdrawn when the light brown gum arabic begins to turn grey. Silver-enhancement for ~1 min, as an average, doubles the size of the gold particles.
21. Certain types of background staining can be avoided by dipping the sections on glass slides in a 0.5% gelatin solution before development.
22. Excess silver grains on the section surface can be removed by dipping the sections on a glass slide for 30 sec in a 2% Farmer's solution (9 parts of 2% sodium thiosulfate

and 1 part of 2% potassium ferricyanide), followed by rinsing in double distilled water.

23. Cryostat sections should be dried onto the adhesive-coated glass slides for at least 1 h to prevent their dislodging.

24. The problem of "fogging" on the surface of sections on glass slides, due to autocatalytic precipitations formed in the developing solution, can be reduced or avoided by coating the sections with 0.05% aqueous gelatin solution; this coat prevents the precipitates from the developer reaching the section surface. After development, the coat can be removed by subjecting the adhering sections to running tap water at 4°C for at least 45 min.

25. If needed, oxidation of paraffin sections and semithin resin sections with Lugol's iodine increases the staining efficiency. After deparaffinization in xylene, the sections should be immersed for 5 min in iodine solution (1% iodine in 2% potassium iodide). Before treatment with the iodine, Epon sections should be treated for 20 min with saturated sodium ethoxide to soften the resin. Following a brief rinse in distilled water, both types of sections should be reduced by treatment with 2.5% aqueous sodium thiosulfate until they become colorless, usually in ~30 sec.

26. Hydroquinone (1,4-dihydroxybenzene, mol. wt. 110.11) is the best reducing agent, and should be used at a concentration of 0.85%. Its solution should be prepared immediately before use. However, according to Burry et al. (1992), N-propyl-gallate is better than hydroquinone.

27. Crude gum arabic (a plant copolymer of D-galactose, L-rhamnose, L-arabinose, and D-glucoronic acid residues) is the protective colloid of choice to inhibit the autocatalytic interaction and ensure a controlled and symmetrical growth of the silver shell in the silver lactate-containing developers as well as silver acetate-containing developers. It can be used for light and electron microscopy at a concentration of 15 to 30%. The solution is prepared by dissolving 1 part of gum arabic in 2 parts of double glass-distilled water by magnetic stirring and kept for at least 24 h at room temperature.

28. Sections untreated with colloidal gold can be silver-enhanced to find out if gold particles and not other substances in the section have induced the reduction of silver ions.

29. Electron beam illumination tends to cause instability of the silver layer and may lead to loss or redistribution of metallic silver during subsequent storage in the presence of air. This adverse effect can be prevented by storing such specimens under vacuum (see Chapter 6 in this volume).

References

Amikura, R., Kobayashi, S., Endo, K., and Okada, M. 1993. Nonradioactive in situ hybridization methods for *Drosophila* embryos detecting signals by immunogold-silver or immunoperoxidase method for electron microscopy. *Dev. Growth Differ.* 35: 617.

Arai, R., Geffard, M., and Calas, A. 1992. Intensification of labelings of the immunogold silver staining method by gold toning. *Brain Res. Bull.* 28: 343.

Behnke, O., Ammitzboll, T., Jessen, H., Klokker, M., Nilausen, K., Tranum-Jensen, J., and Olsson, L. 1986. Nonspecific binding of protein-stabilized gold soles as a source of error in immunocytochemistry. *Eur. J. Cell Biol.* 41: 326.

Breter, H. and Erdmann, B. 1993. Suitability of different protein A-gold markers for immunogold-silver staining in paraffin sections. *Biotech. Histochem.* 68: 206.

Brigati, D. J., Budgeon, L. R., and Unger, E. R. 1988. Immunocytochemistry is automated: development of a robotic workstation based upon the capillary action principle, *J. Histotechnol.* 11: 165.

Burry, R. W., Vandre, D. D., and Hayes, D. M. 1992. Silver enhancement of gold antibody probes in preembedding electron microscopic immunocytochemistry. *J. Histochem. Cytochem.* 40: 1849.

Danscher, G. 1981. Localization of gold in biological tissue. A photochemical method for light and electron microscopy. *Histochimie* 71: 81.

Danscher, G., Hacker, G. W., Grimelius, L., and Norgaard, J. O. 1993. Autometallographic silver amplification of colloidal gold. *J. Histotechnol.* 16: 201.

Danscher, G. and Norgaard, J. O. R. 1983. Light microscopic visualization of colloidal gold on resin-embedded tissue. *J. Histochem. Cytochem.* 31: 1394.

De Brabander, M., Geuens, G., Meydens, R., Moeremans, M., and De Mey, J. 1985. Probing microtubule-dependent intracellular motility with nanometer particle video ultramicroscopy (nanovid ultramicroscopy). *Cytobios* 43: 273.

De Harven, E. and Soligo, D. 1989. Backscattered electron imaging of the colloidal gold marker on cell surfaces. In: *Colloidal Gold: Principles, Methods, and Applications.* Vol. 1. p. 229. Hayat, M. A. (Ed.), Academic Press, San Diego.

De Waele, M., Renmans, W., Segers, E., Jochmans, K., and Van Camp, B. 1988. Sensitive detection of immunogold-silver staining with darkfield and epi-polarization microscopy. *J. Histochem. Cytochem.* 36: 679.

Geoghehan, W. D., Scillian, J. J., and Ackerman, G. A. 1978. The detection of human B-lymphocytes by both light and electron microscopy utilizing colloidal gold-labeled anti-immunoglobulin. *Immunol. Commun.* 7: 1.

Giffin, B. F., Gao, K., Morris, R. E., and Cardell, R. R. 1993. Enhancement of antigenic site detection with gold labeled secondary and tertiary antibodies using the immunogold-silver staining method. *Biotech. Histochem.* 68: 309.

Goode, D. and Maugel, T. K. 1987. Backscattered electron imaging of immunogold-labeled and silver-enhanced microtubules in cultured mammalian cells. *J. Electron Microsc. Tech.* 5: 263.

Goodman, S. L., Park, K., and Albrecht, R. M. 1991. A correlative approach to colloidal gold labeling with video-enhanced light microscopy, low voltage electron microscopy, and high-voltage electron microscopy. In: *Colloidal Gold: Principles, Methods, and Applications.* Vol. 3. p. 370. Hayat, M. A. (Ed.), Academic Press, San Diego.

Gross, D. K. and De Boni, U. 1990. Colloidal gold labeling of intracellular ligands in dorsal root sensory neurons, visualized by scanning electron microscopy. *J. Histochem. Cytochem.* 38: 775.

Gu, J., D´. Andrea, M., Yu, C.-Z., Forte, M., and McGrath, L. B. 1993. Quantitative evaluation of indirect immunogold-silver electron microscopy. *J. Histotechnol.* 16: 19.

Handley, D. A. 1989. Methods for synthesis of colloidal gold. In: *Colloidal Gold: Principles, Methods, and Applications.* Vol. 1. p. 13. Hayat, M. A. (Ed.), Academic Press, San Diego.

Hayat, M. A. 1989. *Principles and Techniques of Electron Microscopy*, 3rd ed. p. 227. CRC Press, Boca Raton, FL and Macmillan Press, London.

Hayat, M. A. 1992. Quantitation of immunogold labeling. *Micron. Microsc. Acta* 23: 1.

Henneberry, H. P. and Aherne, G. W. 1992. Visualization of doxorubicin in human and animal tissues and in cell cultures by immunogold-silver staining. *Br. J. Cancer* 65: 82.

Hermann, R., Schwarz, H., and Muller, M. 1991. High precision immunoscanning electron microscopy using Fab fragments coupled to ultra-small colloidal gold. *J. Struct. Biol.* 107: 38.

Herter, P., Laube, G., Gronczewski, J., and Minuth, W. W. 1993. Silver-enhanced colloidal gold labeling of rabbit kidney collecting duct cell surfaces imaged by scanning electron microscopy. *J. Microsc.* 171: 107.

Hoefsmit, E. C. M., Korn, C., Blijleven, N., and Ploem, J. S. 1986. Light microscopical detection of single 5 and 20 nm gold particles used for immunolabeling of plasma membrane antigens with silver enhancement and reflection contrast. *J. Microsc.* 143: 161.

Holgate, C. S., Jackson, P., Cowen, P. N., and Bird, C. C. 1983. Immunogold silver-staining. New method of immunostaining with enhanced sensitivity. *J. Histochem. Cytochem.* 31: 938.

Horisberger, M., Rosset, J., and Bauer, H. 1975. Colloidal gold granules as markers for cell surface receptors in the scanning electron microscope. *Experientia* 31: 1147.

Hughes, D. A., Kempson, M. G., Carter, N. P., and Morris, P. J. 1988. Immunogold-silver/ Romanowsky staining: simultaneous immunocytochemical and morphologic analysis of fine-needle aspirate biopsies. *Transplant. Proc.* 20: 575.

Hughes, D. A., Morris, P. J., Fowler, S., and Chaplin, A. J. 1991. Immunogold-silver staining of leucocyte populations in lung sections containing carbon particles requires cautious interpretation. *Histochem. J.* 23: 196.

Kellenberger, E. and Hayat, M. A. 1991. Some basic concepts for the choice of methods. In: *Colloidal Gold: Principles, Methods, and Applications.* p. 1. Hayat, M. A. (Ed.), Academic Press, San Diego.

Lucocq, J. M. and Roth, J. 1984. Applications of immunocolloids in light microscopy. III. Demonstration of antigenic and lectin-binding sites in semithin resin sections. *J. Histochem. Cytochem.* 32: 1075.

Namork, E. and Heier, H. E. 1989. Silver enhancement of gold probes (5-40 nm); single and double labeling of antigenic sites on cell surfaces imaged with backscattered electrons. *J. Electron Microsc. Tech.* 11: 102.

Namork, E. 1991. Double labeling of antigenic sites on cell surfaces imaged with backscattered electrons. In: *Colloidal Gold: Principles, Methods, and Applications.* Vol. 3. p. 188. Hayat, M. A. (Ed.), Academic Press, San Diego.

Neagu, C., van der Werf, K. O., Putman, C. A. J., Kraan, Y. M., De Grooth, B. G., van Hulst, N. F., and Greve, J. 1994. Analysis of immunolabeled cells by atomic force microscopy, optical microscopy, and flow cytometry. *J. Struct. Biol.* 112: 32.

Roth, J., Saremaslani, P., Warhol, M. J., and Heitz, P. U. 1992. Improved accuracy in diagnostic immunohistochemistry, lectin histochemistry and in situ hybridization using gold-labeled horseradish peroxidase antibody and silver intensification. *Lab. Invest.* 67: 263.

Scopsi, L., Larsson, L. I., Bastholm, L., and Nielson, M. H. 1986. Silver enhanced colloidal gold probes as markers for scanning electron microscopy. *Histochemistry* 86: 35.

Scopsi, L. 1989. Silver-enhanced colloidal gold method. In: *Colloidal Gold: Principles, Methods, and Applications.* Vol. 1. p. 252. Hayat, M. A. (Ed.), Academic Press, San Diego.

Slot, J. W. and Geuze, H. J. 1985. A new method of preparing gold probes for multiple labeling cytochemistry. *Eur. J. Cell Biol.* 38: 87.

Stierhof, Y.-D., Humbel, B. M., Hermann, R., Otten, M. T., and Schwarz, H. 1992. Direct visualization and silver enhancement of ultra small antibody-bound gold particles on immunolabeled ultrathin resin sections. *Scanning Microsc.* 6: 1009.

Stump, R. F., Pfeiffer, J. R., Seagrave, J., and Oliver, J. M. 1988. Mapping gold-labeled IgE receptors on mast cells by scanning electron microscopy: receptor distributions revealed by silver enhancement, backscattered electron imaging, and digital image analysis. *J. Histochem. Cytochem.* 36: 493.

Tacha, D. E., Bowman, P. D., and McKinney, L. N. 1993. High resolution light microscopy and immunocytochemistry with glycol methacrylate embedded sections and immunogold-silver staining. *J. Histotechnol.* 16: 13.

Trejdosiewicz, L. K., Smoira, M. A., Hodges, G. M., Goodman, S. L., and Livingston, D. C. 1981. Cell surface distribution of fibronectin in cultures of fibroblasts and bladder derived epithelium: SEM-immunogold localization compared to immunoperoxidase and immunofluorescence. *J. Microsc.* 123: 227.

van de Kant, H. J. G., Boon, M. E., and de Rooij, D. G. 1993. Microwave applications before and during immunogold-silver staining. *J. Histotechnol.* 16: 209.

Wagenaar, F., Kok, G. L., Brockhuijsen-Davies, J. M., and Pol, J. M. A. 1993. Rapid cold fixation of tissue samples by microwave irradiation for use in electron microscopy. *Histochem. J.* 25: 719.

Walther, P. and Muller, M. 1985. Detection of small (5–15 nm) gold-labeled surface antigens using backscattered electrons. In: *The Science of Biological Specimen Preparation for Microscopy and Microanalysis.* p. 195. Muller, M., Becker, R. P., Boyde, A., and Wolosweick, J. J. (Eds.), SEM, Chicago.

Chapter 1

Trends in Autometallographic Silver Amplification of Colloidal Gold Particles

Gorm Danscher, Gerhard W. Hacker, Cornelia Hauser-Kronberger, and Lars Grimelius

Contents

1.1 Introduction

The invisible picture hidden in an exposed but undeveloped strip of film is present as nanometer-sized crystal lattices of metallic silver intimately attached to the silver bromide crystals. The silver atoms are created by reduction of free silver ions in the light-sensitive bromide crystals by the electromagnetic beams.

When the picture is developed, i.e., when the invisible silver particles are silver amplified, it is usually done by dipping the strip into a solution of reducing molecules, a so-called chemical developer. When this solution penetrates into the film it dissolves the silver bromide crystals, and silver ions are added to the solution. The nanometer-sized silver crystal lattices affix silver ions, reducing molecules to their surfaces. This initiates a process of reduction of silver ions to metallic silver atoms to take place. In this way the invisible silver crystals grow to a visible size, and the picture takes form.

Colloidal gold crystals and crystals of certain metal sulfides and selenides (e.g., of silver, mercury, and zinc) have the same ability as silver crystal lattices to bind silver ions, reducing molecules to their surfaces and drain the electrons from the reducing molecules

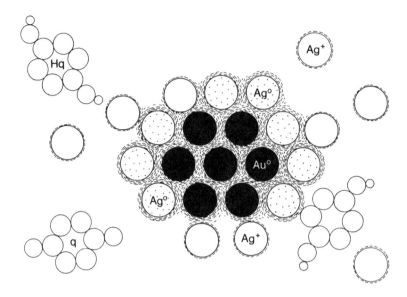

Figure 1 The AMG catalytic crystal lattice when placed in a solution containing reducing molecules — hydroquinone (Hq) — and silver ions (Ag+) will attract both. After being intimately connected to the crystal the hydroquinone molecule will release two electrons into the valence cloud of the crystal lattice. These extra electrons will eventually include two adhering silver ions into the lattice by reducing them to metallic silver atoms. After delivering the electrons the Hq molecules are oxidized to quinone (q), and this molecule is believed to be physically released from the growing crystal by a clip-like movement. Because of this the AMG process can continue as long as reducing molecules and silver ions are present.

to the silver ions; thus turning these into silver atoms. This means that if a section containing such catalytic crystals is exposed to a solution that contains both reducing molecules and silver ions, i.e., an autometallographic (AMG) developer, they will become silver amplified.

As only a few gold atoms or metal sulfide/selenide molecules are sufficient to initiate the AMG silver amplification process, the technique is extremely sensitive but also, being a silver technique, it is highly sensitive to impurities and physical and chemical conditions within and around the tissue sections. It is therefore a precondition for the successful use of AMG that both the general rules and the specific demand for the particular metal/metal molecule in question are strictly followed. The high sensitivity and specificity of the method, and the ease with which the autometallographic technique can be carried out when it becomes a routine, are ample recompense for all the trouble necessary.

1.2 *Light and Electron Microscopical Applications*

Autometallography (Figure 1) can be applied to tissue sections in two principally different ways. Either the sections can be covered with a photographic emulsion and then, after briefly drying, exposed to a chemical developer (Figure 2A), or alternatively, they can be immersed in an AMG developer (Figure 2B). As can be seen from Figure 2C, the two procedures result in the presence of silver ions and reducing molecules in and around the tissue sections, leading to silver amplification of the individual colloidal gold particles present in or on the sections.

The *AMG-emulsion approach* has been little used in general, and in particular for light microscopy. This is undoubtedly because the only silver salt containing emulsions avail-

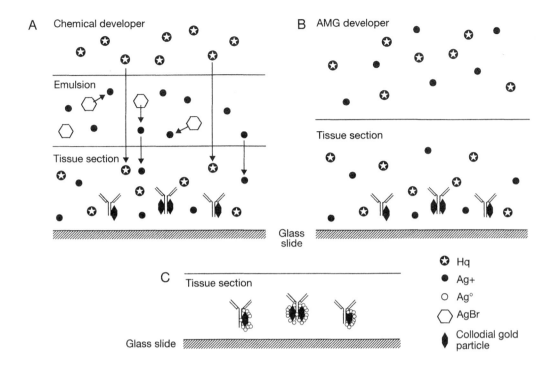

Figure 2 The drawings demonstrate two different ways of performing autometallography. (A) Section containing colloidal gold labeled immunoglobulin molecules covered with an autoradiographic emulsion. The slide is placed in a vial containing a photographic developer (chemical developer). As the reducing molecules move into the emulsion the liquid phase will dissolve the silver bromide crystals and silver ions will mix with the reducing molecules and form an AMG developer. This developer penetrates into the section and causes a silver amplification of the gold crystals, as demonstrated in (C). (B) Schematic presentation of the principle of normal AMG development with a premixed developer.

able are the commercial *autoradiographic* emulsions. These are manufactured for a completely different purpose, i.e., to be maximum sensitive to positron- and beta-rays, to affix permanently to the surface of the sections, etc. Important qualities of a true AMG emulsion, however, include: (1) the silver salt crystals readily dissolve, (2) the crystals have low sensitivity to light, and (3) the emulsion is easy to remove after development. We are presently trying to devise a suitable AMG emulsion, but have not succeeded as yet.

The *AMG-developer-approach* is currently the preferred technique, but a variety of developers have been suggested, both before and since the introduction of silver amplification of colloidal gold particles (Holgate et al., 1983; Danscher and Nørgaard, 1983). These two studies use the finding that metallic gold crystals can be AMG developed (Danscher, 1980, 1981a), and in both presentations the *acid silver lactate developer* devised the same year was used (Danscher, 1981b). The AMG silver lactate developer is still regarded as a suitable choice by virtue of its specificity and sensitivity (Danscher, 1993; Stierhoff et al., 1992; Hacker et al., 1992).

What makes the lactate developer new is (1) its balanced silver release, i.e., silver ions are released intact with their reduction to metallic silver ions on the surface of the colloidal gold particles, and (2) the number of reducing molecules (hydroquinone, Hq) is reduced, compared with the original Liesegang developer. The Liesegang's developer retains the low pH and the protecting colloid *gum arabic* (Danscher, 1981a, 1993).

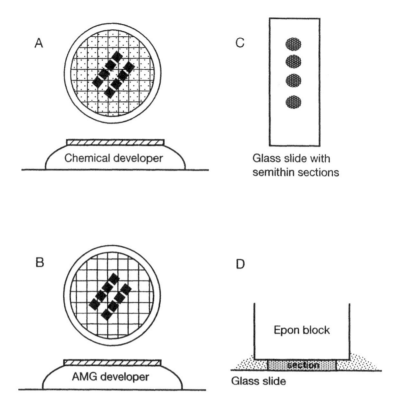

Figure 3 Ultrastructural autometallography can be performed in four separate ways: (1) the grid can be covered by an autoradiographic emulsion and placed upside down on a drop of photographical developer (A); (2) the grid can be placed on top of a drop of AMG developer (B); (3) a semithin section placed on a glass slide can be developed and later glued to a blank Epon block (C, D); or (4) blocks of tissue from AMG developed vibratome sections are stained with osmium and uranyl before being embedded in plastic.

The specificity determined by the controlled quantity of free silver ions in the developer and the sensitivity are probably the result of the acidity of the developer. When the protein-bound colloidal gold particle in or on the section is exposed to the developer, the neutral peptization of the gold particle may decrease and thereby expose a larger part of the particle for AMG silver amplification (Danscher, 1993; John Chandler, personal communication and *vide infra*).

When colloidal gold crystals are used for marking immunoglobulins, as in IGSS (Holgate et al., 1983), it is practical to use a developer that is less sensitive to light. Hacker et al. (1988) have developed such a developer, *the silver acetate developer*, to be used for short-lasting development under continuous observation in the microscope. The acetate developer preserves the qualities of the silver lactate developer and should be chosen for short-term development under direct light microscopic observation.

1.2.1   Ultrastructural Autometallography

Autometallography can be carried out by the emulsion version as well as by the AMG developer version for electron microscopical analysis (Figure 3). One must remember that the grid used should be made of either platinum or nickel. Copper grids cannot resist a low pH and will therefore be partly dissolved in acid developers; gold grids on the other hand will undergo silver-plating and this might disturb the silver amplification of the colloidal gold particles.

We have obtained good results by (1) covering the grid with an autometallographic emulsion, using the loop technique, and 10 min later inverting on a drop of chemical developer (Figure 3A); and (2) placing the grid upside down on top of an acid silver lactate developer or on top of a drop of a commercial AMG developer (Figure 3B).

If one wishes to study the immunocytochemical localization at both the light and electron microscopical level, the following method can be used. Semithin Epon sections are placed on glass slides, covered with gelatin and AMG developed (Figure 3C). After light microscopical analysis, sections to be studied with the electron microscope are covered with unpolymerized Epon and a blank block of Epon is pressed down on the section (Figure 3D). After polymerization, the block is removed by heating the glass slide to 90°C. The block can now be trimmed and thin sections cut and counterstained with salts of uranium and lead (Danscher, 1981a–c).

Another approach is to cut the formaldehyde-glutaraldehyde-fixed tissue into 150-μm sections and expose them to the colloidal gold-marked molecules, after which the sections are AMG developed. Following a careful rinse, the parts of the sections that are to be ultrastructurally analyzed are excised, fixed in OsO_4, stained en bloc with uranyl acetate, and embedded in a resin *lege artis*. This technique gives optimal ultrastructural preservation and localization of the gold-tagged molecules (Danscher and Møller-Madsen, 1986; Danscher, 1991).

1.3 Troubleshooting

1.3.1 Nonspecific Background

1.3.1.1 Clean Tools
Because of the extreme sensitivity of the AMG technique, only two atoms of gold can initiate the silver amplification process (Handley, 1989). The first thing to check is whether the cleaning process of all glass tools and vials has been carried out satisfactorily. Because clean glass tools are imperative for all AMG work, all tools including glass slides must be washed in a 10% Farmer's solution (9 parts of 10% sodium thiosulfate and 1 part of 10% potassium ferricyanide) for 30 min, and all chemicals used must be of analytical grade (Danscher and Montagnese, 1993).

1.3.1.2 Gelatin Cover
If the sections are not gelatinized, two problems can result: small flakes of protective colloid/developer can adhere firmly to the surface, and if these flakes contain autocatalytic silver grains, they will provide a most disturbing "background".

Because of a certain concentration gradient of silver ions and reducing molecules from the surface into the sections, the surface close to AMG sites will grow faster. Especially after longer developing times, this phenomenon can impair quality. Gelatinization of the sections by dipping them in a 0.5% gelatin solution before developing will alleviate the problem or solve it altogether.

Edges on the surface of the sections can also cause unspecific staining. There is to our knowledge no explanation why edges under certain circumstances are able to reduce silver ions when exposed to an AMG developer. Conceivably, a gathering of electrons on protruding edges might reduce the first silver ions and serve as a fuse for the AMG process along the edge. Gelatinization eliminates this problem.

1.3.1.3 Removal of Silver
Excess silver grains on the surface of the sections, whether caused by unspecific staining or by asymmetrical growth of the colloidal gold encapsulating the AMG silver grains, can be removed by treating the sections with a Farmer's solution. If the sections are cryostat or paraffin sections, we use a 2% Farmer's solution (9 parts of 2% sodium thiosulfate and 1

part of 2% potassium ferricyanide). The slides are dipped in this solution for 30 sec and carefully rinsed in distilled water. Epon sections are exposed to a 10% Farmer's for 10 sec (Danscher, 1981a; Danscher and Montagnese, 1993).

1.3.1.4 *Nonspecific and Unique Nonspecific AMG Grains*

AMG silver grains in the sections that do not have a core of colloidal gold would be most disturbing. If the grains are located evenly and completely at random, they belong to the nonspecific background staining and merely impair quality and resolution, but if the fake AMG grains are highly ordered and present only in certain cells or cell organelles, they can be misinterpreted as representing a valid marking.

The unique nonspecific staining can be caused by tissue sections originating from a person who has been exposed to mercury or silver (e.g., amalgam dental fillings, local treatment with silver nitrate, industrial exposure), or gold (e.g., treatment of rheumatism with aurothiomalate). Another confounding factor can result from exposure either of the tissue in situ, or of subsequent tissue blocks or sections to sulfide ions or selenide ions. In such cases, free and loosely bound zinc ions will be bound as zinc sulfide/selenide crystals that will be AMG silver amplified (Danscher, 1991). This is primarily of interest when applying IGSS or other AMG colloidal gold techniques to central nervous system, prostata, pancreas, adrenals, and pituitary tissues.

Whatever kind of sham AMG grains causes the trouble, the problem should be tackled in the same way: the sections should be placed in a 10% Farmer's solution for 30 min or in a 1% aqueous potassium cyanide solution for 1 h. The sections should then be carefully rinsed and *re-exposed to AMG developer*. As gold will not dissolve, the macromolecules are still marked and will be silver amplified, while all other triggering AMG sites, nonspecific as well as unique nonspecific, have been removed (Danscher and Rungby, 1985; Schiønning et al., 1993; Danscher, 1991).

1.4 *Autometallographic Trends*

Two tendencies seem to shape the future in the AMG field. One line of improvement is the introduction of new reducing molecules, better silver ion donors and more suitable protecting molecules. The AMG developer of tomorrow will be able to overcome some of the intrinsic shortcomings of the present ones.

As mentioned earlier, the acidity of the silver lactate and the acetate developers seems not only to support a controlled, reasonably slow AMG process, but also to increase the available surface of the colloidal gold marker crystal. This phenomenon can be explained by changes in an ionic film that surrounds the particle.

Colloidal gold particles can be formed in at least two different ways, but it appears that only the aurochloride technique is used. This technique results in a peptization of the crystal lattices with $AuCl^-$ ions. When, for instance, immunoglobulins bind a colloidal gold particle, part of its surface becomes covered with radicals from the protein, i.e., the macromolecule wraps the crystal in a fold of its own body. The protein-covered part of the crystal has a "neutral" peptization and is believed to be unable to initiate the AMG process (Danscher et al., 1993). As mentioned in the Introduction, the acid AMG developers seem to free the colloidal gold particle from this embrace and, still present in the protein pocket, the gold particle most likely has its total surface available for AMG silver amplification.

A procedure that makes the gold particle maximally visible to the AMG developer will most likely be devised in the future. Not only is it possible to manipulate the state of peptization, it is also possible to exploit a possible increase in the catalytic power of the colloidal gold particle by introducing ions sulfide and selenide in the peptization film (*vide infra*).

At present, the reducing molecule hydroquinone is used in most AMG developers. As can be seen from Figure 2, the molecule is rather small and it is believed that when it is oxidized it changes its form: it folds up and most likely this physical phenomenon releases it effectively from the growing AMG silver grain. New, more powerful reducing molecules could be one way to improve AMG developers.

Silver lactate was the first silver salt introduced into AMG that does not have the quality of being entirely ionized when dissolved (Danscher, 1981). It is of crucial importance that the content of free silver ions in the developer can be controlled by being released from a pool of loosely bound silver ions. If this is not the case the tendency of nonspecific catalytic sites to be created and/or to be AMG amplified is manifestly increased. That is precisely why AMG developers that use silver nitrate are always inferior, for example, to silver lactate or silver acetate developers. In the future we may see more complex silver ion donors used; even organic silver containing molecules might prove to be an ingredient of such developers.

Concerning protective colloids, several types have been tested and earlier gum arabic was found to be superior (Danscher, 1981a). This investigation, however, was just a lifting of the veil, and the field should be expanded. The significance of having the silver ions and the reducing molecules compartmentalized cannot be overestimated. The protective molecules control the potentials of nonspecific AMG sites to be created and seem to suppress important catalytic sites, i.e., the developer becomes more specific but, of course, also takes longer to act.

The other tendency promises the creation of AMG emulsions that suit the different applications for this technique. Advances along these lines include research into determining the right silver halide crystals, the right size of the crystals for the different AMG tasks, and the most ideal chemical developer to match the emulsion determined silver ion concentration.

1.5   Future Autometallographic Markers

Colloidal gold particles were introduced as markers for macromolecules because they bind easily and firmly to most proteins, by virtue of their negative surface, and they are easy to locate in the electron microscope due to their electron density. One problem at that time was that it was impossible to resolve the particles with the light microscope. When the visualization of gold particles became possible, their size could be decreased to less than 1 nm. Recent research has shown that colloidal gold particles less than 2 nm in diameter have so weak an internal coherence that small crystal lattices heated by the electron beam have a tendency to disintegrate. Some of the particles evaporate and the gold atoms liberated condense onto neighboring particles (John Chandler, personal communication). Small particles also seem to be vulnerable to oxidation. Stierhoff et al. (1992) have demonstrated that vacuum protects 1-nm gold particles. These and other shortcomings of colloidal gold make it tempting not only to improve the techniques for silver amplification but also to search for markers that are easier to work with and form powerful catalytic AMG sites.

Several molecules have the same AMG triggering capacity as gold and silver atoms. The known catalytic molecules are either metal sulfide or metal selenide crystals lattices, and the AMG characteristics of such molecules are known only for a limited number of metals, viz., gold, silver, mercury, zinc, and copper (Danscher, 1991). However, the number of metals that can create catalytic sulfide/selenide may be much higher. It has been demonstrated that the quantity of silver atoms necessary to initiate the AMG process is between two to six, of zinc sulfide molecules and gold atoms only two (Hamilton and Logel, 1974; James, 1939, 1977).

1.6   Conclusions

AMG silver amplification is a relatively new and extremely sensitive technique applicable both for tracing endogenous and exogenous metals and for augmentation of AMG catalytic labels of biological molecules in immunological, pharmacological, and toxicological studies. Autometallography therefore makes it possible to achieve advances in many different fields of biology.

References

Danscher, G. 1980. Ultrastructural localization of metals in the CNS by physical development. Gold, mercury and water insoluble metal sulphides. *J. Ultrastruct. Res.* 73: 93.

Danscher, G. 1981a. Histochemical demonstration of heavy metals. A revised version of the sulphide silver method suitable for both light and electronmicroscopy. *Histochemistry* 71: 1.

Danscher, G. 1981b. Localization of gold in biological tissue. A photochemical method for light and electronmicroscopy. *Histochemistry* 71: 81.

Danscher, G. 1981c. Light and electron microscopic localization of silver in biological tissue. *Histochemistry* 71: 177.

Danscher, G. 1991. Applications of autometallography to heavy metal toxicology. *Pharmacol. Toxicol.* 68: 414.

Danscher, G. and Montagnese, C. 1993. Autometallographic localization of synaptic vesicular zinc and lysosomal gold, silver and mercury. *J. Histotechnol.* (in press).

Danscher, G. and Møller-Madsen, B. 1985. Silver amplification of mercury sulfide and selenide: a histochemical method for light and electron microscopic localization of mercury in tissue. *J. Histochem. Cytochem.* 33: 219.

Danscher, G. and Nørgaard, J. O. 1983. Light microscopic visualization of colloidal gold on resin-embedded tissue. *J. Histochem. Cytochem.* 31: 1394.

Danscher, G., Hacker, G. W., Grimelius, L., and Nørgaard, J. O. 1993. Autometallographic silver amplification of colloidal gold. *J. Histotech.* 16: 201.

Hacker, G. W., Grimelius, L., Danscher, G., Bernatzky, G., Muss, W., Adam, H., and Thurner, J. 1988. Silver acetate autometallography: An alternative enhancement technique for immunogold-silver staining (IGSS) and silver amplification of gold, silver, mercury and zinc in tissues. *J. Histotechnol.* 11: 213.

Hamilton, J. F. and Logel, P. C. 1974. The minimum size of silver and gold nuclei for silver physical development. *Photogr. Sci. Eng.* 507: 512.

Handley, D. A. 1989. The development and application of colloidal gold as a microscopic probe. In: *Colloidal Gold: Principles, Methods and Applications*. Vol. 1. pp. 1–32. Hayat, M. A. (Ed.), Academic Press, San Diego.

Holgate, C. S., Jackson, P., Cowen, P. N., and Bird, C. C. 1983. Immunogold-silver staining: new method of immunostaining with enhanced sensitivity. *J. Histochem. Cytochem.* 31: 938.

James, T. H. 1939. The reduction of silver ions by hydroquinone. *Comm. 712 Kodak. Res. Lab.* 61: 648.

James, T. H. 1977. *The Theory of the Photographic Process*, Chapter 13. Macmillan, New York.

Rungby, J., Møller-Madsen, B., and Danscher, G. 1985. Silver intoxication of the CNS. *Nutr. Res. Suppl.* 1: 668.

Schiønning, J. D., Danscher, G., Christensen, M. M., Ernst, E., and Møller-Madsen, B. 1994. Differentiation of silver-enhanced mercury and gold in tissue sections. *J. Histochem.* (in press).

Stierhof, Y.-D., Humbel, B. M., Hermann, R., Otten, M. T., and Schwarz, H. 1992. Direct visualization and silver enhancement of ultra-small antibody-bound gold particles on immunolabeled ultrathin resin sections. *Scanning Microsc.* 4: 1009.

Chapter 2

Silver Staining Techniques, With Special Reference to the Use of Different Silver Salts in Light and Electron Microscopical Immunogold-Silver Staining

Gerhard W. Hacker, Gorm Danscher, Lars Grimelius, Cornelia Hauser-Kronberger, Wolfgang H. Muss, Angelika Schiechl, Jiang Gu, and Otto Dietze

Contents

0-8493-2449-1/95/$0.00+$.50
© 1995 by CRC Press, Inc.

2.1   Introduction

Classical silver stains were widely used in histology and histopathology to visualize different tissue components and cell types (Romeis, 1968; Böck, 1989). Their importance has somewhat declined since the introduction of immunohistochemical techniques, but some of the classical silver staining methods are still useful in both research and routine histopathology. Reticulum, neurons, ganglia, glia, and neuroendocrine cells are examples of tissues and cells identified with silver stains. In addition, substances such as melanin, calcium, and heavy metals have been detected using silver staining techniques. Lately, methods have been introduced allowing the demonstration of mitotic figures (Busch and Vasko, 1988) and some interesting regions of chromatin, which are called the nucleolar organizer regions (NOR) (Goodpasture and Bloom, 1975; Lindner, 1993). Other silver procedures subsumed under the term autometallography (AMG) have been used to detect catalytic tissue metals such as metallic gold and silver, and sulfides or selenides of mercury, silver, and zinc (Danscher, 1991). The AMG setup for silver amplification of metallic gold in tissue sections was introduced in 1981 (Danscher, 1981b) and two years later applied to enhancement of colloidal gold used to label enzymes (Danscher and Norgaard, 1983) and immunoglobulins, the immunogold-silver staining (IGSS) technique (Holgate et al., 1983; Springall et al., 1984; Hacker et al., 1985a). A series of applications of IGSS for related methods and various modifications have been proposed (e.g., Fujimori and Nakamura, 1985; Hacker et al., 1985a, 1994a; Lackie et al., 1985; King et al., 1987; Westermark et al., 1987; Hacker, 1989; Krenács et al., 1989a,b, 1990, 1991; Roth, 1983; Roth et al., 1992a,b; Gu et al., 1993; Tacha et al., 1993), including the introduction of silver acetate AMG as a less light-sensitive alternative to silver lactate and silver nitrate AMG (Skutelsky et al., 1987; Hacker et al., 1988), the use of IGSS for lectin histochemistry (King et al., 1987; Schmidt and Peters, 1987; Skutelsky et al., 1987) and for sensitive nucleic acid detection *in situ* hybridization (Varndell et al., 1984; Liesi et al., 1986; Hacker et al., 1993, 1995) and *in situ* polymerase chain reaction (Zehbe et al., 1992–1994). An excellent review of existing IGSS techniques is given by Hayat and various contributing authors (1993b).

Almost all of the silver methods have been empirically developed, and many of them have been modified to improve their reproducibility. With few exceptions, the chemical background of the classical silver stains is yet unknown. The chemistry and interactions between silver and cellular materials has been discussed by Hayat (1993a).

Silver stains can be divided into three main categories.

2.1.1   Argyrophil Reaction

The actual tissue components of cells retain silver ions from the silver salt solution, but metallic silver becomes visible only after a subsequent reducing process brought about by an external agent(s). The outcome of the staining depends on the type of silver salt, and its concentration, the type and pH of the solvent, the composition of the reducing solution, and fixative used. The more recent argyrophilic methods have the different factors in the staining process strictly controlled. This was not the case for many older methods, however.

2.1.2   Argentaffin Reaction

The cells displaying argentaffin reaction contain chemical substance(s) which can retain silver ions from an ammoniacal silver solution and reduce them to metallic silver (one-step staining procedure).

2.1.3 Autometallography (AMG)

This technique, initially developed during the nineteenth century for photographic processing, was subsequently applied to tissue research by Liesegang (1911, 1928). During the last century the technique has been used to demonstrate certain heavy metals and metal containing molecules (Danscher and Montagnese, 1994) and silver amplified gold particles bound to antibodies and enzymes studied at both LM and EM levels (Holgate et al., 1983; Danscher and Norgaard, 1983). The technique has also been used for *in situ* hybridization (Varndell et al., 1984; Liesi et al., 1985; Hacker et al., 1983). The processing solution called "an AMG developer" contains both a silver salt and a reducing agent. The best developers keep at a low pH; often a protecting agent is added in order to avoid autoreduction (Danscher et al., 1993). The AMG technique is highly specific and extremely sensitive, e.g. gold particles less than 0.5 nm in diameter can be visualized by growing the initial gold lattices into a visible silver crystal.

Some of the silver stains useful for research and histopathological diagnosis are discussed below. Present-day silver staining techniques are used mainly to demonstrate (1) various types of neuroendocrine cells, (2) nucleolar organizer region proteins, (3) heavy metals, and (4) to enhance the visibility of colloidal gold particles used to label immunoglobulins.

2.2 Classical Argyrophil and Argentaffin Techniques

Silver stains have long been used to study the differentiation of endocrine cells, especially those in the pancreatic islets of Langerhans. The Von Gros-Schultz silver method (1912) and the Bodian (1936) silver proteinate (protargol) technique were used for the same purpose. However, the former technique presented a number of technical difficulties, and the latter had practical limitations due to difficulties in obtaining usable silver proteinate. These silver techniques were modified by several authors to achieve better results. The silver staining techniques were initially developed for visualization of the nervous system. Grimelius (1968) developed a silver nitrate stain to obtain the same staining results as the Bodian technique, but was based solely on chemically well-defined substances that were commercially available.

There are four silver techniques that are useful for identifying different neuroendocrine cell types. They are named after Grimelius, Masson, Sevier-Munger, and Hellerström-Hellman. Masson staining is based on the argentaffin reaction, and the other three on argyrophil reactions.

The Grimelius stain visualizes most neuroendocrine cell types and is regarded as a broad-spectrum neuroendocrine marker (Grimelius, 1968; Grimelius and Wilander, 1980; Grimelius et al., 1994). The Masson stain was empirically developed in 1914, and subsequently improved by Hamperl (1927), Singh (1964), and Portela-Gomez and Grimelius (1986; see also Grimelius et al., 1994). The Masson staining facilitates the visualization of the enterochromaffin (EC) cells found in the mucosa of the gastrointestinal tract of various species including human. The Sevier-Munger (1965) technique was also initially developed for neural tissue, but additionally visualizes the enterochromaffin-like (ECL) and D_1 cells of the human gastric mucosa. It also stains EC cells and cells synthesizing gastric inhibitory peptide (GIP). The latter cell type is located in the intestinal mucosa. By combining the staining results of the Sevier-Munger and the Masson methods, one can identify the enterochromaffin-like cells in the gastric mucosa and tumors derived from these cells. Enterochromaffin-like cells cannot yet be detected by immunohistochemical methods in routinely fixed paraffin sections because they contain histamin, which is only identifiable after special tissue processing (Hakanson et al., 1986). Davenport developed an alcoholic silver nitrate method that is useful for neural tissue application. Hellerström and Hellman

(1960) modified this technique so that it could demonstrate the D cells of pancreatic islets. Subsequently, this cell type was shown to contain somatostatin (Polak et al., 1975).

The chemistry of the Grimelius and Masson methods is known to a certain extent. It has been shown by "dot blot" techniques that it is the granular protein chromogranin A that causes the argyrophil reaction in Grimelius staining (Rindi et al., 1986; Lundqvist et al., 1990), but so do dopamine, noradrenalin, and serotonin (Lundqvist et al., 1990). By using the same assay, the argentaffin Masson reaction was found to be positive for Dopa, dopamine, noradrenaline, adrenaline, and 5-hydroxytryptophan (Lundqvist et al., 1990).

2.3   Silver Staining Techniques for Demonstration of Nucleolar Organizer Regions (Ag-NOR)

NORs present in some human chromosomes are believed to be functionally identical to the sites of ribosomal RNA transcription, where RNA transcripts are assembled into functional ribosomes. The silver technique used to demonstrate these regions was empirically developed by Goodpasture and Bloom (1975), and later modified by Howell and Black (1980), Ploton et al. (1986), and Lindner (1993). Initially, the technique was devised for cytogenetic preparation, but subsequently modified for tissue sections. The number and size of these NORs seem, at least for some malignant tumors, to provide useful information about their virulence and prognosis. A new modification of Ag-NOR is being developed by our team that will allow the use of substantially lower silver salt concentrations for the first time under standard conditions and the use of microwave irradiation to speed up the process (Li, Hacker, Danscher and Grimelius, unpublished).

2.4   Autometallography for Detecting Catalytic Tissue Metals

Autometallography allows silver amplification of nanometer size catalytic crystals, i.e., crystals of gold and silver and crystals where the lattices contain both a heavy metal (e.g., Ag, Au, Zn, Hg, Cu) and sulfide or selenide ions. Such autometallographic crystals or crystal lattices have the ability to convey electrons from reducing molecules, which adhere to the surface of the particle, to likewise adhering silver ions, and once started, the autometallographic process will build up a shell of metallic silver around the crystals and reveal their exact position in the tissue. Autometallography is therefore most valuable not only for tracing colloidal gold particles used as labels of immunoglobulin or enzymes at light microscopic (LM) and electron microscopic (EM) levels, but also for tracing crystals initializing autometallography in general (Danscher, 1991). Presently, a method for the demonstration of gold, silver, mercury, and chelatable zinc in tissue sections has been presented for both LM and EM applications (Danscher and Montagnese, 1994).

The silver lactate counterstaining method for resin-embedded semithin sections of osmium-fixed tissue is also an autometallographic method. During a period of 60 min, the semithin sections are exposed to a 0.05% silver lactate solution in distilled water at 60°C. After autometallographic development, silver bound in the tissue either as silver sulfide or metallic silver will be amplified to visible dimensions. This method results in a highly contrasted monochrome photography. The staining may reflect several independent chemical processes (Danscher, 1983).

2.4.1   Autometallographic Demonstration of Certain Metals

2.4.1.1   Gold

A study by Roberts (1935) demonstrated that metallic gold can be silver amplified. Tissue sections from animals exposed to aurothiomalate showed no precipitate after autometallographic development unless the sections had been radiated with UV light before development, demonstrating that the chemically bound gold has to be reduced to metallic gold before being autometallographically catalyzed (Danscher, 1981a,b). It was shown later by Pixe that each gold crystal in the tissues was covered by silver (Danscher, 1984).

2.4.1.2 Silver

Autometallography of tissue sections from a person or an animal exposed to silver will result in a detailed pattern of silver amplified silver in all organs including the brain (Danscher, 1981c, 1991).

2.4.1.3 Zinc

The epithelial cells in several different glands contain secretory vesicles that contain a pool of zinc that can easily be chelated. Neurons containing chelatable zinc in their synaptic vesicles are readily identified in various anatomical structures throughout the brain. The mother cells of these vesicles are called zinc-enriched neurons (ZEN). By treating laboratory animals with either sulfide or selenide ions the chelatable zinc pool can be bound in the vesicles as zinc sulfide/selenide crystal lattices. These crystals have the same catalytic capacity as, for example, colloidal gold particles to initiate an autometallographic silver amplification (Danscher, 1981a–c, 1991).

2.4.1.4 Mercury

When organisms are exposed to mercury (through release from amalgam fillings) the tissues will, after a while, contain mercury bound as mercury sulfide or selenide crystal lattices. It seems as if this process takes place in several different cells including neurons and that such crystals represent a nontoxic form of mercury. After autometallographic development of light microscopical, semithin and thin sections of tissue from a mercury exposed person this particular mercury pool is constantly found in lysosomes (Danscher and Møller-Madsen, 1985a; Danscher, 1981; Danscher and Montagnese, 1994).

For ultrastructural studies, autometallography can be implemented in several different ways: (1) Grids with thin sections can be covered with an autoradiographic emulsion and developed with a chemical developer, (2) the grids can be placed upside down on a drop of autometallographic developer, (3) semithin sections can be autometallographically developed, and after LM analysis, selected sections can be reembedded on top of a blank Epon block, or (4) vibratome sections (150 μm thick) can be autometallographically developed. Blocks and tissue are dissected from these thick sections and stained with uranyl acetate and OsO_4, and then dehydrated and embedded (Danscher et al., 1993).

2.5 Immunogold-Silver Staining and Related Techniques

Only two gold atoms are needed to initiate the silver amplification process. This extreme sensitivity of autometallography has made it useful also for immunocytochemistry that detects substances against which antibodies can be made (proteins, peptides, amines, etc.) and lectin histochemistry that detects agglutinin-binding carbohydrates. More recently, applications for *in situ* hybridization and polymerase chain reaction (PCR) *in situ* hybridization (PISH) have been developed, which allow the detection of single copies of specific DNA or mRNA sequences (Zehbe et al., 1992–1994; Zehbe and Hacker, unpublished). In all these methods, the detection step is based on colloidal gold-adsorbed macromolecules (antibodies, protein A, or streptavidin). The actual application of autometallography for immunohistochemistry has been named IGSS (Holgate et al., 1983). The IGSS technique is highly sensitive and detection efficient, provided that an adequate silver intensification method is used (Stierhof et al., 1992). The high amplification potential of autometallography

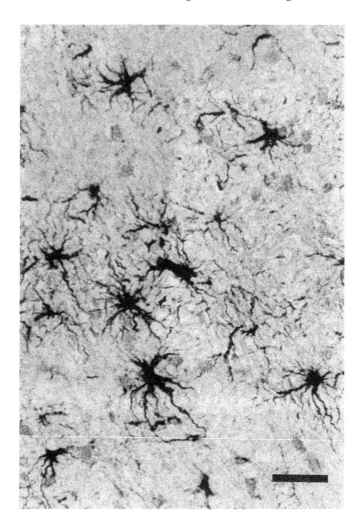

Figure 1   Astrocytes in human brain detected by indirect IGSS with silver acetate autometallography. Sections were incubated overnight at 4°C with primary polyclonal antibodies from rabbit against S-100 proteins diluted 1:2000 in phosphate-buffered saline (pH 7.2) containing 0.1% BSA. Detection was carried out with a 1:1 mixture of goat-antirabbit IgG antibodies each diluted 1:50 with Tris-buffered saline (pH 7.6) containing 0.1% gelatin and incubated for 60 min at room temperature. A very high resolution together with excellent morphological preservation is obtained in this formalin-fixed 5 μm thick paraffin section which was lightly counterstained with hematoxylin, eosin and nuclear fast red. Bar = 25 μm.

makes it possible to obtain dark-gray or black specific staining even if only small quantities of the labeled substance are present (Figures 1 and 2), as is the case in semithin sections or overfixed paraffin sections. Gold particles of 0.8 to 5 nm in diameter should be used in order to achieve high labeling densities and good penetration (Lackie et al., 1985; Gu et al., 1993). Silver amplification makes these initially tiny gold particles visible even at low magnifications. This is a distinct advantage for light and also electron microscopic studies, also including confocal laser microscopy (Figures 3 and 4) (Van Den Pol, 1985; Bastholm et al., 1986; Slater, 1991; Burry et al., 1992; Shimizu et al., 1992; Hacker et al., 1995).

　　Apart from the indirect method, direct, bridge, streptavidin-biotin, protein A-gold-silver staining methods and various other combinations have been described (e.g., Roth, 1982, 1983; Fujimori and Nakamura, 1985; Scopsi and Larsson, 1985; Lah et al., 1990; De Valck et al., 1991;

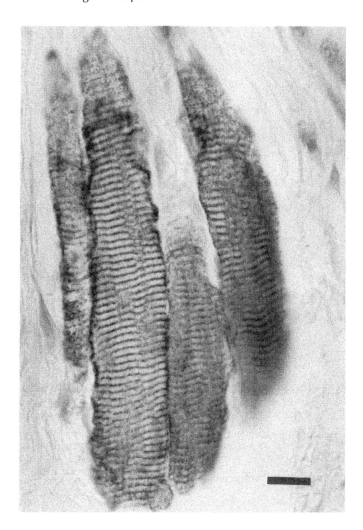

Figure 2 Detection of skeletal muscle fibers with indirect IGSS with silver acetate autometallography. The preparation was treated as described in Figure 1 and *Protocol 1*. Striation can be readily seen at high magnification without any detectable background staining. Mouse monoclonal antibodies against desmin were used in a dilution of 1:100 in phosphate-buffered saline containing 0.1% BSA. Formalin-fixed 5 µm thick paraffin section lightly counterstained with hematoxilin, eosin, and nuclear fast red. Bar = 10 µm.

Roth et al., 1992a,b). Silver lactate and silver acetate have been widely used as the ion source, forming shells of metallic silver around small gold particles (Figure 4a). The IGSS application of autometallography resulted in substantially higher sensitivity and detection efficiency compared with unenhanced immunogold-staining and other immunocytochemical techniques (Holgate et al., 1983; Springall et al., 1984; Hacker et al., 1985; Scopsi and Larsson 1985). In paraffin sections and semithin resin sections, IGSS techniques show a number of advantages when compared to other methods. They may give positive immunostainings where other methods have failed. Therefore, they may greatly facilitate the demonstration of substances present in only small quantities (Springall et al., 1984; Hacker et al., 1985). Using microwave irradiation during the incubation of the primary antibody and the immunogold layer, the whole procedure can be carried out in ~30 min (Gu, 1994).

Hazardous reagents such as potentially carcinogenic DAB commonly used in peroxidase-based methods, or radioisotopes, are avoided in IGSS. Positive reactions in LM

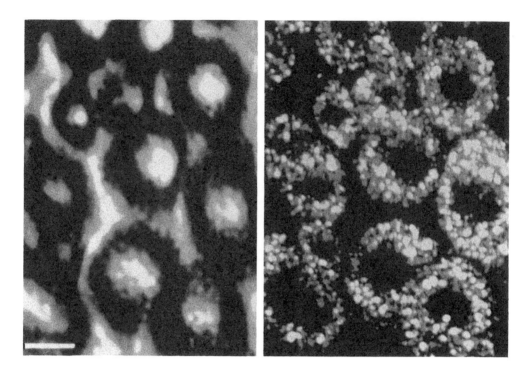

Figure 3   Images of indirect IGSS with silver acetate autometallography given by confocal laser microscopy. Courtesy of Prof. A. Hermann (Salzburg). Transverse section of buccal muscle fibers from *Aplysia californica*. Rabbit polyclonal antibodies to sarcoplasmic Ca-binding protein (SCP) II were applied; SCP II immunoreactivity is localized at the periphery but not in the center of muscle fibers, where mitochondria are located. SCP II is exclusively present in muscle and not found in neurons, which would contain SCP I. From *Eur. J. Neurosci.* 5: 549, 1993. The photo was taken from the TV screen of the laser microscope. Bar = 1 μm.

applications can be easily identified due to the very intense signal, which facilitates the screening of sections even at low magnifications. A very high light and electron microscopic resolution can be obtained (Figures 1–4). It also allows the use of conventional counterstaining such as hematoxylin and eosin and/or nuclear fast red on cryostat or paraffin sections, or azure II, methylene blue and basic fuchsin on semithin resin sections. This improves the morphology for assessment (Springall et al., 1984; Hacker et al., 1988).

As demonstrated by Lackie et al. (1984) and Gu et al. (1993), immunolabeling is maximum when gold particles of small size (2 to 5 nm) are used. Such immunogold reagents penetrate sections better and achieve particle densities higher than those obtained with larger ones (Figure 4). These findings require gold particles of 5 nm or smaller for LM and EM applications. In our hands, 1 nm size particles appear not to give additional labeling compared to 2 to 5 nm size gold. In addition, "ultrasmall" immunogold-reagents appear to be less stable than larger gold particles. This phenomenon was also noticed by De Valck et al. (1991) and Gu et al. (1993).

A novel approach called *nanogold* has been described to be superior to traditional colloidal gold probes which are supposed to be coated with large, bulky proteins for stabilization. Nanogold probes use 1.4 nm diameter gold particles covalently bound to Fab' fragments and do not require protein coating. Possibly due to the smaller size, penetration properties seem to be better than those of colloidal gold probes (Hainfeld et al., 1992 and in this volume; Vandre et al., 1992). When we tried this new reagent in our test systems, the results obtained were comparable to high quality 5 nm colloidal gold probes.

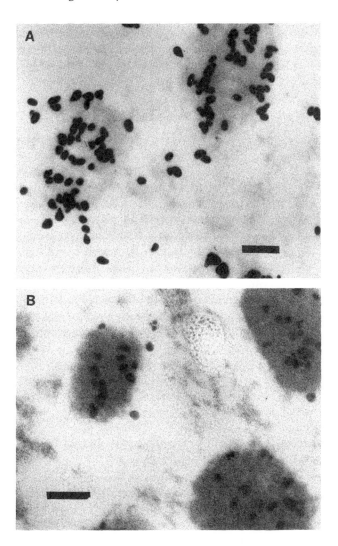

Figure 4   Electron micrographs of silver amplificated immunogold preparations. In both preparations, a relatively uneven amplification of the gold label is obtained. (A) shows the detection of atrial
natriuretic peptide (ANP) in rat heart atrium cardiocytic granules obtained with rabbit polyclonal
primary antibodies and a 5 nm gold-labeled second antibody, amplified for 10 min in dark with the
silver lactate autometallography solution (Danscher, 1981a) but without protective colloid. LR-White
thin section of glutaraldehyde-paraformaldehyde fixed tissue. Courtesy of Dr. R. Rufner (Browns
Mills, NJ). (B) is a similar preparation obtained from glutaraldehyde-paraformaldehyde-fixed human pancreas. B-cell granules are stained with monoclonal mouse primary antibodies to insulin and
a 16-nm gold-adsorbed secondary antibody, and amplified with the silver acetate developer (Hacker
et al., 1988) without gum arabic for 6 min in dark. Bars = 100 nm.

Quality and prices of immunogold reagents available on the market differ greatly, and
for that reason, it is advisable to test different gold-reagents. At present, we recommend
5 nm EM-grade colloidal gold probes which can be used for LM and EM applications
obtained from BioCell (Cardiff, UK), Amersham (Amersham, UK), Aurion (Wageningen,
NL), or Nanogold (Stony Brook, NY). They all yield high labeling densities and show good
penetration properties. In our laboratories we commonly use mixtures of several optimally
diluted immunogold-batches with different gold sizes (LM) or only one gold size (EM)
from different suppliers, each mixed in equal amounts. For LM, indirect IGSS methods

appear to be preferable to direct, protein A-gold or streptavidin-gold-silver methods. The former process can be completed in a shorter time and has the highest detection efficiency. Guidelines for LM and EM working schemes currently used in our laboratories are given in *Protocols 1* and *2*.

2.5.1   Technical Hints for Immunogold-Silver Staining

Various tissue fixatives can be used. Light microscopical IGSS as outlined in *Protocol 1* works well with 4% neutral phosphate-buffered formaldehyde-paraffin sections, Stefanini's/ Zamboni's-fixed paraffin sections (Stefanini et al., 1967), or Bouin's-fixed paraffin sections. Cryostat sections for IGSS should be dried onto the adhesive-coated glass slides for more than 1 h and then subjected to dehydration and rehydration processes by going through increasing concentrations of alcohols to xylene and then back to water. Sometimes, we have postfixed frozen sections from tissues which were not, or only mildly, fixed before freezing, after the dehydration and rehydration treatment using formalin or Stefanini's solution. This step, which may also be carried out before the dehydration and rehydration sequence, may further improve staining and reduce background reactions.

It is also possible to use semithin epoxy sections for IGSS. They should be pretreated with sodium ethoxide or methoxide to soften the resin. Pre-embedding IGSS can be carried out using small gold particles on Stefanini's-fixed cryostat sections which are further processed for EM examination (Lackie et al., 1984; Burry et al., 1992; Zehbe et al., 1993; Kummer et al., 1994). For postembedding ultrastructural studies, it is advisable to use a mild fixation with a buffered mixture of glutaraldehyde and paraformaldehyde, or Stefanini's/Zamboni's solution (Stefanini et al., 1967). Osmium tetroxide fixation should be avoided before autometallography, but can be performed after silver amplification together with uranyl *en bloc* staining (Danscher, 1981a–c). The embedding process includ-ing the use of LR-White, Lowicryl, Epon, or Araldite should include low temperature polymerization (for some antigens not higher than 40°C).

Most IGSS protocols involve oxidation of the sections with Lugol's iodine followed by sodium thiosulfate reduction. In most cases this treatment increases the staining efficiency. If iodine treatment is excluded, the autometallographic enhancement process in most cases needs to be prolonged. For thin sections, Lugol's iodine may be recommended, but is not necessary. As for the washing buffers, most protocols in literature use Tris-buffered saline (TBS) or phosphate-buffered saline (PBS) with a pH of ~7.2. High salt concentrations and the addition of Triton X-100 or Tween 80 to the buffer used before applying the primary antibody may improve the staining. The increased NaCl concentration seems to reduce background staining. The buffer system used before immunogold incubation may be adjusted to pH 8.2 to stabilize the gold reagent (Springall et al., 1984). In order to make the staining protocol easier to handle, it is also feasible to use a buffer with a pH of ~7.6 for all buffer washes. To avoid unspecific reactions, addition of 0.1% gelatin (e.g., cold water fish gelatin) is very effective (Hacker, 1989; Hacker et al., 1994, 1995).

Polyclonal rabbit or guinea pig antisera and monoclonal mouse or rat antibodies have been applied to detect various substances at the LM and EM levels. A primary antibody incubation for 60 to 90 min at room temperature is convenient and efficient, and this time span can be drastically reduced by microwave irradiation (Boon and Kok, 1988). Prolonged incubation and/or double application of the same antibody may further increase the detection efficiency and may allow higher antibody dilutions (Brandtzaeg, 1981; Hacker et al., 1985b). Optimal dilutions of antibodies should be evaluated individually. Dilutions of immunogold reagents should also be optimized by titration and are usually between 1/25 and 1/200. Tris-buffered saline (pH 7.6 to 8.2) is used as diluent for the gold reagent and should contain 0.8% bovine serum albumin (BSA) and 0.1% gelatin. This process helps to prevent aggregation of gold particles and results in less background staining. Postfixation in 2% glutaraldehyde, following the washing steps used to remove unbound immunogold reagent, prevents release of gold reagent from its binding sites in the low pH environment of autometallographic developer. Glutaraldehyde should be diluted in PBS (pH 7.2) which can be kept for at least 2 weeks at 4°C and can be reused several times.

Before autometallographic gold-silver amplification is carried out, semithin and thin sections are washed carefully in glass double-distilled water. For optimal staining results, the purity of water is crucial. Only glassware and plastic or teflonized forceps should be used. If necessary, they should also be cleaned for 30 min in a 10% Farmer's solution (one part sodium thiosulphate and nine parts 10% potassium ferricyanide) (Danscher et al., 1993).

Semithin sections are developed vertically in a glass container, e.g., in a slide container according to Schiefferdecker containing 80 ml of autometallographic developer. If a silver lactate developer is applied, the container should be covered with cardboard box (Danscher, 1981a–c; Hacker et al., 1988). Both autometallographic developers permit monitoring of the staining intensity visually. When the silver lactate developer is used, sections rinsed in double-distilled water can be checked under the microscope, and then further developed if necessary. If one uses silver acetate autometallography, rinsing in distilled water before microscopical observation is unnecessary (Hacker et al., 1988). For silver amplification of immunogold-labeled thin sections on EM grids, both silver lactate and silver acetate autometallographic developers should be covered with a dark box. In this case, development is carried out by floating the grids face-down on top of drops of freshly prepared developer, placed on Parafilm or dental wax. The grids must not be dipped completely in the autometallographic solution, otherwise "crunchy preparations" may result. Development of EM sections should be carried out in dark for 3 to 15 min at room temperature, depending on the gold-size and autometallographic developer used (Hacker et al., 1995).

2.6 Use of Different Silver Salts for AMG Amplification of IGSS

The detection efficiency of IGSS is closely related to the type of autometallographic enhancement used. A number of autometallographic solutions have been developed since Danscher's (1981a–c) first description of silver lactate autometallography. The key points associated with autometallography include prevention of nonspecific silver precipitation, control of reproducibility and homogeneity, and improving efficiency (meaning a high ratio of amplified particles to unstained background) (Stierhof et al., 1992).

2.6.1 Silver Nitrate

Originally, Liesegang (1911, 1928) and Timm (1962) used silver nitrate ($AgNO_3$) in their approach to autometallography. Silver nitrate has a molecular weight of 169.89 D and is composed of colorless, odorless, transparent large crystals or white small crystals (Windholz, 1983). The silver content in silver nitrate is 63.50%. It is a poisonous substance and not photosensitive when pure. Presence of trace amounts of organic material promotes photoreduction. Its solubility in water is very high: one gram of silver nitrate dissolves in 0.4 ml water or 0.1 ml boiling water. Aqueous silver nitrate solutions are neutral to litmus (~pH 6). Silver nitrate is caustic and irritating to human skin and mucous membranes. Swallowing can cause severe gastroenteritis and may be fatal.

In autometallography, silver nitrate was replaced by silver lactate to ensure greater specificity (Danscher, 1981a). Due to earlier political problems in countries of East Europe resulting in the nonavailability of other silver salts, some authors have also recently used a silver nitrate autometallographic developer applied for IGSS by Krenács et al. (1989a). The Krenács' developer consisted of 6 mM silver nitrate and 90 mM hydroquinone in 100 ml of 0.1 M citrate buffer (pH 3.5).

2.6.2 Silver Lactate

Silver lactate ($C_3H_5AgO_3$), the 2-hydroxypropanoic acid silver salt, has a molecular weight of 214.97 D, and is available as monohydrate with a silver content of 54.78%. It is a white or slightly grayish crystalline powder readily affected by light, and can be dissolved in 15 parts of water (Windholz, 1983). Danscher has replaced silver nitrate used in older developers (Liesegang, 1911, 1927; Danscher, 1981a–c) with silver lactate because of a better specificity of the reaction. His autometallographic solutions contain 5.6 mM silver lactate, 77 mM hydroquinone, and 19.8% gum arabic, reacted in low pH citrate buffer (pH 3.8) (Figure 4a). Stierhof et al. (1992) also tested replacement of the acidic pH by a near neutral 0.2 M HEPES buffer (pH 6.8). A slight modification of Danscher's silver lactate developer has been described by Springall et al. (1984) who omitted the use of gum arabic for light microscopical IGSS in order to speed up the process.

2.6.3 Silver Acetate

Danscher's silver lactate developer proved to be both extremely efficient and sensitive for autometallographic detection of catalytic metals and for IGSS. However, for the IGSS method, the development is not easily controllable because of the higher light sensitivity of silver lactate. Silver acetate is less light sensitive than silver lactate (Figure 5), and its use in autometallographic lectin histochemistry was introduced by Skutelsky et al. (1987) and by Hacker et al. (1988) for immunocytochemical use. Silver acetate ($C_2H_3AgO_2$) has a molecular weight of 166.92 D. It is available as slightly grayish, lustrous needles or crystalline powder with a silver content of 64.63%. Silver acetate can be dissolved in 100 parts of cold water or 35 parts of boiling water and is therefore less water soluble than silver lactate (Windholz, 1983). Because of the improved solubility in water, we recommend using crystalline silver acetate powder (Fluka, code no. 85140; Buchs, CH) in a concentration of ~6 mM together with 22 mM hydroquinone for autometallographic purposes. The silver acetate amount needed for autometallography can be dissolved in distilled water at room temperature under continuous stirring within a few minutes. For ultrastructural studies, gum arabic should be added as a protective colloid, giving a more controlled, even, and specific amplification of gold particles (Hacker et al., 1988; Stierhof et al., 1992).

2.6.4 Silver Bromide

Silver bromide (AgBr) has a molecular weight of 187.80 D and is a yellowish, odorless powder which darkens on exposure to light. A maximum of 0.135 mg per liter of water can be dissolved at 25°C (Windholz, 1983). It is used in photography, and in histology it is widely employed for autoradiographic emulsions. Danscher (1983) used autoradiographic emulsions containing silver bromide for the first time for autometallography. We were the first to use this setup for IGSS (Hacker and Grimelius, experiments carried out in 1985, unpublished). Our results did not show any advantage to the emulsion procedure for IGSS. In fact, it turned out to be too complicated and costly for routine use.

2.6.5 Comparison of Silver Lactate, Silver Nitrate, and Silver Acetate

We have thoroughly compared silver lactate, silver acetate, and silver nitrate developers with equivalent molarities or comparable silver ion concentrations under standardized molarity conditions in acidic citrate buffer (pH 3.5 to 3.8) without including gum arabic colloid. Because silver lactate, silver acetate and silver nitrate show relatively good comparable silver ion concentrations, a comparison of equal molarities appears to be justified.

Figure 5 A comparison of the light sensitivities of silver lactate and silver acetate autometallography developers as originally described by Danscher (1981a) (right) and Hacker et al. (1988) (left) without the addition of protective colloid. Both solutions have been mixed freshly and at the same time under comparable conditions and were exposed to strong light of a large neon light tube located 1 m above the unprotected solutions for 15 min at room temperature. The silver lactate developer (right) changed its initial transparency into a black and nontransparent solution within a few minutes, whereas the silver acetate developer (left) stayed clear and transparent even after 15 min of heavy light exposure. This behavior is only reproducible if the glassware is cleaned with Farmer's solution and then thoroughly rinsed.

For these tests, we used sections from formalin-fixed and paraffin-embedded tissues, including rat and human stomach, and rat heart, small intestine, skin, and pancreas. Primary antibodies (mouse monoclonal and rabbit polyclonal) against substances contained in these tissues (gastrin in antrum, cytokeratins and serotonin in small intestine, atrial natriuretic peptide in heart, and insulin and glucagon in pancreas) were utilized in a standard IGSS procedure as given in *Protocol 1*, including Lugol's iodine pretreatment. Negative controls included replacement of the primary antibody with antibody diluent, and also additional omission of the second (gold antibody) layer.

Our experiments (unpublished results) showed that the processes described in the literature for both silver lactate (Danscher, 1981a–c) and silver acetate autometallography (Hacker et al., 1988) both gave optimal silver amplification of gold label without unspecific staining, provided that the intensification process was performed for the desired duration and optimally diluted antibodies were applied (Figure 4). For these experiments, durations of 5 to 30 min of silver amplification at room temperature (RT) were used, and the developers were shielded from strong daylight by covering the slide containers with a black box. The silver salts were used in a molarity of 6 mM, and hydroquinone with 22 and 66 mM. The lower hydroquinone concentration of 22 mM slowed down the speed of autometallography and gave a more controllable staining. In contrast to silver lactate and silver acetate, in both hydroquinone molarities tested, silver nitrate often gave an unacceptably high background staining (Figure 6A), especially with prolonged amplification. With all silver salts in combination with low (22 mM) hydroquinone concentration, specific (silver acetate, silver lactate) or relatively specific (silver nitrate) immunostaining was obtained if the autometallographic process took less than 10 min.

Watching the mixed autometallographic solutions under bright illumination of a neon light tube in the laboratory, we observed that the 6 mM silver nitrate and lactate solutions with a hydroquinone molarity of 77 mM turned black within seconds to minutes. The setup with silver lactate showed a relatively transparent precipitate, whereas silver nitrate gave a darker and nontransparent precipitation. At the start of development, silver acetate was highly transparent, thereafter turned to a shade of gray, and then darkened to a black much later than the other tested developers (Figure 5). Darkening of the developers was much stronger and faster when less than thoroughly cleaned glassware was used. Treatment of the glass containers with Farmer's reducer and subsequent thorough rinsing with double-distilled water prolonged the precipitation process.

The staining controls that omitted the primary and secondary antibodies and that were carried out with shielding of the autometallographic solutions from light gave a very light and even background staining after 30 min of 6 mM silver acetate and silver lactate in the developers with 22 mM of hydroquinone. The same concentrations applied to silver nitrate gave unspecific staining in the connective tissue as early as 10 min after the start of the autometallographic process. Judging from the location of the "unspecific staining", it appeared that collagens were stained by the silver nitrate developer, which occurred as an argyrophilic-type reaction (Figure 5A–C). This reaction was not seen with the silver acetate developer, not even after 60 min of the autometallographic reaction. However, with the silver lactate developer, the argyrophilic-type reaction started after ~30 min of autometallography. This result is not a disadvantage of silver lactate, because 30 min is much longer than the duration of the normal IGSS protocol. It was also found that the appearance of the argyrophilic reaction is related to the Lugol's iodine treatment used in IGSS setups. In the staining controls omitting a primary rabbit antibody but keeping incubation with a 5 nm gold-adsorbed antirabbit IgG reagent, unspecific reactions on the glass slide surface but not in the sections were observed (Figure 7).

Under near-neutral conditions (Hepes buffer pH 6.8 or distilled water), nonspecificity of the staining became much more abundant. Unspecific staining of endocrine cell types sometimes became apparent. Neutral conditions accelerated the reaction and made it much less controllable and reproducible. Those experiments were therefore not performed further.

Stierhof et al. (1992) compared the originally published autometallographic protocols for silver lactate (Danscher, 1981a–c), neutral silver lactate (Lah et al., 1990), silver acetate (Hacker et al., 1988) and two commercially obtainable autometallography kits (IntenSE M from Amersham, UK, and R-Gent from Aurion, Wageningen NL). However, these setups may not be fully comparable, as the conditions were very different (different molarities of the silver salts, concentrations of hydroquinone, buffer systems). Nevertheless, the results of their comparisons are highly interesting and show comparable results to those obtained in our own studies. They found that the use of gum arabic gives superior results especially for EM as compared to those results obtained without a protective colloid, which was also seen in our earlier experiments (Danscher, 1981a; Hacker et al., 1988; Danscher et al., 1993). Silver lactate and silver acetate containing enhancers produced a considerably higher number of amplified particles than the commercial ones (Stierhof et al., 1992). Tacha et al. (1993) described that the staining intensity given by commercially available developers may be increased by also applying gold chloride toning. In EM work, all developers tested by Stierhof et al. (1992) were well suited for nickel grids, but not for copper grids, which strongly influenced the amplification process. These authors also reported that the components of the silver lactate and the silver acetate autometallography solutions may be stored in aliquots at –20°C.

2.7   Use of Autometallography for DNA and RNA Detection

For many years, *in situ* hybridization (ISH) has been the basis for molecular biological analysis of nucleic acid sequences at the LM and the EM levels. DNA or RNA hybridization probes are often used to detect viral genomes in infected cells, to investigate biosynthesis

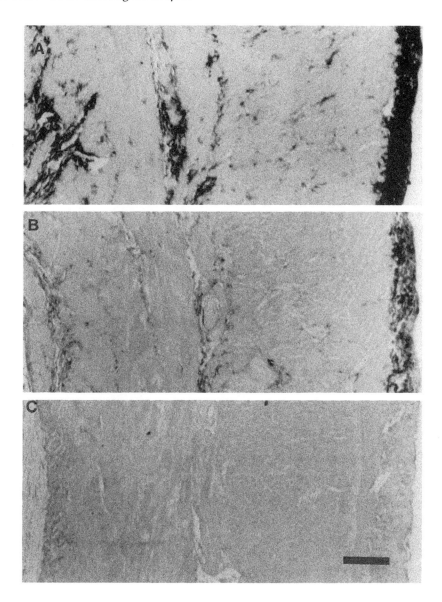

Figure 6 A comparison of silver nitrate (A), silver lactate (B), and silver acetate (C) developers and their effects on IGSS-treated formaldehyde-fixed semiconsecutive 5 μm thick paraffin sections of rat stomach. Indirect IGSS as in *Protocol 1*, but omitting the primary and secondary antibodies, has been employed. Both primary and secondary antibodies were replaced by antibody diluent alone, which were incubated for the desired time (overnight or 60 min, respectively). All three photo micrographs were taken after 1 h of silver amplification with comparable molarities of silver salt (6 mM) and hydroquinone (22 mM). Judging from the location of the "unspecific" background staining obtained, it appears that collagens are strongly stained with the silver nitrate developer (A), lightly stained with silver lactate autometallography (B), and not stained with silver acetate autometallography (C). This staining effect is related to the use of Lugol's iodine treatment used in IGSS. Bars = 50 μm.

of peptides and/or proteins, or to study genetic disease. For nonisotopic labeling, biotin or digoxigenin have been used with success (Varndell et al., 1984; Liesi et al., 1986; Breitschopf et al., 1992; Hacker et al., 1993). These labels avoid the hazardous use of radioactivity and therefore satisfy the requirements of most pathological laboratories, since the probes can be stored and handled much more easily. Nonradioactive labels are cheaper, easier to

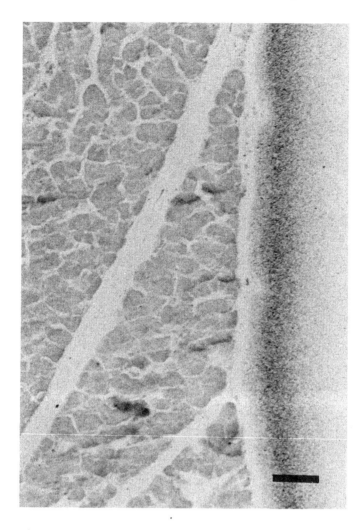

Figure 7 Unspecific staining of the glass surface adjacent to the tissue section was obtained with either PLL- or APES-coated glass slides, in the controls where the primary antibody incubation was replaced by antibody diluent but where the gold-labeled secondary antibody was incubated as usual. This effect was only seen with goat-antirabbit immunogold reagents. In this case, GAR-G5 from Amersham (UK) was used in a dilution of 1:50 in Tris-buffered saline (pH 7.6) containing 0.1% gelatin for 1 h and incubation at room temperature. Bar = 50 μm.

handle and give a higher resolution than radioactive probes. Biotin, digoxigenin, or other labels such as FITC used for nucleic acid detection can be easily visualized by using direct or indirect IGSS methods (Figures 8–10) (Hacker et al., 1993; Zehbe et al., 1992–1994).

Optimized protocols for *in situ* DNA hybridization with biotin-labeled probes and IGSS techniques have been described (e.g., Hacker et al., 1993, 1994, 1995). *In situ* hybridization methods using IGSS detection are very specific and extremely sensitive. Most recently, applications of autometallography for PISH (Figures 9 and 10), or *in situ* self-sustained sequence replication based amplification (*in situ* 3SR) have been or are being presented (Zehbe et al., 1982–1984, and submitted). These methods allow, for the first time, detection of single copies of DNA or mRNA at the cellular level with autometallography. Using these techniques in pre-embedding methods on formalin- or paraformaldehyde-fixed cells followed by a "pop-off" procedure, then osmium tetroxide postfixation and embedding in

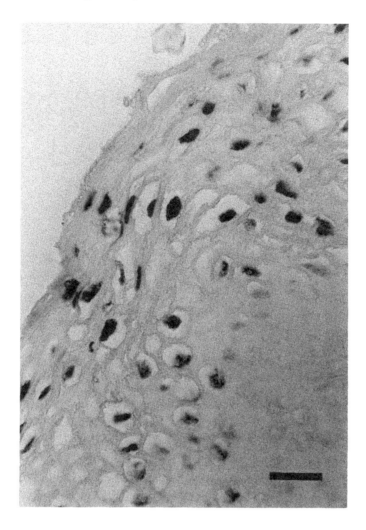

Figure 8 *In situ* hybridization with direct IGSS detection. A formalin-fixed 5 µm thick paraffin section of human condylomata acuminata was hybridized with a biotinylated DNA-probe binding to human papillomavirus (HPV) 6/11 under stringent conditions as described in *Protocol 3*. For the detection of the hybridization sites, 1 nm gold-adsorbed antibiotin antibodies (Amersham, UK) were used in combination with silver acetate autometallography. Bar = 25 µm.

resin, investigation of autometallography-achieved DNA staining in the electron microscope becomes possible (Zehbe et al., 1993). Protocols for *in situ* PCR and *in situ* 3SR have recently been published and are still undergoing further improvements and optimization (Zehbe et al., 1994, and unpublished).

2.8 Perspectives

Protocols for the use of autometallography in the detection of catalytic tissue metals have been extensively described in literature. Autometallography and its application to immunocytochemistry and other IGSS-related methods have many advantages over conventional methods, which is also true for ultrastructural studies. By using small gold particles (2 to 5 nm), high labeling densities and good penetration properties are obtained. This is particularly advantageous for preparations where only scanty immunoreactive structures

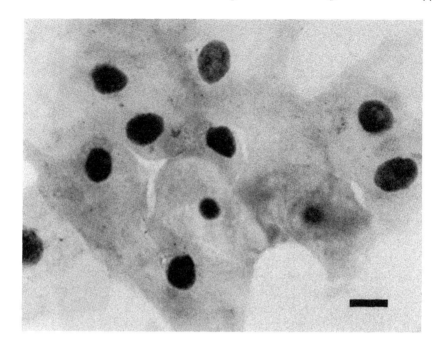

Figure 9   *In situ* polymerase chain reaction (*in situ* PCR) with a combination of IGSS detection using digoxigenin as the marker molecule. Protocol as described in *Anticancer Res.* 12: 2165 (1992). The reaction which is sensitive enough to detect single copies of DNA sequences has been applied to a formalin-postfixed imprint preparation of cervical dysplasia, detecting human papillomavirus (HPV) type 16. Strong nuclear staining of infected cells was obtained, negative controls were upheld; counterstaining with nuclear fast red. *In situ* PCR was performed by Ingeborg Zehbe (Uppsala, Sweden) and one of us (G. W. Hacker). Bar = 10 μm.

are present. Many commercial sources appear unable to produce high quality immunogold reagents. In our tests, some immunogold reagents available on the market gave neither a high enough labeling density nor an acceptable low level background, both of which are required for such a sensitive technique. Limitations include poor penetration of some reagents into the nucleus when demonstrating intranuclear antigens, probably due to problems of charging. "Crunchy" looking background stainings are often due to the nonoptimization of each single step in the protocol. Glassware which has not been properly cleaned, poor quality water, and low quality and relatively costly, insensitive, and unspecific commercial "silver-enhancement kits" present further problems. Such problems can be easily solved, however. Today, IGSS and autometallographic procedures can be used in a variety of applications. Its future is very bright particularly in light of the recently developed *in situ* molecular biological techniques.

2.9   Summary

Autometallography, i.e., the detection of catalytic crystal lattices of metallic gold and silver, and sulfides or selenides of mercury, silver, and zinc, has its roots in "physical development" transplanted from photography to histology by Liesegang at the beginning of this century. In 1981, a series of papers were published by G. Danscher with the purpose of introducing a reliable and easy-to-handle technique for light microscopical and ultrastructural studies. Autometallography has a multitude of applications apart from its use to detect tissue metals. These uses include the highly sensitive and efficient

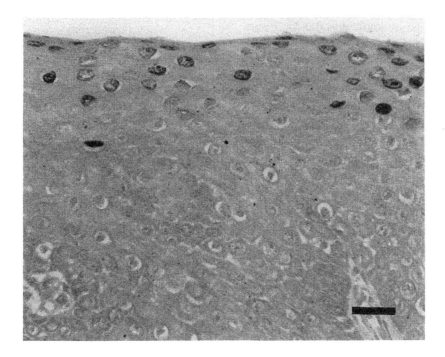

Figure 10 *In situ* PCR as in Figure 9, applied on a formalin-fixed 7 µm thick tissue section. Strong nuclear staining for HPV is obtained using consensus HPV primers and digoxigenin-IGSS-silver acetate autometallographic detection. This case of condylomata acuminata showed no staining with *in situ* hybridization alone, but was strikingly positive with *in situ* PCR, which was performed by Ingeborg Zehbe (Uppsala, Sweden) and one of us (G. W. Hacker). Bar = 50 µm.

in situ colloidal gold tracing of peptides, proteins and amines by immunocytochemistry, carbohydrates by lectin histochemistry, and nucleic acids by *in situ* hybridization and *in situ* polymerase chain reaction techniques. We have briefly reviewed the silver staining techniques used in histology. Different silver salts widely applied for IGSS methods are compared under acidic pH conditions. We also suggest latest protocols for light- and electron-microscopical applications of autometallography for immunocytochemistry and nucleic acid detection.

2.10 Appendix: Protocols

2.10.1 Protocol 1: Indirect Immunogold-Silver Staining Method

2.10.1.1 Immunocytochemistry

1. Mount sections on poly-L-lysine (PLL)- or aminopropyl-triethoxisilane (APES)-coated glass slides (Maddox and Jenkins, 1987) and dry them for 1 h at 60°C. Deparaffinize the sections in xylene and take them to water through a series of graded alcohols. Treat semithin sections with saturated sodium ethoxide (~20 min) and wash in ethanol 3 times for 2 min each.
2. Wash in distilled water for 3 min.
3. Immerse for 5 min in Lugol's iodine (1% iodine in 2% potassium iodide; ready mixed from Merck no. 9261, Darmstadt, Germany).
4. Rinse briefly in distilled water.
5. Treat with 2.5% aqueous sodium thiosulfate until sections become colorless (up to 30 sec).

6. Wash in distilled water for 2 min.
7. Immerse in TBS-gelatin (Tris-buffered saline, pH 7.6, containing 0.1% cold water fish gelatin) for 10 min. In some cases superior results are obtained if the buffer in this step also contains 0.1% Triton X-100 or Tween-80, and 2.5% NaCl.
8. Apply normal serum of the species providing the secondary antibody (1/10 in TBS-gelatin) for 5 min and drain off.
9. Incubate with primary antibody for 90 min at room temperature or overnight at 4°C. The dilution should be tested carefully. The suggested antibody diluent is 0.1 M phosphate- or Tris-buffered saline (TBS or PBS, pH 7.2 to 7.6) containing 0.1% bovine serum albumin and 0.1% sodium azide.
10. Wash in TBS-gelatin 3 times for 3 min each.
11. Apply normal serum 1/10 as in Step 8.
12. Incubate with gold-adsorbed second layer antibodies for 1 h at room temperature. Optimum dilution is usually between 1/25 and 1/200 and should be determined by titration.
13. Wash in TBS-gelatin 3 times for 3 min each.
14. Postfix in 2% glutaraldehyde in PBS (pH 7.2) for 2 min.
15. Rinse 5 times in distilled water for 30 sec each, followed by three washes for 3 min each in double glass-distilled water.
16. Perform silver acetate autometallography.

2.10.1.2 Silver Acetate Autometallography

1. Solutions A and B should be freshly prepared for each experiment. Solution A: Dissolve 80 mg silver acetate (code 85140, Fluka, Switzerland) in 40 ml of glass double-distilled water. Silver acetate crystals can be dissolved within ~15 min by continuous stirring.
2. Citrate buffer: Dissolve 23.5 g of trisodium citrate dihydrate and 25.5 g citric acid monohydrate in 850 ml of deionized or distilled water. This buffer can be kept at 4°C for at least 2 to 3 weeks. Before using, adjust the pH to 3.8 with citric acid solution.
3. Solution B: Dissolve 200 mg of hydroquinone in 40 ml of citrate buffer.
4. Immediately before use, mix solutions A and B.
5. Silver amplification: Place the slides vertically in a glass container (preferably with ≈80 ml volume, up to 19 slides) and cover them by the above mixture. Staining intensity may be checked in the light microscope during the amplification process.
6. Photographic fixer (e.g., Agefix, Agfa Gevaert, FRG, diluted 1:20) may be used to stop the enhancement process immediately. (This solution can be re-used many times). Leave the slides in this solution for ~1 min. Alternatively, a 2.5% aqueous solution of sodium thiosulfate may be used.
7. Rinse the slides carefully in tap water for at least 3 min. After silver amplification, sections can be counterstained with hematoxylin and eosin or nuclear fast red, dehydrated, and mounted in DPX (BDH Chemicals, UK).

2.10.2 Protocol 2: Indirect Immunogold-Silver Staining for Thin Sections

1. Mount thin sections on nickel grids (300–400 mesh) and dry them for 1 h at room temperature.
2. In case of Epon sections, etching with H_2O_2 and/or ethanolic KOH is recommended. Then, wash in distilled water twice for 3 min each.
3. Immerse in TBS or PBS (pH 7.6) containing 0.1% cold water fish gelatin.

4. Apply normal serum of the species providing the secondary antibody (1:20 in TBS-gelatin plus 1% BSA) for 20 min and drain off.
5. Incubate overnight at 4°C with primary antibody using a microtiterplate. The dilution should be tested carefully. The suggested antibody diluent is 0.1 M TBS or PBS containing 1% BSA.
6. Wash 3 times for 3 min each in TBS or PBS containing 1% BSA and 0.1% gelatin.
7. Incubate overnight at 4°C with gold-adsorbed second antibodies. Optimum dilution is usually between 1/20 and 1/100 and should be determined by titration.
8. Wash in PBS or TBS containing 1% BSA twice for 3 min each.
9. Wash in PBS twice for 3 min each.
10. Postfix in 2% glutaraldehyde in PBS (pH 7.2) for 2 min.
11. Rinse 3 times for 30 sec each in double glass-distilled water, followed by three washes for 3 min each in the same.
12. Perform silver acetate autometallography (see *Protocol 1*). On top of dental wax or parafilm, place drops of autometallographic developer and develop for 3 to 10 min. Cover this with a darkbox to protect from harmful daylight. Use plastic or teflonized forceps and avoid any impurity.
13. Rinse in distilled water twice for 3 min each.
14. Dry thin sections at room temperature for 15 min.
15. Contrast the sections as usual with lead citrate and uranyl acetate.
16. Examine in the electron microscope.

2.10.3 Protocol 3: In Situ DNA Hybridization

Note: For *in situ* hybridization, a buffer system called "standard sodium citrate buffer" (SSC) is used. In the staining protocol, dilutions of a concentrated SSC stock solution are used. Preparation of 20x concentrated SSC stock solution (in literature also called 20xSSC): 175.32 g NaCl, 88.23 g Na-citrate in 1 liter H_2O; adjust to pH 7.0 with HCl or citric acid; premixed concentrate is available from Sigma (no. S-6639). To obtain, for example, 2xSSC, the 20xSSC stock solution is diluted 10 times with double distilled water.

1. Paraffin sections (7 to 12 μm thick) are used and deparaffinized in xylene and taken to water via a descending series of ethanols. In case of frozen tissue, cryostat sections are cut. All sections and also cytologic preparations (e.g., cytospins) are mounted on aminopropyl-triethoxysilane-coated glass slides (Maddox and Jenkins, 1987). Cryostat sections and cytological material may be dehydrated in graded ascending alcohols placed into xylene for 30 min, and then rehydrated in graded descending alcohols, in order to improve staining and to overcome penetration problems.
2. Postfix cryostat sections or cytological material in neutral 5% phosphate-buffered formaldehyde or 2% buffered paraformaldehyde for 10 min. Wash sections 3 times for 2 min each in 20 mM PBS (pH 7.2). This first postfixation is not necessary for paraffin sections.
3. Soak in 0.3% Triton X-100 for 15 min to permeabilize sections.
4. Proteolytic treatment: Briefly digest the tissue sections with 0.1% proteinase K in PBS for ~15 min (time and concentration depend on the strength of prefixation of the tissue and the size of the section) at 37°C using a water bath.
5. Wash 3 times for 2 min each in PBS.
6. Wash in distilled water for 2 min, immerse in 50, 70, and 98% isopropanol for 1 min each, and air dry at room temperature.
7. *Optional.* Prehybridization: Incubate with 50% deionized formamide and 10% dext-

ran sulfate in 2xSSC (standard sodium citrate buffer), at 50°C for 5 min. Drain off excess. Special care must be taken that sections never dry in and after the hybridization step.

8. Place a small drop of probe mix (~20 µl of ready to use-probes, or 20 ng of nick-translated probe) onto the section, and then cover it with a 22 × 22 mm coverslip. In this setup, biotinylated or digoxigenin-labeled cDNA probes have been successfully used.

9. Place the slides on a 92°C heating block and incubate for 5 to 10 min.

10. Move the slides into a 37°C oven and incubate for further 2 h.

11. Remove coverslips by soaking with 4xSSC.

12. Wash under stringent conditions in 4xSSC, 2xSSC, 0.1xSSC, 0.05xSSC, and distilled water (each >5 min) at room temperature or preferably at 37°C.

13. Immerse in Lugol's iodine (1% iodine in 2% potassium iodide) for 5 min and briefly rinse in distilled water.

14. Treat with 2.5% aqueous sodium thiosulfate until sections become colorless, and wash in distilled water 3 times for 2 min each.

15. Immerse in TBS-gelatin (TBS [pH 7.6] containing 0.1% cold water fish gelatin) twice for 3 min each.

16. Incubate with gold-adsorbed antibiotin (Amersham, UK) or antidigoxigenin (Boehringer Mannheim, FRG, or Aurion, NL) antibodies for at least 1 h at room temperature or overnight at 4°C. Optimum dilution is between 1/25 and 1/50. Antibody diluent is TBS-gelatin containing 0.8% BSA.

17. Wash in TBS-gelatin 3 times for 3 min each.

18. Postfix in 2% glutaraldehyde in PBS (pH 7.2) for 2 min.

19. Apply silver acetate autometallography as in *Protocol 1*. Counterstain and mount.

Acknowledgments

For the joint collaboration in IGSS related projects, we sincerely thank J.M. Polak and D. Springall (London, UK); P. Lackie (Southampton, UK); I. Zehbe (Uppsala, Sweden); N.E. Carson, M. Forte, R. Rufner, and C. Xenachis (Browns Mills, NJ); S. Dechet and A.-H. Graf (Salzburg, Austria). Comparisons of different silver salts were gratefully accomplished during a visiting professorship of Dr. G.W. Hacker at the Deborah Research Institute, Browns Mills, NJ, and also sponsored by the Swedish Medical Research Council (Stockholm; project no. 00102).

References

Aaseth, L., Olsen, A., Halse, J., and Hovig, T. 1981. Argyria-tissue deposition of silver as selenide. *Scand. J. Clin. Lab. Invest.* 41: 247.

Bastholm, L., Scopsi, L., and Nielsen, M. H. 1986. Silver-enhanced immunogold staining of semithin and ultrathin cryosections. *J. Electr. Microsc. Tech.* 4: 175.

Bodian, A. 1936. A new method for staining nerve fibers and nerve endings in mounted paraffin sections. *Anat. Rec. (Basel)* 65: 89.

Böck, P. (Ed.) 1989. *Romeis Mikroskopische Technik,* 17th ed. Urban and Schwarzenberg Verlag, Munich.

Boon, M. E. and Kok, L. P. 1988. *Microwave Cookbook of Pathology. The Art of Microscopic Visualization,* 2nd ed. p. 161. Coulomb Press, Leyden, The Netherlands.

Brandtzaeg, P. 1981. Prolonged incubation staining of immunoglobulins and epithelial components in ethanol- and formaldehyde-fixed paraffin-embedded tissues. *J. Histochem. Cytochem.* 29: 1302.

Breitschopf, H., Suchanek, G., Gould, R. M., Colman, D. R., and Lassmann, H. 1992. *In situ* hybridization with digoxigenin-labeled probes: sensitive and reliable detection method applied to myelinating rat brain. *Acta Neuropathol.* 84: 581.

Burry, R. W., Vandre, D. D., and Hayes, D. M. 1992. Silver enhancement of gold antibody probes in pre-embedding electron microscopic immunocytochemistry. *J. Histochem. Cytochem.* 40: 1849.

Busch, C. and Vasko, J. 1988. Differential staining of mitoses in tissue sections and cultured cells by a modified methenamine-silver method. *Lab. Invest.* 59: 876.

Danscher, G. 1981a. Histochemical demonstration of heavy metals. A revised version of the sulphide silver method suitable for both light and electron microscopy. *Histochemistry* 71: 1.

Danscher, G. 1981b. Localization of gold in biological tissue. A photochemical method for light and electronmicroscopy. *Histochemistry* 71: 81.

Danscher, G. 1981c. Light and electron microscopic localisation of silver in biological tissue. *Histochemistry* 71: 177.

Danscher, G. 1982. Exogenous selenium in the brain. A histochemical technique for light and electron microscopical localization of catalytic selenium bonds. *Histochemistry* 76: 281.

Danscher, G. 1983. A method for counterstaining plastic embedded tissue. *Stain Technol.* 58: 365.

Danscher, G. and Norgaard, J. O. 1983. Light microscopic visualization of colloidal gold on resin-embedded tissue. *J. Histochem. Cytochem.* 31: 394.

Danscher, G. 1984. Autometallography. A new technique for light and electron microscopical visualization of metals in biological tissue (gold, silver, metal sulphides and metal selenides). *Histochemistry* 81: 331.

Danscher, G. and Møller-Madsen, B. 1985a. Silver amplification of mercury sulfide and selenide: a histochemical method for light and electron microscopic localization of mercury in tissue. *J. Histochem. Cytochem.* 33: 219.

Danscher, G. and Norgaard, J. O. 1985b. Ultrastructural autometallography: a method for silver amplification of catalytic metals. *J. Histochem. Cytochem.* 33: 706.

Danscher, G. and Rungby, J. 1986. Differentiation of histochemically visualized mercury and silver. *Histochem. J.* 18: 109.

Danscher, G. 1988. Can aluminium be visualized in CNS by silver amplification? *Acta Neuropathol.* 76: 107.

Danscher, G. 1991. Applications of autometallography to heavy metal toxicology. *Pharmacol. Toxicol.* 68: 414.

Danscher, G., Hacker, G. W., Norgaard, J. O., and Grimelius, L. 1993. Autometallographic silver amplification of colloidal gold. *J. Histotechnol.* 16: 201.

Danscher, G. and Montagnese, C. 1994. Autometallographic localization of synaptic vesicular zinc and lysosomal gold, silver and mercury. *J. Histotechnol.* (in press).

Faulk, W. P. and Taylor, G. M. 1971. An immunocolloid method for the electron microscope. *Immunochemistry* 8: 1081.

Frederickson, C. J. and Danscher, G. 1990. Zinc-containing neurons in hippocampus and related CNS structures. *Prog. Brain Res.* 83: 71.

Fujimori, O. and Nakamura, M. 1985. Protein A gold-silver staining method for light microscopic immunohistochemistry. *Arch. Histol. Jpn.* 48: 449.

Geoghegan, W. D., Scillian, J. J., and Ackerman, G. A. 1978. The detection of human B-lymphocytes by both light and electron microscopy utilizing colloidal gold-labeled anti-immunoglobulin. *Immunol. Commun.* 7: 1.

Goodpasture, C. and Bloom, S. E. 1975. Visualization of nuclear organizer regions in mammalian chromosomes using silver staining. *Chromosoma* 53: 37.

Grimelius, L. 1968. The argyrophil reaction in islet cells of adult human pancreas studied with a new silver nitrate procedure. *Acta Soc. Med. Upsal.* 73: 271.

Grimelius, L. and Wilander, E. 1980. Silver stains in the study of endocrine cells of the gut and pancreas. *Invest. Cell Pathol.* 3: 3.

Grimelius, L., Su, H., and Hacker, G. W. 1994. The use of silver stains in the identification of neuroendocrine cell types. In: *Modern Methods in Analytical Morphology.* p. 1. chapter 1. Gu, J. and Hacker, G. W. (Eds.), Plenum Press, New York.

Von Gros-Schultz, O. 1918. Neues zur mikroskopischen Untersuchung des Zentralnervensystems. *Sitz.-Ber. Physikal. Med. Ges. Würzburg* 9: 1.

Gu, J., De Mey, J., Moeremans, M., and Polak, J. M. 1981. Sequential use of the PAP and immunogold-staining methods for the light microscopical double staining of tissue antigens. Its application to the study of regulatory peptides in the gut. *Reg. Peptides* 1: 465.

Gu, J., D'Andrea, M., Yu, C.-Z., Forte, M., and McGrath, L. B. 1993. Quantitative evaluation of indirect immunogold-silver electron microscopy. *J. Histotechnol.* 16: 19.

Gu, J. 1994. Microwave immunocytochemistry. In: *Modern Methods in Analytical Morphology.* p. 67. chapter 6. Gu, J. and Hacker, G. W. (Eds.), Plenum Press, New York.

Hacker, G. W., Springall, D. R., Van Noorden, S., Bishop, A. E., Grimelius, L., and Polak, J. M. 1985a. The immunogold-silver staining method. A powerful tool in histopathology. *Virchows. Arch. A* 406: 449.

Hacker, G. W., Polak, J. M., Springall, D., Ballesta, J., Cadieux, A., Gu, J., Trojanowski, J. Q., Dahl, D., and Marangos, P. J. 1985b. Antibodies to neurofilament protein and other brain proteins reveal the innervation of peripheral organs. *Histochemistry* 82: 581.

Hacker, G. W., Grimelius, L., Danscher, G., Bernatzky, G., Muss, W., Adam, H., and Thurner, J. 1988. Silver acetate autometallography: an alternative enhancement technique for immunogold-silver staining (IGSS) and silver amplification of gold, silver, mercury and zinc in tissues. *J. Histotechnol.* 11: 213.

Hacker, G. W. 1989. Silver-enhanced colloidal gold for light microscopy, In: *Colloidal Gold — Principles, Methods, and Applications.* Vol. 1. p. 297. Hayat, M. A. (Ed.), Academic Press, San Diego.

Hacker, G. W., Graf, A.-H., Hauser-Kronberger, C., Wirnsberger, G., Schiechl, A., Bernatzky, G., Wittauer, U., Su, H., Adam, H., Thurner, J., Danscher, G., and Grimelius, L. 1993. Application of silver acetate autometallography and gold-silver staining methods for *in situ* DNA hybridization. *Chinese Med. J.* 106: 83.

Hacker, G. W., Hauser-Kronberger, C., Graf, A.-H., Danscher, G., Gu, J., and Grimelius, L. 1994a. Immunogold-silver staining (IGSS) for detection of antigenic sites and DNA sequences. In: *Modern Methods in Analytical Morphology.* p. 19. chapter 3. Gu, J. and Hacker, G. W. (Eds.), Plenum Press, New York.

Hacker, G. W., Danscher, G., Gu, J., Hauser-Kronberger, C., Muss, W., Sonnleitner-Wittauer, U., Zehbe, I., Rufner, R., Carson, N. E., Xenachis, Chr., Forte, M., Juhl, S., Stoltenberg, M., and Dietze, O. 1995. Autometallography and its use in electron microscopy. In: *Beiträge zur Elektronenmikroskopischen Direktabbildung un Analyse von Oberflächen.* Vol. 26. Ehrenwerth, U. (Ed.), Verlag R.A. Remy, Münster.

Hainfeld, J. F. and Furuya, F. R. 1992. A 1.4-nm gold cluster covalently attached to antibodies improves immunolabeling. *J. Histochem. Cytochem.* 40: 177.

Hakanson, R., Böttcher, G., Ekblad, E., Panula, P., Simonsson, M., Dahlsten, M., Hallberg, T., and Sundler, F. 1986. Histamine in endocrine cells in the stomach. A survey of several species using a panel of histamine antibodies. *Histochemistry* 86: 5.

Hamperl, U. 1927. Über die 'gelben (chromaffinen)' Zellen im gesunden und kranken Magenschlauch. *Virchows Arch.* 226: 509.

Hayat, M. A. 1993a. *Stains and Cytochemical Methods,* pp. 318–336. Plenum, New York.

Hayat, M. A. (Ed.) 1993b. Immunogold-silver staining for light and electron microscopy. *J. Histotechnol.* 16: 197.

Hellerström, C. and Hellman, B. 1960. Some aspects of silver impregnation of the islets of Langerhans in the rat. *Acta Endocrinol. (Kbh.)* 35: 518.

Holgate, C. S., Jackson, P., Cowen, P. N., and Bird, C. C. 1983. Immunogold-silver staining: new method of immunostaining with enhanced sensitivity. *J. Histochem. Cytochem.* 31: 938.

Howell, W. M. and Black, D. A. 1980. Controlled silver-staining of nucleolus organizer regions with a protective colloidal developer: a 1-step method. *Experimentia* 36: 1014.

Huang, W.-M., Gibson, S. J., Facer, P., Gu, J., and Polak, J. M. 1983. Improved section adhesion for immunocytochemistry using high molecular weight polymers of L-lysine as a slide coating. *Histochemistry* 77: 275.

Krenács, T., Molnár, E., Dobó, E., and Dux, L. 1989a. Fibre typing using sarcoplasmic reticulum Ca^{2+}-ATPase and myoglobin immunohistochemistry in rat gastrocnemius muscle. *Histochem. J.* 21: 145.

Krenács, T., Láslik, Z., and Dobo, E. 1989b. Application of immunogold-silver staining and immunoenzymatic methods in multiple labeling of human pancreatic Langerhans islet cells. *Acta Histochem.* 85: 79.

Krenács, T., Krenács, L., Bozóky, B., and Iványi, B. 1990. Double and triple immunocytochemical labeling at the light microscopical level in histopathology. *Histochem. J.* 22: 530.

Krenács, T., Iványi, B., Bozóky, B., Lászik, Z., Krenács, L., Rázga, Z., and Ormos, J. 1991. Postembedding immunoelectron microscopy with immunogold-silver staining (IGSS) in Epon 812, Durcupan ACM and LR-White resin embedded tissues. *J. Histotechnol.* 14: 75.

Kummer, W., Hauser-Kronberger, C., and Muss, W. H. 1994. Pre-embedding immunohistochemistry in transmission electron microscopy. In: *Modern Methods in Analytical Morphology.* p. 187. chapter 12. Gu, J. and Hacker, G. W. (Eds), Plenum Press, New York.

Lah, J. J., Hayes, D. M., and Burry, R. W. 1990. A neutral pH silver development method for the visualization of 1-nanometer gold particles in pre-embedding electron microscopic immunocytochemistry. *J. Histochem. Cytochem.* 38: 503.

Lackie, P. M., Hennessy, R. J., Hacker, G. W., and Polak, J. M. 1985. Investigation of immunogold-silver staining by electron microscopy. *Histochemistry* 83: 545.

Liesegang, R. E. 1911. Die Kolloidchemie der histologischen Silberfärbungen. In: *Kolloidchemische Beihefte (Ergänzungshefte zur Kolloid-Zeitschrift) — Monographien zur reinen und angewandten Kolloidchemie.* p. 1. Oswald, W. (Ed.), Theodor Steinkopff Verlag, Dresden.

Liesegang, R. 1928. Histologische Versilberung. *Z. Wiss. Mikrosk.* 45: 273.

Liesi, P. Julien, J.-P., Vilja, P., Grosveld, F., and Rechardt, L. 1986. Specific detection of neuronal cell bodies: *in situ* hybridisation with a biotin-labeled neurofilament cDNA probe. *J. Histochem. Cytochem.* 34: 923.

Lindner, L. E. 1993. Improvements in the silver staining technique for nuclear organizer regions (AgNOR). *J. Histochem. Cytochem.* 41: 439.

Lundqvist, M., Arnberg, H., Candell, J., Malmgren, M., Wilander, E., Grimelius, L., and Öberg, K. 1990. Silver stains for identification of neuroendocrine cells. A study of the chemical background. *Histochem. J.* 22: 615.

King, T. P., Brydon, L., Gooday, G. W., and Chappell, L. H. 1987. Silver enhancement of lectin-gold and enzyme-gold cytochemical labeling of eggs of the nematode *Onchocerca gibsoni. Histochem. J.* 19: 281.

Maddox, P. H. and Jenkins, D. 1987. 3-Aminopropyltriethoxysilane (APES): a new advance in section adhesion. *J. Clin. Pathol.* 40: 1256.

Norgaard, J. O., Møller-Madsen, B., Hertel, N., and Danscher, G. 1989. Silver enhancement of tissue mercury: demonstration of mercury in autometallographic silver grains from rat kidneys. *J. Histochem. Cytochem.* 37: 1545.

Ploton, D., Menager, M., Jeannesson, P., Himber, G., Pigeon, F., and Adnet, J. J. 1986. Improvement in the staining and in the visualization of the argyrophilic proteins of the nucleolar organizer region at the optical level. *Histochemistry* 18: 5.

Polak, J. M., Pearse, A. G. E., Grimelius, L., Bloom, S. R., and Arimura, A. 1975. Growth-hormone release-inhibiting hormone in gastrointestinal and pancreatic D cells. *Lancet* i: 1220.

Portela-Gomes, G. M. and Grimelius, L. 1986. Identification and characterization of enterochromaffin cells with different staining techniques. *Acta Histochem.* 79: 161.

Rindi, G., Buffa, R., Sessa, F., Tortora, O., and Solcia, E. 1986. Chromogranin A, B and C immunoreactivities of mammalian endocrine cells. Distribution, distinction from costored hormones/prohormones and relationship with the argyrophil component of secretory granules. *Histochemistry* 85: 18.

Romeis, B. 1968. *Mikroskopische Technik.* Oldenbour Verlag, Wien.

Roth, J. 1982. Applications of immunocolloids in light microscopy. Preparation of protein A-silver and protein A-gold complexes and their application for the localization of single and multiple antigens in paraffin sections. *J. Histochem. Cytochem.* 30: 691.

Roth, J. 1983. The colloidal gold marker system for light and electron microscopic cytochemistry. In: *Immunocytochemistry.* Vol. 2. p. 217. Bullok, G. R. and Petrusz, P. (Eds.), Academic Press, London.

Roth, J., Saremaslani, P., and Zuber, C. 1992a. Versatility of anti-horseradish peroxidase antibody-gold complexes for cytochemistry and *in situ* hybridization: preparation and application of soluble complexes with streptavidin-peroxidase conjugates and biotinylated antibodies. *Histochemistry* 98: 229.

Roth, J., Saremaslani, P., Warhol, M. J., and Heitz, P. U. 1992b. Improved accuracy in diagnostic immunohistochemistry, lectin histochemistry and *in situ* hybridization using a gold-labeled horse-radish peroxidase antibody and silver intensification. *Lab. Invest.* 67: 263.

Scopsi, L. and Larsson, L.-I. 1985. Increased sensitivity in immunocytochemistry. Effects of double application of antibodies and of silver intensification on immunogold and peroxidase-antiperoxidase staining techniques. *Histochemistry* 82: 321.

Schmidt, J. and Peters, W. 1987. Localization of glycoconjugates at the tegument of the tapeworms Hymenolepis nana and H. microstoma with gold labeled lectins. *Parasitol. Res.* 73: 80.

Sevier, A. C. and Munger, B. L. 1965. A silver method for paraffin sections of neural tissue. *J. Neuropath. Exp. Neurol.* 24: 130.

Shimizu, H., Ishida-Yamamoto, A., and Eady, R. A. 1992. The use of silver-enhanced 1-nm gold probes for light and electron microscopic localization of intra- and extracellular antigens in skin. *J. Histochem. Cytochem.* 40: 883.

Shirabe, T. 1979. Identification of mercury in the brain of Minamata disease victims by electron microscopic X-ray microanalysis. *Neurotoxicology* 1: 349.

Singh, I. 1964. A modification of the Masson-Hamperl method for staining argentaffin cells. *Anat. Anz.* 115: 81.

Skutelsky, E., Goyal, V., and Alroy, J. 1987. The use of avidin-gold complex for light microscopic localization of lectin receptors. *Histochemistry* 86: 291.

Slater, M. 1991. Differential silver enhanced double labeling in immunoelectron microscopy. *Biotech. Histochem.* 66: 153.

Slomianka, L., Danscher, G., and Frederickson, C. J. 1990. Labeling of the neurons of origin of zinc-containing pathways by intraperitoneal injections of sodium selenite. *Neuroscience* 38: 843.

Springall, D. R., Hacker, G. W., Grimelius, L., and Polak, J. M. 1984. The potential of the immunogold-silver staining method for paraffin sections. *Histochemistry* 81: 603.

Stefanini, M., De Martino, C., and Zamboni, L. 1967. Fixation of ejaculated spermatozoa for electron microscopy. *Nature* 216: 173.

Stierhof, Y.-D., Humbel, B. M., Hermann R., Otten, M. T., and Schwarz, H. 1992. Direct visualization and silver enhancement of ultra-small antibody-bound gold particles on immunolabeled ultrathin resin sections. *Scanning Microsc.* 6: 1009.

Tacha, D. E., Bowman, Ph.D., and McKinney L. A. 1993. High resolution light microscopy and immunocytochemistry with glycol methacrylate embedded sections and immunogold-silver staining. *J. Histotechnol.* 16: 13.

Timm, F. 1962. Der histochemische Nachweis der Sublimatvergiftung. *Beitr. Gerichtl. Med.* 21: 195.

Umeda, M., Saito, K., and Saito, M. 1969. Cytotoxic effects of inorganic, phenyl and alkyl mercuric compounds on HeLa cells. *Jpn. Exp. Med.* 39: 15.

De Valck, V., Renmans, W., and Segers, E. 1991. Light microscopical detection of leukocyte cell surface antigens with a one-nanometer gold probe. *Histochemistry* 95: 483.

Van Den Pol, A. N. 1985. Silver-intensified gold and peroxidase as dual immunolabels for pre- and postsynaptic neurotransmitters. *Science* 228: 332.

Vandre, D. D. and Burry, R. W. 1992. Immunoelectron microscopic localization of phosphoproteins associated with the mitotic spindle. *J. Histochem. Cytochem.* 40: 1837.

Varndell, I. M., Polak, J. M., Sikri, K. L., Minth, C. D., Bloom, S. R., and Dixon, J. E. 1984. Visualisation of messenger RNA directing peptide synthesis by in situ hybridisation using a novel single-stranded cDNA probe. *Histochemistry* 81: 597.

Westbrook, G. L. and Mayer, M. L. 1987. Micromolar concentrations of Zn^{2+} antagonize NMDA and GABA responses of hippocampal neurons. *Nature* 328: 640.

Westermark, K., Lundqvist, M., Hacker, G. W., Karlsson, A., and Westermark, B., 1987. Growth factor receptors in thyroid follicle cells. *Acta Endocrinol. (Copenh.) Suppl.* 281: 252.

Windholz, M. (Ed.) 1983. *The Merck Index*. Merck & Co., Rahway, NJ.

Zehbe, I., Hacker, G. W., Sällström, J., Rylander, E., and Wilander, E. 1992. *In situ* polymerase chain reaction (*in situ* PCR) combined with immunoperoxidase staining and immunogold-silver staining (IGSS) techniques. Detection of single copies of HPV in SiHa cells. *Anticancer Res.* 12: 2165.

Zehbe, I., Hacker, G. W., Muss, W. H., Sällström, J., Rylander, E., Graf, A.-H., Prömer, H., and Wilander, E. 1993. An improved protocol of *in situ* polymerase chain reaction (PCR) for the detection of human papillomavirus (HPV). *J. Cancer Res. Clin. Oncol. Suppl.* 119 :22.

Zehbe, I., Sällström, J., Hacker, G. W., Rylander, E., Strand, A., Graf, A.-H., and Wilander, E. 1994. Polymerase chain reaction (PCR) *in situ* hybridization: detection of human papillomavirus (HPV) DNA in SiHa cell monolayers. In: *Modern Methods in Analytical Morphology.* p. 297. chap. 18. Gu, J. and Hacker, G. W. (Eds.), Plenum Press, New York, in press.

Zieger, K. 1938. Physikochemische Grundlagen der histologischen Methodik. *Wiss. Forschungsber.* 48: 55.

Chapter 3

A Method for Monitoring the Extent of Silver Intensification of Colloidal Gold Staining

A. E. Kalyuzhny, R. Elde and H. H. Loh

Contents

3.1 Introduction

The detection of an antigen using immunohistochemical techniques based on the intensification of staining of colloidal gold labeled probes has been termed as immunogold-silver staining (IGSS). These methods are reported to possess greater sensitivity than those utilizing the enzymatic activity of peroxidase (Danscher, 1981a–d). Immunogold-silver staining is widely used in diagnostic studies of HIV and rubella (Rocks et al., 1991; Patel et al., 1991) in revealing regulatory peptides (Hauser-Kronberger et al., 1993a, b; Albegger et al., 1991) and has been also suggested as a detection system for *in situ* hybridization (Zehbe et al., 1992; Hacker et al., 1993). The high sensitivity of IGSS is attributed to the ability of silver ions in the presence of a reducing agent (hydroquinone) to create a shell of metallic silver on the surface of colloidal gold, a process known as physical development (Liesegang, 1911).

0-8493-2449-1/95/$0.00+$.50
© 1995 by CRC Press, Inc.

It has been shown that the longer the time of development the larger the size of silver deposits (Lackie et al., 1985). To make the staining more controllable it has been recommended to retard the physical development by including gum arabic into the solution. This process enables one to assess staining intensity by removing slides from the enhancement solution and observing them under the optical microscope. However, the use of gum arabic significantly complicates the staining procedure since it requires time for its solubilization and washing of the preparations after silver staining (Hacker et al., 1988). Without addition of gum arabic silver staining proceeds much faster making visual assessment of enhancement extremely difficult. Thus the only possibility for monitoring the staining is to refer to the time of development.

Our experiments suggest that the final contrast of silver intensified gold particles may vary significantly under conditions of standardized temperature, concentration, and developing time; thus, giving the risk of over- or under-intensification. These differences may be due to variations in the catalytic activity of the physical developer. To examine this possibility, we monitored the rate of change in the absorbance of the physical developer in parallel with intensification of stained tissue sections with an antipeptide antibody to an opioid binding cell adhesion molecule (OBCAM) (Lane et al., 1992).

3.2   Materials and Methods

3.2.1   Tissue Preparation

Four- to six-week-old male mice of the ICR strain were sacrificed by cervical dislocation. Brains were frozen with gaseous CO_2, cut at $-11°C$ into 16-mm thick coronal sections, and thaw-mounted onto Probe-On (Fisher Scientific) slides. They were dried by vacuum desiccation at room temperature, and stored at $-20°C$ until use.

3.2.2   Immunohistochemistry

All incubation and washing steps were done at room temperature in 0.01 M phosphate-buffered saline (PBS, Sigma) (pH 7.4) unless otherwise indicated. Rabbit serum raised against the peptide MRIENKGHISTL (OBCAM 270-281) (Lane et al., 1992) was used at a dilution of 1:600 in buffer containing 1% normal goat serum (NGS, Sigma). Sections were preincubated for 10 min with 10% NGS/PBS, and then with antiserum to OBCAM 270-281 at 4°C overnight. They were washed for 1 h with buffer and fixed with freshly prepared 0.5% glutaraldehyde for 2 min. After rinsing with PBS they were incubated with 0.1% glycine to block unreacted aldehyde groups, rinsed with PBS, and incubated with goat-antirabbit biotinylated IgG (Dako, Code No: E353), diluted 1:300 with PBS. Slides were washed for 30 min with buffer and incubated with 1 µg/ml horseradish peroxidase-conjugated streptavidin (Jackson Immunoresearch Laboratories, Code No. 016-030-084) for 20 min. Slides were washed for 20 min with buffer and incubated for 90 min with 4-, 6-, 12-, or 18-nm colloidal gold labeled goat anti-horseradish peroxidase antibodies (Jackson Immunoresearch Laboratories, Code No: 123-185-021; 123-195-021; 123-205-021 and 123-215-021, respectively), diluted with PBS containing 0.01% Triton X-100 and 0.01% Tween 20 to $OD_{525} = 0.08$. Sections were washed for 20 min with buffer and fixed with 2% glutaraldehyde for 10 min. Finally, sections were washed twice for 5 min each with buffer, and then five times for 2 min each with deionized water.

3.2.3   Silver-Enhancement

Unless otherwise indicated sections were preincubated for 5 min in hydroquinone (Kodak, 250 mg/50 ml 0.1 M citrate buffer (pH 3.79) diluted 1:1 with deionized water. Sections were immersed in physical developer freshly prepared by mixing 50 ml of silver acetate (Fisher Scientific; 2 mg/ml) in deionized water and 50 ml of hydroquinone (5 mg/ml) in citrate buffer. Immediately after mixing and filtering, a portion of the physical developer was monitored for absorbance at $\lambda = 360$ nm in a spectrophotometer (4049 Novaspec, LKB). Silver intensification of histological sections was terminated at times corresponding to changes in absorbance of developer within the range 0.003 to 0.008 units. Slides were rinsed with distilled water and placed in a Rapid (Kodak) photographic fixer at 1:4 dilution for 2 min followed by subsequent washing for 5 min with running tap water. They were cleared with xylene, and mounted under a coverslip with Permount (Fisher Scientific).

3.2.4 Cytochemical Controls

Sections were incubated with (1) pre-immune serum, (2) OBCAM 270–281 antiserum preabsorbed with the peptide OBCAM 270–281, and (3) without colloidal gold anti-horseradish peroxidase.

3.2.5 Data Recording and Image Analysis

Tissue sections were observed and photographed with a Nikon Diaphot equipped with a Nikon F2000 camera using manually fixed exposure and black-and-white Techpan film (Kodak). The image analysis system (Quantimet 500; Leica-Cambridge, UK) was attached to the microscope via a 4910 series CCD camera (Cohu Inc., USA). Images were averaged (12 frames) prior to analysis. A measuring frame equal to 1357 pixels was selected to determine the optical density of six randomly chosen areas in the granular layer of the cerebellum. For studies of the contrast of silver-enhanced gold particles of different sizes we analyzed the gigantocellular reticular nuclei in adjacent sections of the medulla oblongata using a measuring frame of 11316 pixels. After measuring the gray level of the background and that of the stained features the detection window was set to the boundary between specific staining and background so that images were collected with intensities ranging from 0 to 173 out of a possible 256 gray level.

3.3 Results

OBCAM 270–281 immunoreactivity (ir) was found to be widely distributed in mouse brain. As has been previously reported, staining was found in neuronal perikarya in many mouse brain regions (Kalyuzhny et al., 1993).

3.3.1 Absorbance of Physical Developers as a Function of Time

The absorbance of physical developers was monitored in six consecutive experiments (Figure 1) until OD_{360} increased to 0.008 units. Despite identical concentration of reagents and identical reaction conditions the kinetics of absorbance changes differed among all physical developers; the time interval between the endpoints (0.008 units) ranged from several seconds to 7 min.

3.3.2 Correlation Between the Absorbance of Physical Developer and Optical Density of Histological Sections

Silver-enhancement of histological sections continued until absorbance of the developer reached 0.008 units. Figure 2A corresponds to curve No. 1 and Figure 2B to curve No. 6 of Figure 1, respectively. Figure 2B represents the section through nucleus ambiguus which

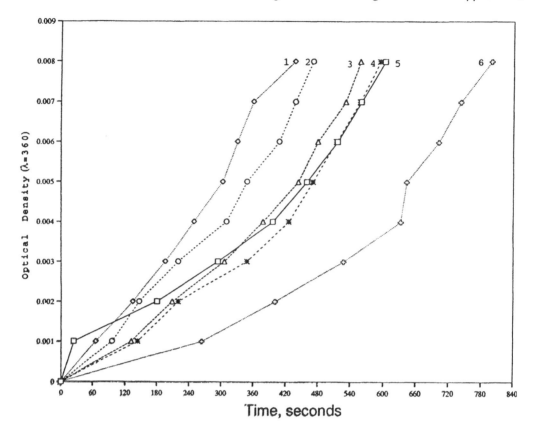

Figure 1   Absorbance dynamics of physical developers. Absorbance of developer solutions has been monitored using the spectrophotometer in six independent experiments. Each point of absorbance is plotted against the time required to reach that absorbance.

was developed ~7 min longer than that of Figure 2A. Image analysis revealed that both images had similar optical density of stained neurons (Figure 3).

3.3.3   *Relationship Between Optical Density of Physical Developer and That of Immunogold-Silver Stained Histological Sections*

At certain time points silver-enhancement of adjacent sections of cerebellum was terminated and the optical density of developer and of the tissue sections measured. Photomicrographs correspond to termination of development at the time at which absorbance of the developer increased to 0.003, 0.004, 0.005, 0.006, 0.007, 0.008 (Figure 4A–F, respectively).

3.3.4   *Correlation Between the Size of Gold Probes and Contrast of Silver-Enhancement*

Adjacent coronal sections through gigantocellular nuclei were incubated with 4-, 6-, 12-, or 18-nm gold probes (Figure 5A–D, respectively). Dilution of all anti-peroxidase gold labeled antibodies was done in such a way that density of colloidal gold was identical. Sections were incubated in the same developer until its absorbance reached 0.008. Figure 5 clearly demonstrates that the intensity of staining decreased as the diameter of colloidal gold increased. Image analysis confirmed this finding (Table 1).

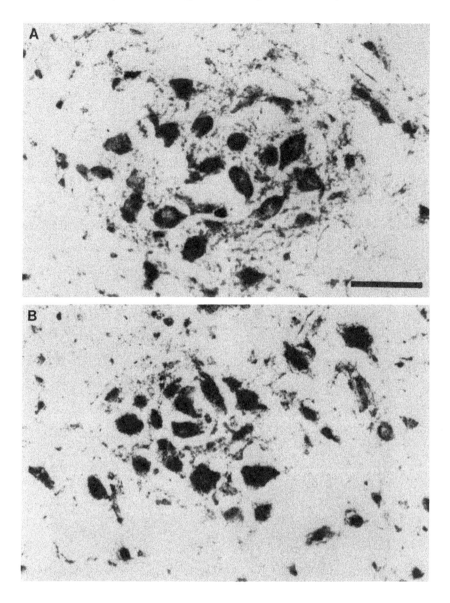

Figure 2 Immunogold-silver stained neurons in nucleus ambiguus. Silver-enhancement allowed to continue until absorbance of the developer reached 0.008 units. (A) curve No. 1 (435 sec of silver enhancement); (B) curve No. 6 (815 sec of silver enhancement) of Figure 1. Scale bar = 150 μm.

3.4 Discussion

The intensification of colloidal gold staining by silver is a time-dependent process, i.e., the longer the developmental time the larger the size of silver precipitates (Lackie et al., 1985). Without including gum arabic into the developer the process is fast and optimum enhancement could be obtained in 5 to 7 min (Hacker et al., 1988). In our study we further confirmed the observation by Hacker and colleagues that formation of silver precipitates occurs not only in the presence of colloidal gold absorbed by tissue but also spontaneously in physical developer itself (Hacker et al., 1988). We reasoned that the absorbance of this solution may predict the density of precipitates.

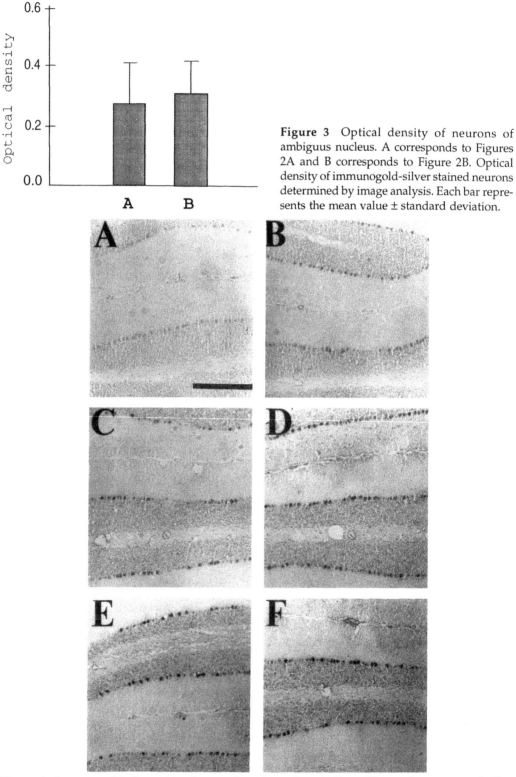

Figure 3 Optical density of neurons of ambiguus nucleus. A corresponds to Figures 2A and B corresponds to Figure 2B. Optical density of immunogold-silver stained neurons determined by image analysis. Each bar represents the mean value ± standard deviation.

Figure 4 Immunogold-silver stained coronal sections of cerebellum. Silver-enhancement of adjacent sections was subsequently terminated at points in absorbance of physical developer ranging from 0.003 to 0.008. Pictures correspond to termination at absorbances of: (A) 0.003; (B) 0.004; (C) 0.005; (D) 0.006; (E) 0.007; (F) 0.008. Scale bar = 400 μm.

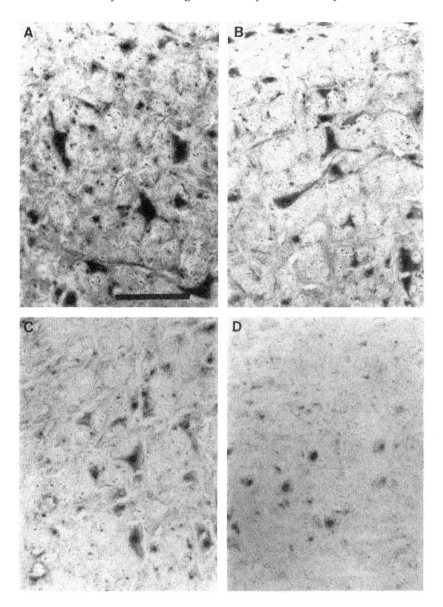

Figure 5 Correlation between the size of gold probes and contrast of silver-enhancement. Coronal sections through gigantocellular nuclei incubated with 4-nm (A), 6-nm (B), 12-nm (C), and 18-nm (D) gold probes and silver enhanced until optical density of developer solution reached .008 units. Scale bar = 150 μm.

As shown in Figure 1, similar developer solutions reached corresponding points of absorbance although at different time intervals. Absorbance curves had irregular profiles which indicates that the intensity of formation of silver precipitates could vary in an unpredictable manner. This observance could be the explanation for irreproducible staining intensity when the criterion for enhancement is based on the duration of that process. On the other hand, similarity in image contrast (Figure 2) and optical density (Figure 3) of sections enhanced for different times but to equal points of absorbance of the developer solution suggests an advantage in measuring optical density of the developer instead of the time.

Data presented in Figure 4 and Table 1 demonstrate that silver more readily enhances smaller gold particles rather than large ones, thus confirming the results obtained by

Table 1 Image Analysis of Neurons in the
Gigantocellular Nuclei Incubated with
Gold Probes of Different Size

| GOLD PROBES 4 nm | | |
Area (pixels)	Int.Gray	MeanGray
Total	11316.000	3519.000	90.884
Mean	35.363	10.997	0.284
Std Dev	66.413	23.48	0.029
Max	559.000	184.000	0.424
Min	4.000	1.000	0.254
Features	320		

| GOLD PROBES 6 nm | | |
Area (pixels)	Int.Gray	MeanGray
Total	6885.000	1319.000	50.826
Mean	28.333	5.428	0.209
Std Dev	46.888	9.994	0.044
Max	337.000	78.000	0.532
Min	4.000	0.000	0.156
Features	243		

| GOLD PROBES 12 nm | | |
Area (pixels)	Int.Gray	MeanGray
Total	2239.000	392.000	24.108
Mean	18.815	3.294	0.203
Std Dev	18.468	3.870	0.034
Max	100.000	21.000	0.433
Min	4.000	0.000	0.149
Features	119		

| GOLD PROBES 18 nm | | |
Area (pixels)	Int.Gray	MeanGray
Total	833.000	134.000	13.421
Mean	12.621	2.030	0.203
Std Dev	7.908	1.678	0.035
Max	42.000	8.000	0.348
Min	4.000	0.000	0.158
Features	66		

Lackie et al. (1985). The physical mechanism responsible for this phenomenon remains to be established, but may be due to the greater penetration of small gold particles into the tissue. Alternatively, it may be due to the higher quantity of gold particles per unit volume. It has been found that the higher the quantity of metal nuclei the higher the density of deposited silver (Arens and Eggert, 1929). In our pilot studies using immunoperoxidase staining protocol with 3-amino-9-ethylcarbazol (AEC) as a dye, we demonstrated that OBCAM-ir in spinal cord is found in nuclei, the cytoplasm, cell membranes, and axons of neurons (Kalyuzhny et al., 1993). In the present study, a relationship was found between the diameter of the probe and the cellular compartments that were stained, such that 18-nm gold revealed cell nuclei, 12-nm revealed nuclei and cell membranes, 6- and 4-nm revealed nuclei, membranes, and axons. This phenomenon is probably related to the relative abundance of the antigen in each of these compartments.

It is not yet clear why the formation of silver precipitates in developers containing equal proportions of silver acetate and hydroquinone may vary. We suggest that minor variations in concentrations of hydroquinone and silver acetate, and shifts in pH and temperature may have major influence on its autocatalytic activity. Thus, we suggest that monitoring the optical density of the physical developer is a relevant predictor of its ability to induce silver-enhancement while providing a simple and accurate method for standardizing silver intensification of colloidal gold.

The only reliable method for monitoring the extent of silver amplification known at present appears to be the examination of samples under the microscope during development. However, exposure to light can increase background staining (Hacker et al., 1988) and visual assessment, because of its subjectivity, may not be desirable. In contrast, this method, which is based on the monitoring of the optical density of physical developer, does not have those disadvantages. It is simple and enables one to obtain reproducible silver-enhancement of colloidal gold.

3.5   Summary

Silver intensification has proven to be a very sensitive method for amplifying immunostaining when using colloidal gold detection systems. However, the signal-to-noise ratio varies greatly, even when standardized conditions are used. We have determined that optical density of the physical developer solution may vary significantly within a given time interval. Using image analysis we found that the optical density in tissue sections stained with immunogold and enhanced with silver correlates with the optical density of the physical developer solution. An increase in the diameter of the gold probe decreased the contrast of preparations enhanced with silver. Thus, we suggest that monitoring the optical density of the physical developer is a relevant predictor of its ability to induce silver-enhancement while providing a simple and accurate method for standardizing silver intensification of colloidal gold.

Acknowledgment

The authors express their appreciation to Dr. N. M. Lee for many helpful discussions. This work was supported in part by NIH grants DA 00564, DA 01583, DA 705504, DA 05695 and the F. Stark Fund of the Minnesota Medical Foundation.

References

Albegger, K., Hauser-Kronberger, C. E., Saria, A., Graf, A. H., Bernatzky, G., and Hacker, G. W. 1991. Regulatory peptides and general neuroendocrine markers in human nasal mucosa, soft palate and larynx. *Acta Otolaryngol. (Stockh.)* 111: 373.

Arens, H. and Eggert, J. 1929. Das Wachstum des kolloiden Silbers in Gelatineschichten. *Z. Electrochem.* 35: 728.

Danscher, G. 1981a. Histochemical demonstration of heavy metals. A revised version of the sulfide method suitable for both light and electron microscopy. *Histochemistry* 71: 1.

Danscher, G. 1981b. Localisation of gold in biological tissue. A photochemical method for light and electron microscopy. *Histochemistry* 71: 81.

Danscher, G. 1981c. Light and electron microscopical localisation of silver in biological tissue. *Histochemistry* 71: 177.

Danscher, G. 1981d. Autometallography. A new technique for light and electron microscopical visualisation of metals in biological tissue (gold, silver, metal sulphides and metal selenides). *Histochemistry* 81: 331.

Hacker, G. W., Graf, A. H., Hauser-Kronberger, C., Wirnsberger, G., Schiechl, A., Bernatzky, G., Wittauer, U., Su, H. C., Adam, H., Thurner, J., et al. 1993. Application of silver acetate autometallography and gold-silver staining methods for in situ DNA hybridization. *Chin. Med. J.* 106: 83.

Hacker, G. W., Grimelius, L., Danscher, G., Bernatzky, G., Muss, W., Adam, H., and Thurner, J. 1988. Silver acetate autometallography: An alternate enhancement technique for immunogold-silver staining (IGSS) and silver amplification of gold, silver, mercury and zinc in tissues. *J. Histotechnol.* 11:213.

Hauser-Kronberger, C., Albegger, K., Saria, A., and Hacker, G. W. 1993. Regulatory peptides in the human larynx and recurrent nerves. *Acta Otolaryngol. (Stockh.)* 113: 409.

Hauser-Kronberger, C., Hacker, G. W., Muss, W., Saria, A., and Albegger, K. 1993. Autonomic and peptidergic innervation of human nasal mucosa. *Acta Otolaryngol. (Stockh.)* 113: 387.

Kalyuzhny, A. E., Elde, R., Lee, M. N., and Loh, H. H. 1993. Localization of OBCAM-immunoreactivity in mouse brain. *International Narcotics Research Conference, Skovde, Sweden,* July 10–15, P14 (abstract).

Lackie, P. M., Hennessy, R. J., Hacker, G. W., and Polak, J. M. 1985. Investigation of immunogold-silver staining by electron microscopy. *Histochemistry* 83: 545.

Lane, C. M., Elde, R., Loh, H. H., and Lee, N. M. 1992. Regulation of an opioid-binding protein in NG 108-15 cells parallels regulation of delta-opioid receptors. *Proc. Natl. Acad. Sci.* 89: 11234.

Liesegang, R. 1911. Die Kolloidchemie der histologischen Silber-Farbungen. *Kolloid. Beihefte* 3: 1.

Patel, N., Rocks, B. F., and Bailey, M. P. 1991. A silver enhanced, gold labeled, immunosorbent assay for detecting antibodies to rubella virus. *J. Clin. Pathol.* 44: 334.

Rocks, B. F., Patel, N., and Bailey, M. P. 1991. Use of a silver-enhanced gold-labeled immunoassay for detection of antibodies to the human immunodefficiency virus in whole blood samples. *Ann. Clin. Biochem.* 28: 155.

Zehbe, I., Hacker, G. W., Rylander, E., Sallstrom, J., and Wilander, E. 1992. Detection of single HPV copies in SiHa cells by in situ polymerase chain reaction (in situ PCR) combined with immunoperoxidase and immunogold-silver staining (IGSS) techniques. *Anticancer Res.* 12: 2165.

Chapter 4

Comparison of Embedding Media for Immunogold-Silver Staining

Tibor Krenács and László Krenács

Contents

4.1   Introduction

Catalytic reduction of silver ions by heavy metals found in tissues or used as labeling particles in affinity cytochemistry has been widely used for both research and diagnostic purposes. RNA, sugar residues, antigens and complementary nucleic acid sequences can be detected by silver-enhancement of gold colloid coupled to specific proteins or nucleic acid sequences. (For more details, see Chapter 1.) For immunocytochemistry, immunogold-silver staining (IGSS; Holgate et al., 1983) has been widely used, e.g., in unembedded tissues or cells (de Waele et al., 1986), routinely processed paraffin or resin-embedded sections for light microscopy (Hacker et al., 1986), free floating sections (Van den Pol, 1985), thin frozen sections (Atherton et al., 1992) and resin-embedded thin sections (e.g., Bienz et al., 1986; Krenács et al., 1989, 1991) for subcellular antigen detection.

Its increasing popularity for immunoelectron microscopy is due to the commercial availability of high-quality small (5 nm) and ultrasmall (1 to 1.4 nm) gold probes (Amersham Pcl. and Nanogold Inc.) (Hainfield and Furuya, 1992), which can hardly be seen without silver amplification in counterstained thin sections. With the ultrasmall gold probes, the drawback of poor probe penetration into the unembedded tissues seems to be overcome, which makes IGSS a technique of promise in pre-embedding immunoelectron microscopy (Burry et al., 1992; Vandré and Burry, 1992; see also Chapter 14). The IGSS method can also be expected to gain increasing attention from researchers in the near future because of the relative simplicity of the postembedding approach, supplemented with the advantages of clear recognition of the ultrastructure when a particulate label versus the diffuse DAB precipitate is used, and because of the high sensitivity provided by the ultrasmall gold probes.

Several variables can affect the sensitivity and the quality of an IGSS immunoreaction in resin-embedded tissues: (1) the affinity of the reagents; (2) the antigen loss or masking resulting from chemical fixation, dehydration, resin monomer action, and the conditions of resin curing (Causton, 1984); (3) the degree of hydrophilicity of the polymerized resin (Newman and Hobot, 1987); (4) the size of the gold probe (Lackie et al., 1985); and (5) the conditions of silver amplification. Most of the above factors have already been studied to some extent for postembedding immunogold methods (see Merighi, 1992), but the effect of different resins, used for electron microscopy, on silver-enhancement has not yet been examined. In previous work, we observed that larger particles were formed in Durcupan ACM than in LR-White when the same silver-enhancement protocol was used (Krenács et al., 1991). The question of whether this relationship holds true between other hydrophobic and hydrophilic resins, and how reliable silver-enhancement can be in different resin-embedded tissues seemed to deserve attention. This contribution focuses on postembedding antigen detection with the IGSS method when different embedding polymers for electron microscopy are applied.

4.2   Postembedding Immunoelectron Microscopy with Immunogold-Silver Staining

For postembedding immunocytochemistry in resin-embedded thin sections, the immunogold method has been a general approach for more than a decade (Merighi, 1992). Relatively large colloidal gold particles (10 to 40 nm) have permitted visualization of the antigenic sites on the counterstained ultrastructure with acceptable sensitivity in most cases. Silver-enhancement of the immunogold signal for ultrastructural studies was first applied in the pre-embedding method (Van den Pol, 1985). This was soon being followed by the description of a postembedding double antigen detection, where the same immunoconjugate was used with or without enhancement in subsequent reactions (Bienz et al., 1986). The recognition of the inverse relationship between the size of the coupled gold particles and the particle density (sensitivity) of the IGSS reaction (Lackie et al., 1985; Van den Pol, 1986) provided a further basis for the use of IGSS in electron microscopy. However, this approach of bringing together the high sensitivity provided by the small conjugate with good visualization resulting from the particle enlargement had been exploited in only a few postembedding studies until recently (e.g., Krenács et al., 1989, 1991; Bozóky et al., 1993). Not only has silver-enhancement been very useful when small (5 nm) gold conjugates are used, but it has also become an inevitable aid for immunoelectron microscopy since ultrasmall (1 to 1.4 nm) gold conjugates have recently become commercially available and have been favorably received (Burry et al., 1992; Hainfield and Furuya, 1992; Shimizu et al., 1992; Vandré and Burry, 1992). Besides the quality of the reagents, the result of the silver-enhanced immunogold method in resin sections greatly depends on the physicochemical features of the polymerized resin lattice surrounding the antigen, which governs antigen access and antibody penetration and provides the microenvironment for the silver reduction step.

4.3   Physicochemical Character of Resins

There are basically two kinds of tissue embedding resins in electron microscopy: the polymers of aliphatic or aromatic epoxides and acrylates. Epoxy resins comprised an almost exclusive basis for transmission electron microscopy for decades (Hayat, 1989), until the introduction of the low-temperature-cured polyhydroxy acrylates in the early

1980s (Carlemalm et al., 1982; Newman et al., 1983). This emergence coincided with the first boom in postembedding antigen detection (see Merighi, 1992). The general considerations relating to the physicochemical properties of the commercial resins and their influence on structural preservation, antigenicity and electron beam stability have subsequently been determined and reviewed (Causton, 1984; Newman and Hobot, 1987, 1989; Newman, 1989; Hobot, 1989). We provide only a brief account of these considerations with an introduction to the different resin types used in this study.

Epoxy monomers and polymerized epoxy resins (Araldite equivalent Durcupan ACM, FLUKA Chem., Buchs, Switzerland; Epon 812, SERVA, Heidelberg, Germany) have very low affinity for water and consequently exhibit a rather hydrophobic character. This feature necessitates complete tissue dehydration before embedding. This effect, coupled with the possible binding of the reactive epoxy groups to tissue proteins (Lee and Neville, 1967), may lead to massive alterations in the antigen conformation. Etching with an oxidizing agent (e.g., hydrogen peroxide), or possible cleavage of the ester bonds in the polymerized epoxides by sodium (m)ethoxide, may produce a gel layer on the surface region of the tissue section, which may allow an antigen-antibody interaction (Causton, 1984). Harsh etching for 5 to 30 min with sodium (m)ethoxide will completely eliminate the epoxy resin from the semithin section before immunostaining (Lackie et al., 1985; Krenács et al., 1990). Araldite (Durcupan ACM) monomers have a low rate of diffusion into the tissue, but the aromatic side-groups provide the polymerized resin with good beam stability (Causton, 1984). The small degree of cross-link density observed in Araldite-embedded sections renders them somewhat more useful for immunoelectron microscopy than the low-viscosity, but highly cross-linked aliphatic Epon 812.

The new generation of polyhydroxy acrylates (i.e., Lowicryls, Lowi GmbH, Waldkraiburg, Germany; LR-White and LR-Gold, London Resin Co., Woking, Surrey, UK) can be polymerized with some 5 to 30% of water in the tissue, which may help maintain the tissue molecules in a state closer to their native conformation and allows better antigen access for antibodies (Carlemalm, 1982; Newman and Hobot, 1987). The aliphatic Lowicryls and the aromatic LR-Gold can be used for tissue infiltration and polymerized at low temperature, –35 to –70°C, as catalyzed by ultraviolet or blue light. These conditions minimize the antigen loss by extraction and the probability of resin monomer reaction with tissue molecules (Causton, 1984). Aromatic hydrophilic acrylate LR-White can best be polymerized at 50°C, or at room temperature with an accelerator (Newman and Hobot, 1987).

4.4 *The Silver-Enhancement Protocol*

According to the principle of physical silver development (autometallography), metallic gold bound to the antigen catalyzes the reduction of silver ions to a metallic silver precipitate in the presence of a reducing agent (Danscher and Nörgaard, 1983). The end-product is seen as dark-brown to black in the light microscope, or appears as electronedense dots, gradually increasing with the duration of amplification in the electron microscope. Since the introduction of an acidic silver nitrate solution containing hydroquinone by Liesegang (1928), several modifications of the protocol have appeared, involving the use of other silver sources (silver lactate or silver acetate), reducing agents (formaldehyde, N-propyl-gallate), pH (alkaline, neutral), and protecting colloid (gum acacia) (see Chapter 2). Several companies have elaborated and marketed stabilized enhancer kits which are recommended for application with good results for several months if stored in the dark at 4°C. However, in our experience, more standard results can be achieved if fresh enhancer solution is made on each occasion. We achieved acceptable results with a method practically similar to that of Liesegang (1928) at both light microscopic and ultrastructural levels

(Krenács et al., 1989a,b, 1990, 1991; Bozóky et al., 1993). However, the use of silver nitrate with a relatively high concentration of hydroquinone and without protecting colloid resulted in a fast catalysis and inhomogeneity of the particles, and it was difficult to establish the point just before individual particles joined, particularly when the particle density was high. More reliable results were obtained with the silver acetate enhancer (Krenács and Krenács, in press), introduced originally for light microscopy by Hacker et al. (1988), which was exclusively used in the present study.

4.5   Materials and Protocols Utilized

Tissues fixed in 4% formaldehyde (made freshly of paraformaldehyde) containing 0.1% glutaraldehyde and dehydrated in a graded ethanol series was routinely embedded from 100% ethanol in Epon 812, Durcupan ACM, (at 56°C) and Lowicryl K4M (with UV light at –35°C), from 96% ethanol in LR-White (at 50°C), or from 90% ethanol in LR-Gold (with 0.1% initiator in UV light) resins. Semithin sections of 1 μm were collected on 3-aminopropyltriethoxysilane (SIGMA, St. Louis, MO) coated glass slides. The Durcupan resin was removed by immersing the sections for 20 min in a mixture of methanol/benzene (1:1) saturated with metallic sodium, followed by washing with methanol/benzene. Before the immunoreactions, the acrylate semithin sections were microwave cooked twice for 5 min each in 0.01 M citrate buffer (pH 6.0) at 750 W.

Yellow/gold thin sections mounted on Formvar-coated nickel grids were etched with 5% hydrogen peroxide for 10 min, followed by blocking treatment with a 5% normal goat serum-1% bovine serum albumin-PBS buffer for 30 min. Routine antigen detection as described elsewhere (Krenács et al., 1989) involved incubation of the semithin sections and thin sections ("on-grid") with (1) monoclonal (mouse anti-actin, 1:100, Enzo Diagn., NY) and rabbit polyclonal antibodies (anti-ws keratin, 1:100, DAKO A/S, Glostrup, Denmark; anti-SR-Ca-ATPase, 1:100, Krenács et al., 1989; anti-*E. coli* wall, 1:100, Iványi et al., 1990) overnight at 4°C, (2) mouse anti-rabbit IgG (DAKO, 1:50, only for polyclonals) and (3) a goat anti-mouse IgG-5 nm gold (GAM-G5, 1:30, Amersham Plc, Buckinghamshire, UK), or a GAM(Fab')-1.4 nm gold probe (1:30, Nanogold Inc., Stony Brook, NY). In semithin sections, a GAM-G10 probe (1:30, Amersham) was also used. After washing with 1% BSA/PBS solution, the sections were postfixed in 1% glutaraldehyde for 5 min, to avoid detachment of the labeling particles. The washing out of the fixative and ions from the sections with double-distilled water was followed by silver-enhancement by means of the acetate method of Hacker et al. (1988) described below.

Solutions A (250 mg hydroquinone in 50 ml 0.1 M citrate buffer, pH 3.5) and B (100 mg silver acetate in 50 ml double-distilled water) were prepared by magnetic stirring for ~15 min each, and after dissolution equal parts of the two were mixed. To keep the process standard, the enhancement was started 60 sec after the solutions A and B were mixed. After appropriate silver development and washing with distilled water, treatment with 5% sodium thiosulfate for 2 min was applied. Etching with sodium methoxide, or hydrogen peroxide (the same as in epoxy sections) had no effect on the immunoreactions in acrylate-embedded semithin sections. Likewise, no reactions were observed in the control experiments where the primary antibodies were substituted with normal serum. As a final step, the sections were routinely counterstained with uranyl acetate and lead citrate for 5 min each, and studied with a Philips CM10 electron microscope at 60 kV.

4.6   Immunogold-Silver Staining in Semithin and Thin Sections

Following resin elimination and immunogold staining, silver-enhancement for ~7 min resulted in a compact jet-black reaction product in the Araldite semithin sections (Figure 1A). Without pretreatment, only weak IGSS reactions with a moderate background were

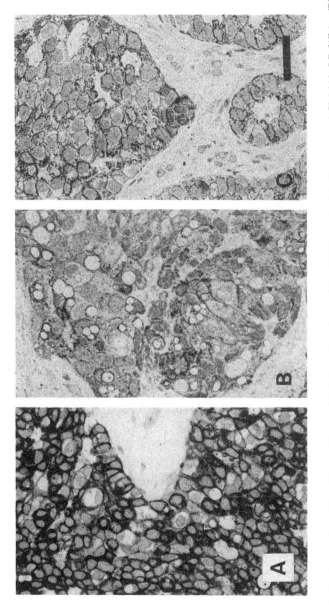

Figure 1 Immunostaining for ws-keratin with the Nanogold (1.4 nm) probe and the IGSS method in Durcupan ACM (A), LR-White (B), and Lowicryl K4M (C) resin-embedded semithin sections (1 μm) of breast cancer. (A) Elimination of Durcupan with sodium methoxide allowed complete penetration of the tissue by the reagents and strong labeling. (B, C) More delicate perinuclear, periluminar, and cell membrane reactions are found in the hydrophilic resins with higher intensity and lower background in Lowicryl (C) than in LR-White (B). Microwave cooking of the sections was inevitable for reagent penetration and getting acceptable reactions in hydrophilic resins. Bar = 20 μm.

detected in both types of acrylates with the GAM-G5 and Nanogold (1.4 nm) probes. GAM-G10 gave no appreciable reaction on acrylate sections, even after pretreatment with hydrogen peroxide, or sodium methoxide. Poor sensitivity was also observed in parallel reactions made with the labeled streptavidin-biotin immunoperoxidase method (data not shown). Microwave pretreatment substantially improved the staining intensity with both the 5- and 1.4-nm probes, resulting in delicate but strong staining in Lowicryl (Figure 1C) and a somewhat weaker reaction in LR-White (Figure 1B).

In order to assess and compare the effects of the embedding resins in the IGSS reaction ultrastructurally, several immunoreactions were performed by following the same protocol including the silver-enhancement step. The results are shown at the same magnification (see Figures 2–4). In a double staining reaction with the same gold probe, the silver-enhanced particles obtained in the first sequence can easily be differentiated from the 5-nm unenhanced particles used for the second antigen (Figure 2A). These 5-nm particles provide a standard for particle size assessment in the subsequent reactions. Immunostaining with the Nanogold probe (1.4 nm) for muscle actin in skeletal muscle, followed by the same silver-enhancement, resulted in somewhat heterodisperse particles with about the same diameter in LR-White (Figure 2B) and LR-Gold (Figure 2D), but gave significantly larger particles in Durcupan (Figure 2C).

Both the particle size heterogeneity within a section and the particle size differences observed in LR-White and Durcupan became gradually reduced after a longer period of silver-enhancement (Figure 2B inset). Under the same conditions as for actin, immunodetection for E. coli wall antigen was made in two other resins (Figure 3). Although there was no significant difference, a somewhat more pronounced particle heterogeneity with some larger dots was observed in Epon 812 (Figure 3) than in LR-White (Figure 3, inset). Quite uniform particles similar in size to those observed in LR-White and LR-Gold (Figure 2B and 2D) were formed in Lowicryl K4M when either a 5 nm gold containing immunoprobe was used (Figure 4A), or the same reaction as shown in Figure 1C was performed "on grid" with the Nanogold probe (Figure 4B).

4.7 Discussion

Not many examples have yet been published on the application of IGSS in resin-embedded semithin sections (Al-Nawab and Davies, 1989; Bowdler et al., 1989; Krenács et al., 1990; Velde and Prins, 1990) or thin sections (Bienz et al., 1986; Bozóky et al., 1993; Krenács et al., 1989, 1991; Penschow and Coglan, 1993; Pettitt and Humphris, 1991; Shimizu et al., 1992; Slater, 1991). However, a revival of this approach is expected, due to the recent availability of ultrasmall (1.0 to 1.4 nm) gold probes and the improvements achieved in acrylate embedding technology. For IGSS, the demands imposed on the embedding media in postembedding immunogold electron microscopy, as reviewed by Causton (1984), must be supplemented with a compatibility with the silver-enhancement protocol. The properties of polyhydroxy acrylates are closer to those theoretically demanded from an ideal resin, and usually allow better morphological preservation and immunoreactions in the tissues than with epoxides (Roth et al., 1981; Newman and Hobot, 1987).

At both light microscopic and ultrastructural levels, the degree of antigen access to immunoreagents in the resins is the cornerstone of successful labeling. The cross-link density, defining a theoretical "poresize" within the resins, can be adjusted within a limited range by changing the relative amount of cross-linking (curing) agent, or by chemical etching of the polymerized resins (Causton, 1984). The latter may not only increase the "poresize", but also make the resin and tissue molecules more hydrated, while having deleterious effects on the ultrastructural morphology. If the resin can be eliminated, both the immunoenzyme and IGSS methods can be almost as successful in semithin

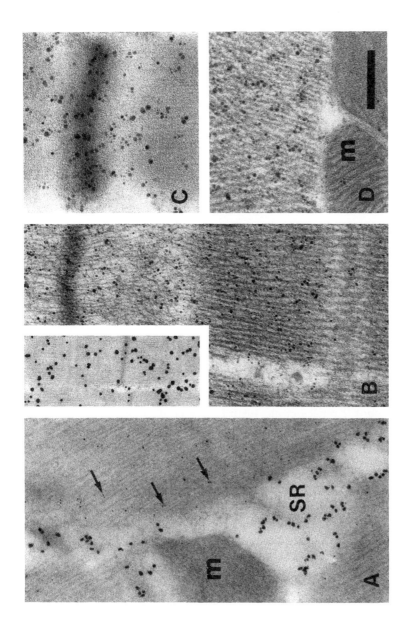

Figure 2 Actin (A to D) and SR-Ca-ATPase (A, large dots) reactions with the IGSS method in skeletal muscle embedded in LR-White (A, B), Durcupan ACM (C) and LR-Gold (D) resins. (A) Double labeling with a RAM-G5 probe used without (small particles for actin, arrows) and with silver-enhancement (large particles for SR-Ca-ATPase). Note that enhancement for 3 min enlarged the 5-nm particles to about triple their diameter. Compare particle sizes in B to D to these. (B to D) Most of the particles in hydrophilic resins (B, D) are about 5 to 10 nm, whereas most of them in Durcupan (C) are about 10 to 15 nm. Inset in (B) shows that longer intensification reduces particle size inhomogeneity except where neighboring particles aggregate. M: mitochondria, SR: sarcoplasmic membrane-system. Bar = 200 nm.

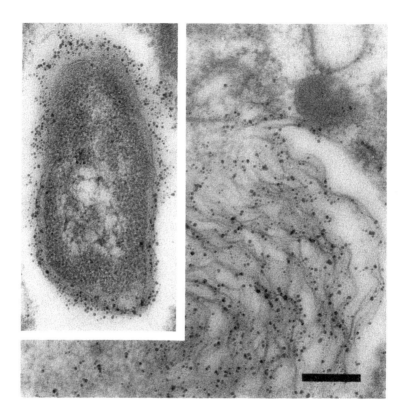

Figure 3   Bacterial wall antigen detected with the Nanogold (1.4 nm) probe followed by a silver-enhancement for 3 min in *E. coli* pyelonephritis. The particles are more homogeneous on the bacterium embedded in LR-White (inset), than in the residual myelin figures of a macrophage (right) embedded in Epon resin. Most of the particles are about 5 to 8 nm in LR-White (inset) and about 3 to 10 nm with some larger ones in Epon (right) when compared to those in Figure 2A. Bar = 200 nm.

sections as in dewaxed paraffin sections, which is the case with epoxides after sodium (m)ethoxide pretreatment (Krenács et al., 1990; Lackie et al., 1985). However, not every antibody that works well in paraffin sections can recognize antigens in deplasticized epoxy sections (unpublished observation), probably because of the effects of resin monomers on the antigen determinants, which can not be eliminated by etching.

The penetration of larger (10 to 40 nm) gold probes into intact resin sections of either epoxides (Lackie et al., 1985), or even partially dehydrated acrylates (Newman and Hobot, 1987; Timms, 1986) was found to be fully hindered. We also observed serious penetration problems with the 10-nm gold probe. In addition, it was shown that complete dehydration may completely prevent immunodetection in LR-White, despite the same reaction giving impressive results in partially dehydrated tissues (Herken et al., 1988). Newman et al. (1983) likewise stressed the importance of partial dehydration with LR-White embedding, which also allowed easy antigen detection with 5-nm gold probes and IGSS in Lowicryl K4M embedded semithin sections (Al-Nawab and Davies, 1989; Velde and Prins, 1990). Although etching of the acrylate resins is not thought to be needed for immunoelectron microscopy (Causton, 1984), pretreatment of the conventionally fixed, highly dehydrated and acrylate-embedded semithin sections does seem to be necessary. In our hands, complete (for Lowicryl K4M) and almost complete tissue dehydration (up to 96% ethanol for LR-White) gave the best ultrastructural preservation. At the same time, however, the high degree of dehydration must account for the poor immunoreactions and background

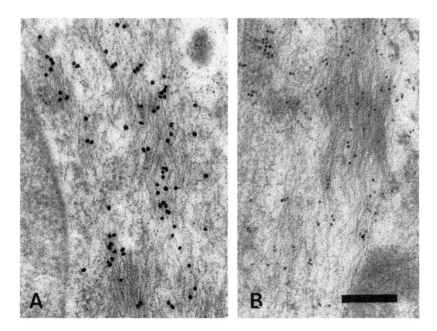

Figure 4 Ws-keratin detected with a GAM-G5 reagent (A) and the Nanogold (1.4 nm) probe (B) followed by a silver-enhancement for 3 min in the same breast cancer tissue as in Figure 1C embedded in Lowicryl K4M resin. The quite homogeneous particles are about 8 to 10 nm after the Nanogold labeling (B) and 15 to 20 nm following the use of RAM-G5 (A) when compared to the particles in Figure 2A. Bar = 200 nm.

staining in intact acrylate semithin sections with both IGSS and immunoperoxidase methods. Here we have described an effective means of achieving a high-standard IGSS reaction in strongly dehydrated acrylate-embedded tissues.

Microwave cooking has recently become a booming field of antigen retrieval in aldehyde-fixed, dewaxed tissue sections (Cattoretti et al., 1993). Boiling of the sections in mildly acidic buffer before the immunoreaction must have a mild hydrolytic effect on the resin lattice, similarly as presumed for aldehyde cross-links and tissue molecules (Cattoretti et al., 1993), which probably allows better reagent penetration and some degree of renaturation (hydration) of the antigens. Exploitation of these benefits in postembedding immunoelectron microscopy deserves further examination.

The acetate-based silver-enhancement protocol of Hacker et al. (1988) was found to be compatible with every kind of resin used in this study, which resulted in round and countable, though somewhat heterodisperse particles. The destructive effect of acidic pH developers which was observed on the ultrastructure in a pre-embedding study by Lah et al. (1990), was not seen in resin-embedded tissues. The silver amplification step permitted enlargement of the unrecognizable (1.4 nm) gold particles to a predetermined, recognizable size in the contrasted ultrastructure, lending high resolution and sensitivity to the method. We also confirmed our preliminary observation that the gold particles bound to tissue antigens in Durcupan led to somewhat faster catalyzes of the silver precipitation, resulting in larger particles than those in LR-White with either the silver nitrate or acetate developer (Krenács et al., 1991, and in press). In this respect, there was no significant difference between the polyhydroxy acrylates, LR-White, LR-Gold, and Lowicryl K4M.

In Epon 812-embedded sections, the particle size heterogeneity was more pronounced as compared with that in LR-White. The differences between the two basic resin types in particle enlargement kinetics, which seems gradually reduced as the enhancement period is elongated, may be explained in at least two ways. On the one hand, the acrylates, being more hydrophilic than the epoxides, may allow a somewhat better penetration of small and ultrasmall (1 to 5 nm) gold probes. As a result, the access of the enhancer should be somewhat delayed in comparison with that of the epoxides (Krenács et al., 1991). However, pretreatment of the sections with hydrogen peroxide renders the epoxides rather permeable in their surface layer (Lackie et al., 1985), whereas such an effect on acrylates has neither been published by others nor found in our experiments with semithin sections. This second possibility also suggests that the better access of the particles by the enhancer in pretreated epoxy resin may result in faster growing particles in epoxides, and particularly in Durcupan, than in acrylates. The reactive groups remaining from aldehyde fixation and in the resin components, or created by treatment such as etching, do not seem to make differences in effect on the silver-enhancement between the basic resin types, since both types were treated in the same way during the reactions. The effect of these groups can not be completely disregarded, however.

In conclusion, highly sensitive postembedding ultrastructural antigen detection can be achieved with small (5 nm) and ultrasmall gold (1.4 nm) probes if the labeling particles are catalytically enlarged to a predictable size. Silver-enhancement can be reliably performed in both epoxides and acrylates with a silver-acetate based protocol, although some initial differences in the kinetics of the catalyzes must be considered. Standard, predictable particle size may have a great impact, particularly in quantitative morphometric studies based on counting of the label particles (for details, see Chapter 7.) It must be stressed that the relationships described here may be slightly modified if different tissue handling protocols are followed, e.g., the use of freeze substitution or other dehydrating agents than ethanol, or when the tissue blocks contain a higher amount of water during acrylate embedding.

4.8   Summary

Immunogold-silver staining, which has been used for a decade mainly in the light microscopic antigen detection, has recently had promising applications to ultrastructural immunostaining. Small (5 nm) and particularly ultrasmall (1 to 1.4 nm) gold particles covered or conjugated with high affinity antibodies can be applied in pre-embedding methods and to thin sections of resin-embedded specimens. However, these particles can hardly be recognized on the contrasted ultrastructure. With the enlargement of the labeling particles to a predetermined size by a catalytic silver reduction, one can achieve the highest available sensitivity and resolution in a particulate immunolabeling combined with a reliable signal recognition. A silver-enhancement protocol utilizing silver acetate, hydroquinone, and no protecting colloid was found to be compatible with a wide range of electron microscopic resin types of either epoxides (Durcupan ACM and Epon 812) or polyhydroxy acrylates (Lowicryl K4M, LR-White and LR-Gold) in postembedding immunoelectron microscopy. For good results in semithin sections, epoxides have to be eliminated with sodium (m)ethoxide. Likewise, pretreatment of the routinely fixed and highly dehydrated acrylate-embedded semithin sections by microwave cooking was found to be necessary. Ultrastructurally, some differences may be observed in particle enlargement kinetics between the hydrogen peroxide etched epoxides and acrylates. This fact may lose its significance with prolonged silver-enhancement, however. It should be remembered that a high density reaction permits only limited particle enlargement in order to avoid fusion of the neighbouring particles.

Acknowledgments

We are grateful to J. M. Hainfield (Nanogold Inc., Stony Brook, NY) for Nanogold probe, to Zsolt Rázga for help in setting Lowicryl embedding protocol, to Mrs. Mária Bakacsi and

Mrs. Katalin Pajor for skillful technical assistance, and to Mihály Dezsô for help in photographic work.

References

Al-Nawab, M. D. and Davies, D. R. 1989. Light and electron microscopic demonstration of extracellular immunoglobulin deposition in renal tissue. *J. Clin. Pathol.* 42: 1104.

Atherton, A. J., Monaghan, P., Warburton, M. J., and Gusterson, B. A. 1992. Immunocytochemical localization of the ectoenzyme aminopeptidase N in the human breast. *J. Histochem. Cytochem.* 40: 705.

Bienz, K., Egger, D., and Pasamontes, L. 1986. Electron microscopic immunocytochemistry. Silver enhancement of colloidal gold marker allows double labeling with the same primary antibody. *J. Histochem. Cytochem.* 34: 1337.

Bowdler, A. L., Griffits, D. F. R., and Newman, G. R. 1989. The morphological and immunohistochemical analysis of renal biopsies by light and electron microscopy using single processing method. *Histochem. J.* 21: 393.

Bozóky, B., Krenács, T., Rázga, Zs., and Erdôs, A. 1993. Ultrastructural characteristics of glial fibrillary acidic protein expression in epoxy-embedded human brain tumors. *Acta Neuropathol.* 86: 295.

Burry, R. W., Vandré, D. D., and Hayes, D. M. 1992. Silver enhancement of gold-conjugated antibody probes in pre-embedding electron microscopic immunocytochemistry. *J. Histochem. Cytochem.* 40: 1849.

Carlemalm, E., Garavito, R. M., and Villiger, W. 1982. Resin development for electron microscopy and an analysis of embedding at low temperature. *J. Microsc.* 126: 123.

Cattoretti, G., Pileri, S., Parravicini, C., Becker, M. H. G., Poggi, S., Bifulco, C., Key, G., D'Amato, L., Sabattini, E., Feudale, E., Reynolds, F., Gerdes, J., and Rilke, F. 1993. Antigen unmasking on formalin-fixed paraffin-embedded tissue sections. *J. Pathol.* 171: 83.

Causton, B. E. 1984. The choice of resins for electron immunocytochemistry. In: *Immunolabeling for Electron Microscopy.* p. 29. Polak, J. M. and Varndell, I. M. (Eds.), Elsevier, Amsterdam.

Danscher, G. and Nörgaard, J. O. R. 1983. Light microscopic visualization of colloidal gold on resin embedded tissue. *J. Histochem. Cytochem.* 31: 1394.

De Waele, M., De Mey, J., Remnans, W., Labeur, C., Reynaert, P. H., and Van Camp, B. 1986. An immunogold-silver staining method for the detection of cell surface antigens in light microscopy. *J. Histochem. Cytochem.* 34: 935.

Hacker, G. W., Grimelius, L., Danscher, G., Bernatzky, G., Muss, W., Adam, H., and Thurner, J. 1988. Silver acetate autometallography: An alternative enhancement technique for immunogold-silver staining (IGSS) and silver amplification of gold, silver, mercury and zinc in tissues. *J. Histotechnol.* 11: 213.

Hacker, G. W., Springall, D. R., Van Noorden, S., Bishop, A. E., Grimelius, L., and Polak, J. M. 1986. The immunogold-silver staining method. A powerful tool in histopathology. *Virchows Arch. (Anat. Pathol.)* 406: 449.

Hainfeld, J. F. and Furuya, F. R. 1992. A 1.4 nm gold cluster covalently attached to antibodies improves immunolabeling. *J. Histochem. Cytochem.* 40: 177.

Hayat, M. A. 1989. *Principles and Techniques of Electron Microscopy,* 3rd. ed. Macmillan Press, London and CRC Press, Boca Raton, FL.

Herken, R., Fussek, M., Barth, S., and Götz, W. 1988. LR-White and LR-Gold resins for postembedding immunofluorescence staining of Laminin in mouse kidney. *Histochem. J.* 20: 427.

Hobot, J. A. 1989. Lowicryls and low temperature embedding for colloidal gold methods. In: *Colloidal Gold: Principles, Methods, and Applications.* Vol. 2. p. 76. Hayat, M. A. (Ed.), Academic Press, San Diego.

Holgate, C. S., Jackson, P., Cowen, P., and Bird, C. 1983. Immunogold-silver staining: new method of immunostaining with enhanced sensitivity. *J. Histochem. Cytochem.* 31: 938.

Iványi, B., Krenács, T., Dobó, E., and Ormos, J. 1990. Demonstration of bacterial antigen in macrophages in experimental pyelonephritis. *Virchows Arch. B* 59: 83.

Krenács, T., Iványi, B., Bozóky, B., Lászik, Z., Krenács, L., Rázga, Zs., and Ormos, J. 1991. Postembedding immunoelectron microscopy with immunogold-silver staining (IGSS) in Epon 812, Durcupan ACM and LR-White resin embedded tissues. *J. Histotechnol.* 14: 75.

Krenács, T. and Krenács, L. Immunogold-silver staining (IGSS) for immunoelectron microscopy and in multiple detection affinity cytochemistry. In: *Modern Analytical Methods in Morphology.* Gu, J. and Hacker, G. W. (Eds.), Plenum, New York, in press.

Krenács, T., Krenács, L., Bozóky, B., and Iványi, B. 1990. Double and triple immunocytochemical labeling at a light microscopic level in histopathology. *Histochem. J.* 22: 530.

Krenács, T., Lászik, Z., and Dobó, E. 1989. Application of immunogold-silver staining and immunoenzymatic methods in multiple labeling of panreatic Langerhans islet cells. *Acta Histochem.* 85: 79.

Krenács, T., Molnár, E., Dobó, E.,and Dux, L. 1989. Fibre typing using sarcoplasmic reticulum Ca-ATP^{2+}ase and myoglobin immunohistochemistry in rat gastrocnemius muscle. *Histochem. J.* 21: 145.

Lackie, P. M., Hennessy, R. J., Hacker, G. W., and Polak, J. M. 1985. Investigation of immunogold-silver staining by electron microscopy. *Histochemistry* 83: 545.

Lah, J. J., Hayes, D. M., and Burry, R. W. 1990. A neutral pH silver development method for the visualization of 1 nanometer gold particles in pre-embedding electron microscopic immunocytochemistry. *J. Histochem. Cytochem.* 38: 503.

Lee, H. and Neville, K. 1967. *Handbook of Epoxy Resins.* McGraw-Hill, New York.

Newman, G. R. 1989. LR White embedding medium for colloidal gold methods. In: *Colloidal Gold: Principles, Methods, and Applications.* Vol. 2. p. 48. Hayat, M. A. (Ed.), Academic Press, San Diego.

Newman, G. R. and Hobot, J. A. 1987. Modern acrylics for post-embedding immunostaining techniques. *J. Histochem. Cytochem.* 35: 971.

Newman, G. R. and Hobot, J. A. 1989. Role of tissue processing in colloidal gold methods. In: *Colloidal Gold: Principles, Methods, and Applications.* Vol. 2. p. 33. Hayat, M. A. (Ed.), Academic Press, San Diego.

Newman, G. R., Jasani, B., and Williams, E. D. 1983. A simple post-embedding system for the rapid demonstration of tissue antigens under the electron microscope. *Histochem. J.* 15: 543.

Penschow, J. D. and Coghlan, J. P. 1993. Secretion of glandular kallikrein and renin from the basolateral pole of mouse submandibular duct cells: An immunocytochemical study. *J. Histochem. Cytochem.* 41: 95.

Pettitt, J. M. and Humphris, D. C. 1991. Double lectin and immunolabeling for transmission electron microscopy: pre- and post-embedding application using the biotin-streptavidin system and colloidal gold-silver. *Histochem. J.* 23: 29.

Roth, J., Bendayan, M., Carlemalm, E., Villinger, E., and Garavita, M. 1981. Enhancement of structural preservation and immunocytochemical staining in low temperature embedded pancreatic tissue. *J. Histochem. Cytochem.* 29: 663.

Slater, M. 1991. Differential silver enhanced double labeling in immunoelectron microscopy. *Biotech. Histochem.* 66: 153.

Shimizu, H., Ishida-Yamamoto, A., and Eady, R. A. J. 1992. The use of silver-enhanced 1-nm gold probes for light and electron microscopic localization of intra- and extracellular antigens in skin. *J. Histochem. Cytochem.* 40: 883.

Timms, B. G. 1986. Postembedding immunogold labeling for electron microscopy using LR-White resin. *Am. J. Anat.* 175: 267.

Van den Pol, A. N. 1985. Silver-intensified gold and peroxidase as dual ultrastructural immunolabels for pre- and postsynaptic neurotransmitters. *Science (N.Y.)* 228: 332.

Van den Pol, A. N. 1986. Tyrosine hydroxylase immunoreactive neuros throughout the hypothalamus receive glutamate decarboxylase immunoreactive synapses: A double pre-embedding immunocytochemical study with particulate silver and HRP. *J. Neurosci.* 6: 877.

Vandré, D. D. and Burry, R. W. 1992. Immunoelectron microscopic localization of phosphoproteins associated with the mitotic spindle. *J. Histochem. Cytochem.* 40: 1837.

Velde, C. I. and Prins, F. A. 1990. New sensitive light microscopical detection of colloidal gold on ultrathin sections by reflection contrast microscopy. Combination of reflection contrast and electron microscopy in post-embedding immunogold histochemistry. *Histochemistry* 94: 61.

Chapter 5

Silver-Enhancement of Nanogold and Undecagold

James F. Hainfeld and Frederic R. Furuya

Contents

5.1 Introduction

Most previous work with immunogold-silver staining (IGSS) has been done with colloidal gold particles. More recently, large gold compounds ("clusters") having a definite number of gold atoms and defined organic shell, have been used frequently with improved results (Vandré and Burry, 1992). These gold clusters, large compared to simple compounds, are, however, at the small end of the colloidal gold scale in size; undecagold is 0.8 nm and Nanogold is 1.4 nm in diameter. They may be used in practically all applications where colloidal gold is used (light and electron microscopy, dot blots, etc.) and in some unique applications, where the use of at least the larger colloidal gold particles is not feasible. This is exemplified by running gold labeled proteins on gels (which are later detected by silver-enhancement). The main differences between gold clusters and colloidal golds are the small size of the clusters and their covalent attachment to antibodies or other molecules.

5.2 Undecagold

The gold core of undecagold, which contains 11 gold atoms, is 0.8 nm in diameter. The structure of this cluster has been solved by X-ray diffraction of crystals (McPartlin et al.,

0-8493-2449-1/95/$0.00+$.50
© 1995 by CRC Press, Inc.

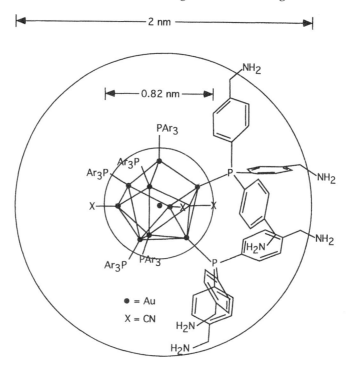

Figure 1 Structure of the undecagold cluster.

1969), and is shown in Figures 1 and 2. The overall diameter, including the organic shell, is 2.0 nm. One or more of the organic moieties around the periphery of this cluster may be derivatized to a functional group, such as an amine (Bartlett et al., 1978), a maleimide (reacts with thiols; Yang et al., 1984; Safer et al., 1986), or N-hydroxysuccinimide ester (reacts with amines; Reardon and Frey, 1984), so that the cluster can be covalently attached to IgG, Fab, or other molecules, including lipids, peptides, proteins, carbohydrates, lectins, and nucleic acids. The attachment is stable (covalent) and specific (coupling to a specific amino acid or site). For example, the undecagold may be coupled to the hinge sulfhydryl of an Fab' fragment (Figure 3; Hainfeld, 1987). The product may be isolated by gel filtration chromatography to remove any unbound gold cluster, and the reaction is virtually quantitative, that is, almost all Fab' molecules are labeled with a gold cluster (Figure 4). The

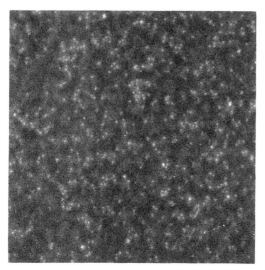

Figure 2 Dark field scanning transmission electron micrograph (STEM) of the undecagold clusters on a thin carbon film. Bright dots are the 11 gold atom cores that have a 0.8 nm diameter. Since it is a compound, there is complete uniformity of size. Full width is 0.128 μm.

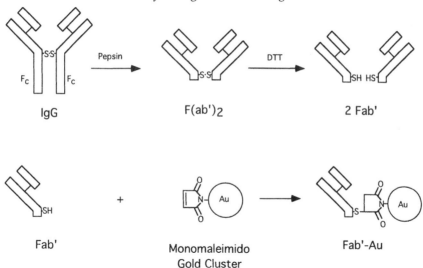

Figure 3 Covalent coupling of undecagold to the hinge thiol of an Fab' fragment.

undecagold cluster is small enough (MW = 5000) to hardly affect the properties of the antibodies, as seen by their identical retention time by gel filtration HPLC and identical or only slightly altered R_f on polyacrylamide electrophoresis gels.

Although the undecagold cluster has many interesting applications, including high resolution single molecule electron microscopy (Hainfeld et al., 1991; Hainfeld, 1992; Milligan et al., 1990), it has a few disadvantages. The cluster is so small that it is difficult to see in the transmission electron microscope (TEM) (although it can be visualized in the high resolution scanning transmission electron microscope (STEM)). It is even more difficult to visualize when it is sitting on ~20 nm of protein (without silver-enhancement). It is sensitive to beam damage by the electron beam and becomes less visible with higher doses (Wall et al., 1982). Moreover, it does not develop with silver robustly. However, upon studying its silver-enhancement properties further for this report, we find that it should be quite useful and unique for many applications.

5.3 Nanogold

Because of the limitations of undecagold, a larger gold cluster with a 1.4-nm gold core was developed (Hainfeld and Furuya, 1992) (Figure 5). This too has an organic shell, giving an

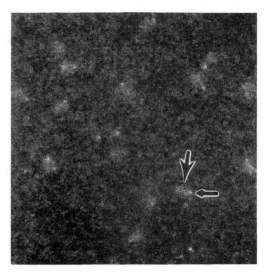

Figure 4 Dark field STEM micrograph of Fab' fragments labeled with undecagold. Large arrow points to whitish area that is the (unstained) Fab' fragment; small arrow points to an undecagold cluster attached to its hinge region. Full width is 0.128 μm.

Figure 5 Darkfield STEM micrograph of Nano-gold 1.4 nm clusters (bright dots). Full width is 0.128 μm.

overall diameter of 2.7 nm. Although crude preparations may contain small amounts of a smaller cluster, it is, on the whole, very uniform, and a major fraction can be purified that is extremely uniform and even forms microcrystals (Figure 6). This cluster, termed $Au_{1.4nm}$, or "Nanogold", has very interesting properties. It can be covalently attached to target molecules in a similar way to undecagold (Figures 7 and 8). Nanogold is much more visible than undecagold in the TEM, does not damage under the electron beam, and develops well with silver (Hainfeld and Furuya, 1992).

5.4   Advantages of Using Gold Clusters

There are several benefits gained by using undecagold or Nanogold over traditional colloidal gold:

1.   Due to their small size, penetration into permeabilized tissue, for example, is significantly better (Burry et al., 1992). In fact, it is up to 40 times that of colloidal gold probes (even when compared to 1 nm colloidal gold probes), and they can diffuse up to 20 μm. Fab' antibody probes may be routinely made, whereas IgG

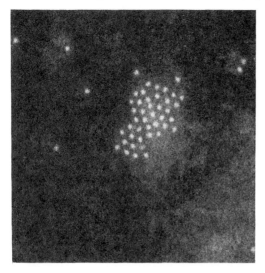

Figure 6 Darkfield STEM micrograph of Nano-gold (bright dots) showing a microcrystal. Full width is 0.064 μm.

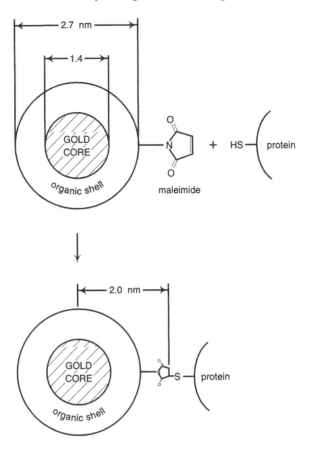

Figure 7 Covalent coupling scheme of Nanogold to a protein thiol showing dimensions of the gold and distance from thiol.

probes are used with most colloidal gold probes. The use of the Fab' fragment (MW = 50,000) already reduces the size of the probe by a factor of ~3 from those made with IgG (MW = 150,000).

2. It has been commonly found that smaller gold particles lead to a greater staining density than larger gold particle probes (Hayat, 1989). Use of the ultrasmall gold

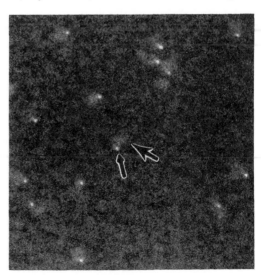

Figure 8 Dark field STEM micrograph of Fab' fragments labeled with Nanogold. Large arrow points to whitish area that is the (unstained) Fab' fragment; small arrow points to a Nanogold cluster attached to its hinge region. Full width is 0.128 µm.

clusters (Nanogold-Fab') has also been shown to effect more quantitative antigen labeling (Vandré and Burry, 1992).

3. The attachment of gold clusters is covalent and stable, whereas colloidal probes may undergo dissociation (Horisberger and Clerc, 1985). Dissociation leads to free antibody which competes for target sites and reduces signal (Kramarcy and Sealock, 1990).

4. Gold cluster-labeled antibodies are not aggregated, as some colloidal gold preparations tend to do. In colloidal gold immunoconjugates, the IgG molecules are adsorbed to the gold particles. Since the gold is "sticky", it tends to aggregate. With the gold cluster approach, the immunoglobulin is coupled only at a specific site, and no aggregation occurs. The gold is not "sticky" and does not attach without chemical activation. In fact, the conjugates are routinely chromatographed using gel filtration, and no aggregates are observed, even upon storage.

5. There is typically 1 antibody per gold cluster, whereas with colloidal gold there may be 0.4 to 25 proteins per gold particle (Horisberger and Clerc, 1985).

6. The size of the gold clusters are very precise since these are compounds; larger colloidal gold particles can be very regular, but smaller ones (<3nm) are heterogeneous. "1 nm" colloidal gold can vary from 1 to 3 nm (Hainfeld, 1990).

7. For immunoconjugates, the gold clusters are usually covalently attached to the hinge sulfhydryl of Fab' or IgG. This is away from the hypervariable binding region of the antibody, so antibody activity is generally completely retained. With colloidal gold, antibodies may attach such that they are no longer active.

8. The small size of the gold clusters improves resolution over larger colloidal gold probes.

9. Gold clusters are available as reagents and can be covalently reacted with antibodies and many other molecules to make a variety of specific molecular probes that are not possible with the colloidal gold approach. For example, the gold clusters may be attached to IgG through their carbohydrate moiety, to nucleic acid bases, to hormones, and potentially to single-chain antibodies. Because of these advantages, generally a much higher density of gold labeling is observed than that obtained with colloidal gold (Vandré and Burry, 1992), and many unique experiments may be designed that are not possible using colloidal gold.

5.5   Disadvantages of Gold Clusters

For particular applications, use of colloidal gold probes may be preferable to gold clusters. The disadvantages of the gold clusters are (1) They are very small, i.e., 0.8 or 1.4 nm. For routine electron microscopy at low magnifications, 10 to 20 nm gold does not require silver-enhancement and may be more convenient. (2) For postembedding work, where diffusion of label is not significant (antibodies do not penetrate the resin), there may be less of an advantage gained by using cluster conjugates compared with colloidal gold conjugates. (3) Although the clusters are regular, after silver-enhancement to 5 to 20 nm, their size is more irregular than colloidal gold of the same size. They are therefore more difficult to use in double labeling schemes. (4) The undecagold develops to give less signal than Nanogold; so even though undecagold is smaller and might seem to be the preferable choice under the criteria "advantages" above, Nanogold may give more satisfactory silver enhanced results. (5) For "do-it-all-yourself" investigators, the gold clusters are much more difficult to make than colloidal gold because they involve multistep organic and inorganic syntheses.

5.6   Silver-Enhancement of Gold Clusters

For high-resolution electron microscopy applications, where the small gold clusters are visible directly, silver-enhancement is not necessary and would detract from those results.

These applications include study of individual molecules either unstained (Hainfeld, 1992) or in a low density stain (e.g., vanadate, where the gold is still visible; Hainfeld et al., 1993), in ice embedded (frozen hydrated) samples (Boisset et al., 1992; Wilkens and Capaldi, 1992), and in crystalline or helical arrays where image processing may be applied (Milligan et al., 1990). Recent work with thin sections by Stierhof et al. (1992), using dark field electron microscopy, demonstrated visibility of 1-nm gold (colloid) directly and should be applicable to gold cluster labeling.

For most other applications, silver-enhancement is required. These applications include standard immunocytochemistry using the TEM or light (or confocal) microscope, and gold staining of gels and blots. Upon the introduction of Nanogold, its excellent response to silver-enhancement was documented (Hainfeld and Furuya, 1992). It was used for light microscopy of cell types, and on immuno-dot blots, where it showed exceptional sensitivity, down to 0.1 pg, or 10^{-18} moles, thus making it more sensitive than most radioactive, fluorescent, and enzymatic probes.

Burry et al. (1992) reported some previous light and electron microscopic studies where they compared Nanogold silver-enhancement to that of 1-nm colloidal gold. They found that upon enhancement, Nanogold gave more uniform particle sizes and a better correlation between enhancement time and particle density. After 6 min of development, the particles had an average size of 20 nm; after 15 min, the sections were optimal for light microscopic localization. These results were with N-propyl-gallate (NPG) enhancer, and differences may be found with alternative silver-enhancer preparations. They also observed a shrinking of silver particles upon posttreatment with OsO_4.

In a study of immunolocalization of proteins associated with the mitotic spindle, Vandré and Burry (1992) found that Nanogold-Fab' was superior to other colloidal gold probes, including a 1 nm one. They compared penetration, label uniformity, and labeling density using various gold probes, followed by silver-enhancement. The tissue in this study was fixed with glutaraldehyde (0.7%), permeabilized with saponin (0.1%), reacted with primary (60 min, 37°) then secondary antibodies (60 min, 37°), postfixed with glutaraldehyde (1.6%), silver enhanced (with NPG, 5 min for electron microscopy, 8 to 10 min for light microscopy), osmicated (30 min, 0.1% OsO_4), then embedded in Epon and sectioned. A tertiary fluorescein conjugated antibody test indicated that an equivalent amount of secondary antibody had reached the microtubules from the Nanogold and 1-nm colloidal probe, but the silver staining was much poorer with the colloidal gold. This implied that free antibody was present in the colloidal gold preparation and degraded its gold staining performance. Nanogold localizations correlated exactly with conventional indirect immunofluorescence, and permitted ultrastructural determination at a high resolution. The use of 10 to 20 nm colloidal gold preparations failed to penetrate adequately into the structures of interest in this study and therefore gave unsatisfactory results. Light microscopy of silver-enhanced samples also showed better staining with Nanogold compared to the 1-nm colloidal gold. For localization of MPM-2 antibody, only the Nanogold conjugates demonstrated specific localization to the centrosome and kinetochores by electron microscopy, confirming immunofluorescence localization studies. This study demonstrated the ability of Nanogold probes to penetrate into dense, structurally complex cytoskeletal organelles, such as the centrosome and the midbody.

Another study with 1-nm colloidal gold immunoconjugates demonstrated significant microaggregation of the gold and antibody, thus making the effective size of the "1 nm" colloidal gold probe much larger (Hainfeld, 1990). This would also explain the significantly lower gold/silver staining compared to Nanogold, where the Fab'-$Au_{1.4nm}$ conjugates are monomolecular, chromatographically purified, and unaggregated.

This chapter attempts to further quantitate the silver-enhancement properties of undecagold and Nanogold, to compare these results with silver-enhancement of colloidal gold, and to show several applications.

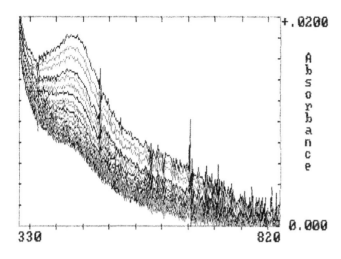

Figure 9 Spectral changes of the silver enhancer (LI Silver, Nanoprobes) alone over time. The spectral range is 330 to 820 nm. The sample was scanned once per minute over 30 min; each scan is represented by a one line spectrum. The first scan is the lowest one with the least absorbance. As can be seen, many scans nearly overlap during the first few minutes, but become further apart as time progresses (the uppermost scans) indicating an acceleration of development.

5.7 Silver-Enhancement Rates

Previously, development rates of colloidal golds have been studied (Gu et al., 1993). Likewise, Nanogold has been compared with 1 nm colloidal gold (Burry et al., 1992). Here we expand the study to include undecagold, Nanogold, 1, 3, and 15 nm colloidal gold. Colloidal golds were first stabilized with bovine serum albumin (BSA) to prevent aggregation upon addition of the enhancer.

Traditionally, electron microscopy samples at various times of development have been used to study and document the rate of silver growth. Here we have also used ultraviolet-visible (uv-vis) spectroscopy to quantify initial silver growth. The experiment involved the incubation of a gold solution with silver developer (LI Silver, Nanoprobes) in a disposable cuvette in a diode array spectrophotometer (Hewlett Packard 8452A), and record spectra every minute. At first, everything is in solution and small silver particle growth can be monitored. Later, large silver particles contribute heavily to light scattering, they eventually fall out of solution, and extensive amounts of silver coats the walls of the cuvette forming a silver mirror. Although there are shortcomings of this method, it appears promising, particularly for early development times.

The first sample is the developer alone. Silver enhancers are known to autonucleate typically after ~30 min and produce some silver particle background. The spectral changes over time are shown in Figure 9. These changes can be plotted versus time if a particular wavelength is followed. The wavelength selected should be one that exhibits the most change. In this case, the range of 430 to 450 nm was averaged (to reduce noise) since a peak at ~440 nm appears strongly over time. Much of the general light scattering and deposition on the walls results in less transmittance, which can be roughly monitored by the absorbance at ~800 nm (an average of 800 to 820 was used). The difference of the 440 absorbance minus this background at 800 nm should better represent the solution behavior. These signals are plotted in Figure 10 for the silver developer.

Similarly, other gold solutions may be studied. Nanogold development is shown in Figures 11 and 12. A final peak at ~440 nm develops similar to the silver developer alone. However, development is ~10,000 times faster, although it is difficult to estimate this factor from these measurements. Also, the absorbance at 810 nm rapidly rises after ~2 min (Figure 12), indicating significant deposition of silver on the walls of the cuvette and light scattering. For comparison, development of a ~tenfold more dilute Nanogold solution is shown in Figure 13. Because the gold still produces a strong signal, it raises the question as to at what level gold could be ultimately detected by this means (uv-vis spectroscopy with silver-enhancement). Therefore, a very dilute Nanogold solution was measured that gave

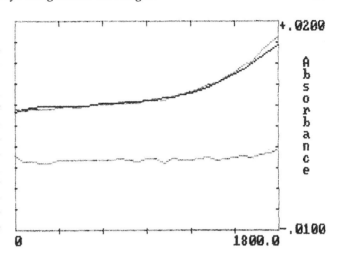

Figure 10 Absorbance at particular wavelengths for the silver enhancer alone plotted vs. time (in seconds; full width is 30 min). Top light trace is the absorbancy of the average of 430 to 450 nm, thus picking up the 440-nm peak that develops (Figure 9). The lower light trace is the absorbance at 810 nm (800 to 820 nm data averaged), showing little change or coating of the cuvette walls with silver over this period. The darker trace is the difference between these curves.

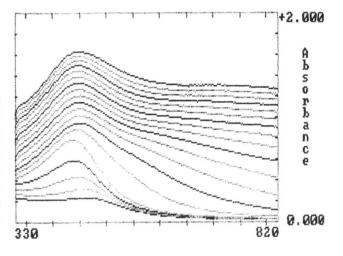

Figure 11 Silver-enhancement of Nanogold tracked by uv-vis spectral changes. Each spectrum is 1 min apart. An initial peak appears at ~460 nm, shifts to ~430 then back to 450 nm. Nanogold concentration was 2.2 × 10⁻⁶ M. Concentration of silver developer (LI Silver, Nanoprobes) was the same as used in the control shown in Figures 9 and 10.

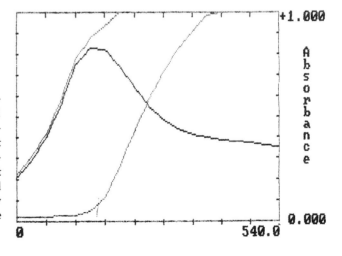

Figure 12 Development of Nanogold (2.2 × 10⁻⁶ M) from Figure 11 replotted using the change at a particular wavelength vs. time. Top left light line is 440 nm (430 to 450 averaged) and rightmost light line is at 810 nm (800 to 820 averaged). Full time scale is 540 sec (9 min). Heavy line is the difference between the other two plots.

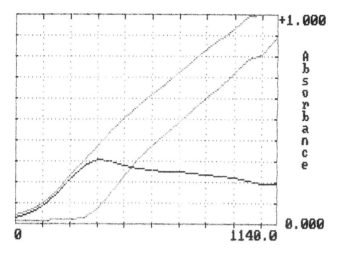

Figure 13 Development of Nanogold (2.9 × 10⁻⁷ M) monitored by
absorbance at 440 nm (top light line),
810 nm (lower light line), and the
difference of these graphs (dark line).
Time scale is in sec, 1140 max (19
min).

a significant reading over the background silver enhancer alone. This is shown in Figures
14 and 15. The absorbance (at 440 nm) vs. time for this dilute gold solution with silver
enhancer compared with the silver enhancer alone is plotted in Figure 16. A significant
difference can be detected in 5 to 10 min. Assuming a 50 μl cuvette cell, this would give a
detection of ~10⁻¹³ moles by this method.

Undecagold silver-enhancement was subjected to the same uv-vis spectral analysis.
This result is shown in Figures 17 and 18. The spectra show a very sharp peak appearing
at 370 nm. Usually sharp peaks indicate a tight size distribution or perhaps a new gold-
silver alloy cluster that is well defined. Further work is needed to characterize this new
product. It is clear that the end-point (after 15 min) is quite different from that of the
Nanogold or silver enhancer alone (they have peaks at 440 nm). A second difference
between the undecagold and Nanogold is the slow and somewhat delayed development
of the undecagold. Undecagold begins with almost a zero rate which then picks up after
a few minutes (Figure 18). Also, the concentration of undecagold was ~5 times higher than
that of the Nanogold shown in Figures 11 and 12, but it gave less overall development.
Silver-enhancement of undecagold is therefore slower and less robust than that of Nanogold.
That is not to say, however, that the use of undecagold should not be considered further.
It is a smaller probe, and the overall performance with silver-enhancement, as shown later,
can be quite useful in most applications. For comparison, the results for Nanogold and
undecagold discussed thus far are plotted on the same scale in Figure 19.

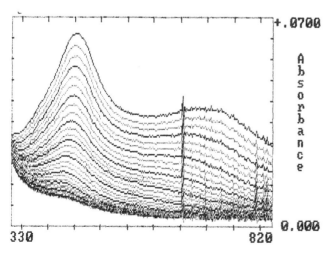

Figure 14 Silver-enhancement of a
very dilute Nanogold solution (1.7 ×
10⁻⁹ M) monitored by uv-vis spectroscopy. Spectra are recorded every
1 min. A major peak at 450 nm and a
minor one at ~710 nm appear.

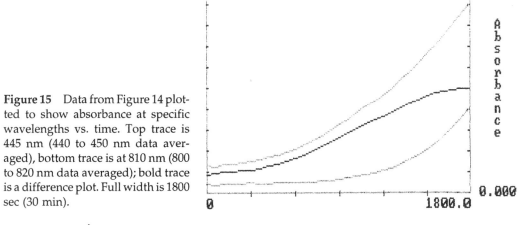

Figure 15 Data from Figure 14 plotted to show absorbance at specific wavelengths vs. time. Top trace is 445 nm (440 to 450 nm data averaged), bottom trace is at 810 nm (800 to 820 nm data averaged); bold trace is a difference plot. Full width is 1800 sec (30 min).

Figure 16 Silver-enhancement of a very dilute Nanogold solution (1.7×10^{-9} M) compared to the silver-enhancement solution (LI Silver) alone. The optical density at 440 nm was monitored over time. Significant signal is seen after 5 to 10 min for the Nanogold, and would correspond to a detection of $\sim 10^{-13}$ moles in a 50 µl cuvette cell.

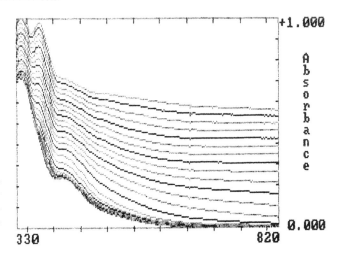

Figure 17 Undecagold spectra taken at 1 sec intervals during silver-enhancement. The undecagold concentration was 1.2×10^{-5} M. A distinct peak at 370 nm appears.

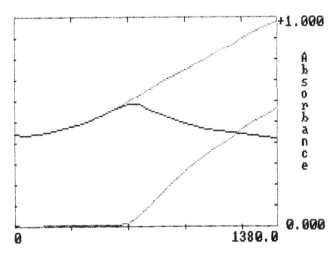

Figure 18 Undecagold silver-enhancement (same as for Figure 17) monitored over time using the absorbance at 372 nm (370 to 374 nm data averaged, top light trace), 810 nm (800 to 820 nm averaged, lower light trace) and their difference (bold line). Time period is 1380 sec (15 min).

For comparison, 3-nm colloidal gold was treated in the same fashion (Figures 20 and 21). The 3-nm gold shows the usual appearance of a peak at 440 nm. Compared to the Nanogold, development at approximately the same gold concentration (Figure 13), there is relatively little coating of the walls with silver or heavy light scattering (as judged by 810 nm absorption) for the 3-nm gold, but an extensive amount with the Nanogold. More silver product is produced by the Nanogold, so it appears more efficient at silver development.

Fifteen nm colloidal gold development is shown in Figures 22 and 23. The usual 530 nm peak for this size colloidal gold is seen in its spectrum at zero time (the lowest spectrum in Figure 22). Development is accompanied by a shift of the 530 nm peak to 500 nm, and an appearance of the usual 440 nm peak. Similar to the 3-nm colloidal gold, there is little change in the 810 absorption. The solution remains clear but colored over this time, and is free of any noticeable gross silver deposition. Much more silver product is observed than with the Nanogold of a similar concentration (Figure 15), where an O.D. value of only 0.07 is reached after 30 min, whereas the 15-nm gold gave 0.9 O.D. after 10 min. There are,

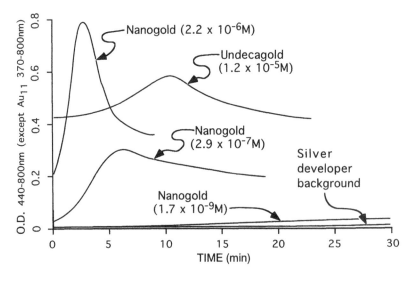

Figure 19 Development of Nanogold and undecagold as assayed by absorbance in solution. The results from Figures 10, 12, 13, 15, and 18 are plotted here on the same scale. The absorbance difference of 440 nm minus 810 nm is used for all samples except the undecagold, which is 370 nm minus 810 nm.

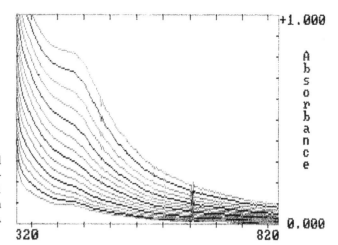

Figure 20   Spectra of 3-nm colloidal gold as it is silver enhanced in solution. Spectra are recorded every 1 min. Colloidal gold concentration was 5×10^{-7} M, and included 1% BSA for stabilization.

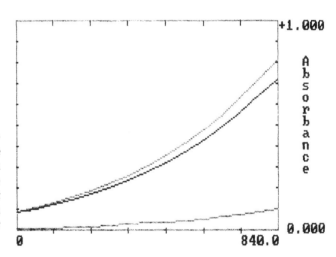

Figure 21   Silver-enhancement of 3-nm colloidal gold (same as Figure 20) replotted to show the change in absorbance at 440 nm with time (top light trace), 810 nm (bottom light trace) and the difference between these two (bold trace). Time is over 840 sec (14 min).

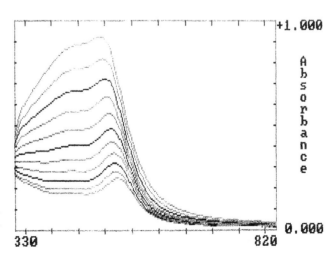

Figure 22   Spectra of 15 nm colloidal gold as it is silver enhanced in solution. Spectra are recorded every 1 min. Colloidal gold concentration was 1.3×10^{-9} M, and included 1% BSA for stabilization.

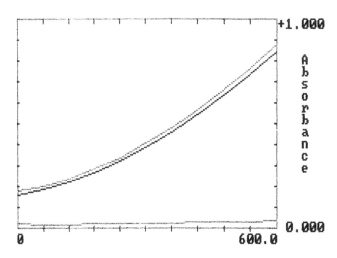

Figure 23 Silver-enhancement of 15-nm colloidal gold (same as Figure 22) replotted to show the change in absorbance at 440 nm with time (top light trace), 810 nm (bottom light trace), and the difference between these two (bold trace). Time is over 600 sec (10 min).

however, ~1000 more gold atoms per 15-nm particle than per Nanogold, and this affects the results. Development of 40 times more dilute 15 nm colloidal gold solution is shown in Figures 24 and 25. A slightly different behavior is observed than for the more concentrated sample. The peak at 530 nm shifts to 500 then 480 nm, but instead of a peak at 440 nm, one at 570 appears. Development is also accompanied by a larger proportional shift in the 810 nm absorbance. On a molar basis, 15-nm colloidal gold is more detectable than Nanogold under these conditions.

5.8 Application to Electron Microscopy

As far as we know, there has not been any previous report of undecagold probes being used with silver-enhancement. As shown above, these probes do develop, albeit slower than other gold particles. The advantage of their smaller size may compensate for their slower development. A TEM micrograph of silver enhanced undecagold is shown in Figure 26. There is some diversity of sizes, but many particles are ~10 nm, a useful range for ultrastructural studies. The use of undecagold probes with silver-enhancement for such uses should be pursued further. Nanogold develops more rapidly than undecagold, as shown above. A typical TEM micrograph of silver enhanced Nanogold is shown in Figure 27.

Gold cluster immunoprobes are the smallest commercially available (they may be obtained from Nanoprobes, Inc., Stony Brook, NY, or such distributors as in the U.S.: E.F.

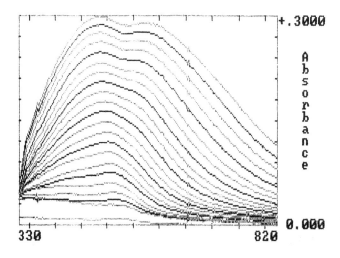

Figure 24 Spectra of 15-nm colloidal gold as it is silver enhanced in solution. Spectra are recorded every 1 min. Colloidal gold concentration was 3.3×10^{-11} M, and included 0.025% BSA for stabilization.

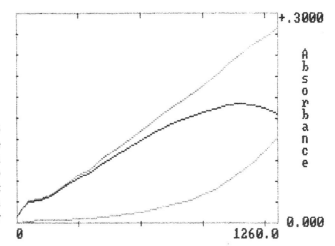

Figure 25 Silver-enhancement of 15 nm colloidal gold (same as Figure 24) replotted to show the change in absorbance at 440 nm with time (top light trace), 810 nm (bottom light trace), and the difference between these two (bold trace). Time is over 1260 sec (21 min).

Fullam, Polysciences; Japan: Cosmo Bio Co., Funokoshi, Ltd., Wako Chemicals; Germany: Biotrend, Ltd.). Use of Fab'-Nanogold ($Au_{1.4nm}$) is considerably smaller than IgG-1 nm colloidal gold because the Fab' fragment is about one third the size of the IgG; additionally, there is no aggregation. Although Fab-colloidal gold probes have been made (Baschong and Wrigley, 1990), they are stabilized by adsorbing BSA typically, which is larger (68 kDa) than the Fab fragment (50 kDa), thus increasing the overall probe size. The most striking improvements in immunolabeling with gold cluster probes over colloidal gold are often observed in pre-embedding localization where penetration of the probe is an important factor. The small probes penetrate and more quantitatively label antigens. Subsequent silver-enhancement leads to a strong signal and improved localizations.

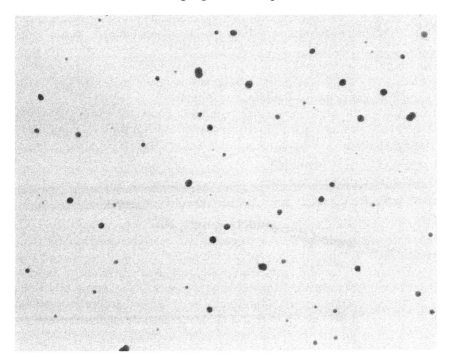

Figure 26 Transmission electron micrograph of undecagold clusters silver enhanced for 15 min. Undecagold concentration was 1.2×10^{-5} M and LI Silver (Nanoprobes) was used as the developer. Many particles are in the 10 nm range. Full width is 700 nm.

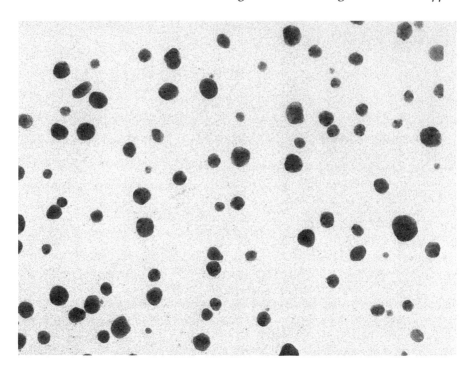

Figure 27 Transmission electron micrograph of Nanogold clusters silver enhanced for 3 min. Nanogold concentration was 7×10^{-6} M and IntenSE M (Amersham) was used as the developer. Many particles are in the 20 nm range. Full width is 700 nm.

An example of the excellent immunostaining obtainable using Nanogold-Fab' followed by silver-enhancement (done by Dr. Susan J.-H. Tao-Cheng), is shown in Figure 28, where synaptophysin was localized in PC12 cell neurites. Other examples of pre-embedding use are found in an article by Vandré and Burry (1992) where they found dramatic improvement in immunolabeling at the ultrastructural level using Nanogold probes compared to 1 nm and other sized colloidal gold probes.

Postembedding immunostaining using antibodies coupled to Nanogold (followed by silver-enhancement) has also given improved results. A beautiful example of this, provided by Dr. Sarah Bacon, is shown in Figure 29, where GABA containing terminals from rat spinal cord were clearly identified.

5.9 *Application to Light and Confocal Microscopy*

The greatly improved penetration of Nanogold-Fab probes over colloidal gold probes can also provide dramatic results for light and confocal microscopy. One example of this is the staining of a nuclear oncogene protein by Dr. Yair Gazitt. A primary monoclonal antibody was used, followed by goat-antimouse-Fab'-Nanogold, then silver-enhancement with LI Silver. This procedure is shown in Figure 30. Intense staining was observed in the nucleus where expected, confirming results with fluorescently labeled secondary antibodies. An interesting aspect of this study was that a wide variety of commercial colloidal gold immunoprobes using gold particles of different sizes were tested but all failed to access the nucleus and stain the antigen.

Another excellent result was obtained by Vandré and Burry (1992). They immunostained cells during mitosis with antitubulin antibodies and probed with Nanogold-Fab', followed by silver-enhancement. Intense and specific staining of the spindle microtubules was

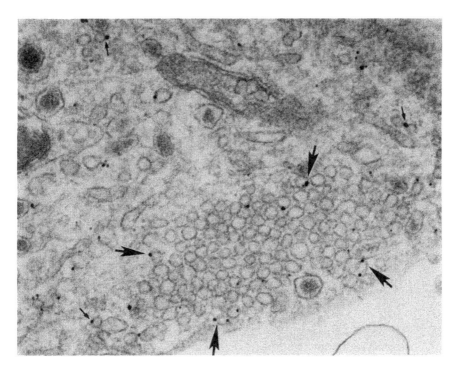

Figure 28 Transmission electron micrograph of Nanogold-Fab' (GAM) targeting a monoclonal antibody to synaptophysin in a PC12 cell neurite. Large arrows indicate silver enhanced gold located on the membranes of a cluster of synaptic vesicles. Small arrows indicate labeling on individual small clear vesicles. Note the excellent structural preservation and specific labeling. PC12 cells were grown in culture in a monolayer and treated with nerve growth factor. After fixing and permeabilizing (with 0.1% saponin), primary and secondary antibody incubations were done for 1 h. each, then silver enhanced with HQ Silver (Nanoprobes) for 4 min. Poststaining was with OsO$_4$ and uranyl acetate, and subsequent embedding was in Epon. Silver spots show a relatively uniform diameter of ~10 nm. Full width is 1.35 μm. This work was done by Dr. Susan J.-H. Tao-Cheng, NINDS EM Facility, Laboratory of Neurobiology, NINDS/NIH.

observed (Figure 31A). A parallel experiment using 1 nm colloidal gold-IgG probe showed only weak staining (Figure 31B). These results indicate that frequently better results can be obtained in immunocytochemistry for light microscopy using the small Nanogold probes and silver-enhancement than with colloidal gold probes, especially when tissue or cell penetration is a factor, e.g., when probing for internal cellular structures.

5.10 Application to Blots

Immunodot blots using colloidal gold labeled secondary probes have been previously described (Moeremans et al., 1984). Silver-enhancement of these dot blots gave excellent sensitivity, and were deemed to be more sensitive than radioactive or fluorescent probes (Moeremans et al., 1984; Hsu, 1984; Holgate et al., 1983; Wu et al., 1990). The gold clusters (undecagold and Nanogold) may also be used in this application. The question is: Are they more sensitive than colloidal gold or are there other advantages/disadvantages to their use? For blots, development is carried out for longer times (20 to 60 min) than for electron or light microscopy to produce a maximal amplification and size of the silver particles. Their size is then generally beyond that useful for electron microscopy. Also, the development is near the longest possible without incurring excessive background (due to developer self nucleation), in order to achieve maximal sensitivity and visibility. Development

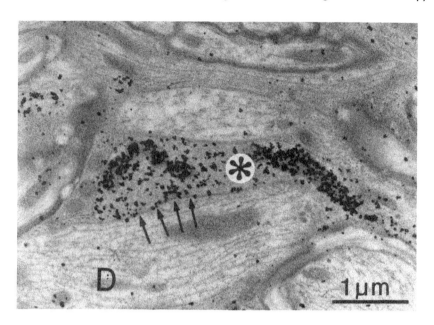

Figure 29   Transmission electron micrograph showing GABA-containing terminals (asterisks) forming symmetrical synaptic specializations (arrows) with dendrites (D) in the thoracic spinal cord of the rat. After embedding in Durcupan (Fluka) and sectioning, immunolocalization was done by incubating with GABA antiserum (Incstar, 1:2000, 4°C, 18 hrs), Nanogold (goat-antirabbit, 1:40, room temperature, 90 min), and intensified with HQ Silver (Nanoprobes) for 6 min. Counterstaining was with lead citrate. There is excellent structural preservation and the silver enhanced gold particles are ~20 nm. This work was done by Dr. Sarah Bacon, Oxford University, Department of Pharmacology, U.K.

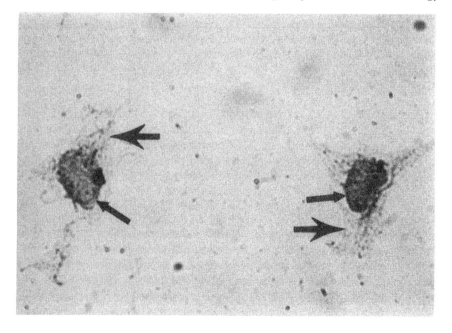

Figure 30   Light micrograph of two neuroblastoma cells (large arrows) immunostained for a nuclear protein using Nanogold-Fab'. The cell boundary and cytoplasm are very light since the cytoplasm is unstained. After incubation with the primary monoclonal and then goat-antimouse Fab'-Nanogold, silver-enhancement was done using LI Silver (Nanoprobes). Small arrows point to the intense nuclear staining. Full width is 290 μm. This work was done by Dr. Yair Gazitt, Bone Marrow Purging Laboratory, University of Florida College of Medicine.

Figure 31A Light micrograph of spindle microtubules labeled with a monoclonal antitubulin antibody, followed by Nanogold-Fab', and silver enhanced. LLC-PK cells were grown in monolayer culture, fixed with glutaraldehyde (0.7%, 15 min), permeabilized with 0.1% saponin, incubated with antitubulin antibodies (1:250, Amersham) for 1 h at 37°C, rinsed, then incubated with antimouse Nanogold-Fab' (1:50, Nanoprobes) for 1 h at 37°C. Samples were postfixed, silver enhanced with a N-propyl-gallate enhancer for ~9 min, osmicated (0.1% OsO_4, 30 min), dehydrated, embedded in Epon, and sectioned. Intense staining was observed. Full width is 95 μm.

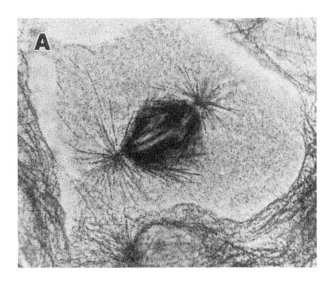

can be extended with fresh enhancer solution every ~20 to 30 min (after an intervening water wash), since the self-nucleation process depends upon the time the enhancer sits once mixed. At some point, however, no further improvement in signal-to-noise ratio is seen (usually at 30 to 60 min), even after applying fresh developer solutions. The silver grains presumably slow in growth or their percentage change in radius decreases.

In order to investigate the sensitivities achieved on blots, several experiments were done. One was to spot known concentrations of gold solutions (without antibody) directly onto nitrocellulose paper. After drying, these were developed with silver enhancer. Typical results showing undecagold and Nanogold are shown in Figure 32. This showed a sensitivity (minimal amount detected by eye) of 10^{-19} moles for Nanogold and 10^{-15} moles for undecagold. Similar blots were done for 1-, 3-, and 15-nm colloidal gold, and the sensitivities for all are summarized in Table 1.

These results show that Nanogold reaches a silver enhanced end-point that is slightly greater than 15-nm colloidal gold on a molar basis. This is different from the solution development results followed by uv-vis spectroscopy reported above, where the 15 nm gave a better signal for dilute solutions. Undecagold seems comparable to 1-nm colloidal gold.

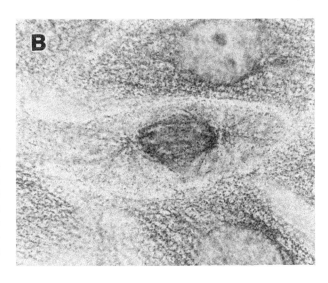

Figure 31B Same preparation as described in (A), except that AuroProbe One goat-antimouse was used (1:50, Amersham). Only weak staining was observed. Full width is 95 μm. This work was done by Drs. Dale Vandré and Richard Burry, Department of Cell Biology, Neurobiology, and Anatomy, The Ohio State University.

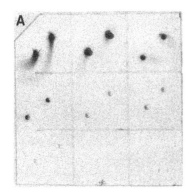 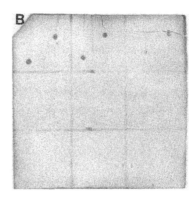

Figure 32   Dilutions of Nanogold (A) and undecagold (B) applied to nitrocellulose paper, dried, and silver enhanced. Amount in upper left box of each blot was 1 picomole per spot (applied in 1 µl). Dilutions were made in water, each box representing a tenfold dilution, and duplicate samples were spotted in each box. Lower right box was diluent only (in this case water). Development was for 30 min. Actual blots were whiter, but overexposed prints were made to improve visibility of spots. Frequently, the faintest spot must be detected with a hand lens. The Nanogold developed in all 8 boxes, corresponding to detection of 10^{-19} mol. The silver enhanced undecagold was visible by eye only in 4 boxes, corresponding to 10^{-15} mole sensitivity.

Gold clusters have an organic shell and are stable in salt solutions; gold colloids made by the usual means have a thin ionic shell of $AuCl_2^-$ and H^+ (Weiser, 1933), but the sols are unstable to the addition of salts or many other molecules which cause the particles to aggregate. Colloidal gold is stabilized by the adsorption of immunoglobulins, BSA, or other polymers, such as Carbowax (a type of polyethylene glycol, PEG). These adsorbents form a coating around the gold. A question is whether these coatings affect the development by silver enhancers. To answer this question, the above blot experiments were repeated with the golds conjugated to antibodies (Nanogold and undecagold), and antibodies followed by the usual BSA stabilization for the colloidal golds. Since the golds absorb in the visible range, the amount of gold may still be measured. These tests gave the same resultant sensitivities as gold without antibodies listed in Table 1. The protein attachment does not therefore appear to interfere with silver-enhancement.

Another level of testing is to put a target molecule on the nitrocellulose, e.g., mouse IgG, block the rest of the paper with the usual BSA blocking solution, then incubate with goat-antimouse antibody attached to the various golds studied. Sensitivity of detection is

Table 1   Sensitivity (Minimal Amount
Detected by Eye) on Blots of Gold
Solutions Only, Spotted on
Nitrocellulose, Dried, Then Silver
Enhanced for 30 min

Gold	Sensitivity (moles)
Undecagold	10^{-15}
Nanogold	10^{-19}
1-nm colloidal	10^{-15}
3-nm colloidal	10^{-16}
15-nm colloidal	10^{-18}

Note: Experimental error, determined from
several runs, is in the order of a factor
of 5 to 10.

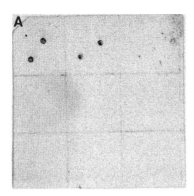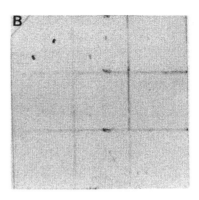

Figure 33   Immunodot blot with Nanogold-IgG (A) and undecagold-IgG (B). First, a mouse IgG was applied in tenfold dilutions to the first 8 boxes (a duplicate spotted in each box), and buffer (the diluent, PBS) applied to the lower right box, dried, then blocked in BSA buffer (Moeremans et al., 1984). The highest concentration, in the upper left box, was 0.1 μg. The blots were then incubated with the goat-antimouse gold conjugate for 1 h at room temperature, washed, and silver enhanced for 30 min. Blots are printed darker than they were to improve visibility of spots. The Nanogold-IgG showed 5 boxes visible (10^{-11} g or ~10^{-16} mol target detected), and the undecagold-IgG showed 2 boxes developing (10^{-8} g or 10^{-13} moles detected).

then measured. This corresponds to the usual immunodot blot procedure and therefore compares the conjugates and should indicate the best one for this application. Examples are shown in Figure 33 for Nanogold-IgG and undecagold-IgG. A compilation of the sensitivities for the various golds is given in Table 2.

These results indicate that Nanogold and 15-nm colloidal gold immunoconjugates give approximately the same sensitivity, and that undecagold conjugates are ~1,000 times less sensitive in this application, and perform slightly worse than 1 nm colloidal gold conjugates.

Immunoblots depend on the quality of the antigen and antibody used, as well as the alterations that may occur upon conjugation to the gold. The above immunodot blots used the same antibody and antigen material for comparison. However, we frequently observe more sensitive detection using the Nanogold conjugates. An example of this is shown in Figure 34 where 0.1 pg mouse IgG target was detected, corresponding to 6.7×10^{-19} moles of target antigen. This is similar to the 5.5×10^{-18} moles (of target DNA) detected using chemiluminescence (Pollard-Knight et al., 1990). However, the chemiluminescent experiment required a 12 h exposure of film and subsequent development, whereas the gold

Table 2   Sensitivity (Minimal Amount Detected by Eye) on
Immunodot Blots of Gold Immunoprobes (Goat-Antimouse)
to a Mouse IgG Target (the Antigen),
After Silver-Enhancement for 30 min

Gold Immunoprobe	Sensitivity (g target)	Sensitivity (moles target)
Undecagold	10^{-8}	10^{-13}
Nanogold	10^{-11}	10^{-16}
1 nm colloidal	10^{-9}	10^{-14}
3 nm colloidal	10^{-7}	10^{-12}
15 nm colloidal	10^{-11}	10^{-16}

Note: More sensitive results have been obtained (10^{-13} g, Figure 34), but these are
reported since the same antibody and antigen preparations were used for all
the golds, giving a better comparison.

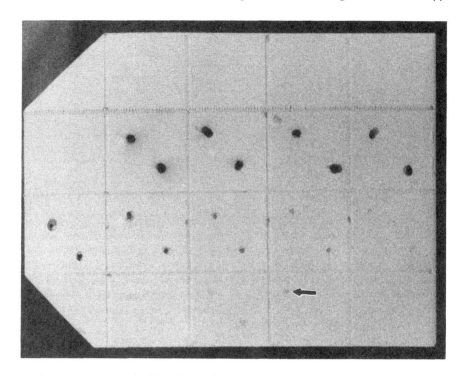

Figure 34   Sensitive immunodot blot obtained with Nanogold antimouse Fab' to a mouse IgG target showing 0.1 pg detection (arrow), corresponding to 6.7×10^{-19} moles of target. One microliter of mouse IgG (antigen target) was spotted onto nitrocellulose containing the following amounts (each box contains a duplicate spot): top row (of spots): 10 ng, 2.5 ng, 1 ng, 0.25 ng; middle row: 100 pg, 25 pg, 10 pg, 2.5 pg; bottom row: 1 pg, 0.25 pg, 0.1 pg, buffer blank. Membrane was blocked with 4% BSA. Goat-antimouse Fab'-Nanogold was incubated for 2 h, washed, and developed 2 × 15 min with LI Silver. Buffer and nonspecific antibody/antigen controls were blank.

detection produces a permanent record directly in ~30 min. It appears that gold detection on blots may rival the other most sensitive detection schemes thus far developed.

5.11 Heating Gold Clusters

One ancillary point is whether the gold clusters are sensitive to heating, since many procedures require a heating step, for example, curing of Epon or other embedding resins at 60°C, or boiling proteins in SDS before electrophoresis. Yang et al. (1984) described "marked, progressive changes in the visible absorption spectrum" detectable within a few minutes when heating undecagold in the presence of air. They reported that most decomposition did not occur under anaerobic conditions, or in the presence of borohydride.

We have therefore measured the sensitivity of the gold clusters to heating at various temperatures for various times in various solvents. They are, on the whole, very resistant to heating. Integrity was assessed by measuring the change of absorbance at 420 nm. At room temperature or 37°C there was no significant loss (<10%) for undecagold or Nanogold after 4 h, either in PBS or 20 mM phosphate (pH 7.4), 0.5 M NaCl. At 60°C the undecagold showed no change after 4 h, but the Nanogold showed ~20% loss. At 100°C, the undecagold was stable for at least 4 h, and the Nanogold showed 42% loss during this period (Figure 35). Behavior was similar in the buffers mentioned. Interestingly, heating to 100°C in 1.3% SDS for 5 min of either undecagold or Nanogold caused no loss, so the SDS may stabilize the structures. The results of Yang et al. (1984) were confirmed by first bubbling air into the undecagold solution for 1 h. After that, heating for 5 min at 100°C caused a 23% loss of cluster.

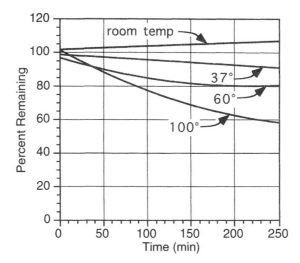

Figure 35 Plot of Nanogold stability upon heating. Degradation was monitored by following the optical density at 420 nm. Solvents of PBS or 20 m*M* phosphate, pH 7.4, 0.5 M NaCl gave virtually the same results and were averaged. Fitted curves are shown, and values >100% indicate experimental error.

Blots of clusters that had been heat treated to 100°C (i.e., the gold was applied to a nitrocellulose paper, dried, then silver-enhanced) showed approximately the same sensitivity as unheated gold. Slight losses in detectability, comparable to the losses detected by UV (420 nm measurements, Figure 35), were observed in the blots. From this study it appears that undecagold may be used in degassed buffers at 100°C over long periods, and that Nanogold may also be heated without terribly significant losses even at 100°C for 1 h.

5.12 Application to Gels

Colloidal gold immunoconjugates have previously been used to detect bands on blot transfers of gels (Moeremans et al., 1984). Most stabilized colloidal gold is negatively charged, but it can also be positively charged by coating with polylysine (Skutelsky and Roth, 1986). These charged gold particles have now been tried as postrun gel stains which presumably work by differential affinity for proteins or nucleic acids vs. affinity for the gel material; sometimes more gold binds to the gel matrix leading to a negative staining of bands. Initial transfer of the gel band to a blot material (e.g., nitrocellulose) is frequently required since diffusion of gold colloids into the gel matrix is poor, whereas staining of a blot is closer to a surface binding situation.

Gold clusters may be used in the same way: immunoconjugates may be used, or positive, negative, and uncharged gold clusters may be synthesized to exploit other staining strategies. In addition, the small size of the gold clusters permits easy diffusion into gels and avoids the necessity of first making blot transfers. Another possibility is to actually electrophorese the clusters on a gel. Undecagold monomers and synthetic undecagold dimers were separated by SDS polyacrylamide gel electrophoresis (PAGE) by Yang et al. (1984).

Furthermore, it should be possible to run gold cluster-labeled proteins on gels. Since gold clusters are relatively small compared to most proteins, and since they are covalently bound, they should run with the protein during PAGE. This process was reported for an undecagold labeled ribosome subunit by Weinstein et al. (1989). The 5 kDa weight of the

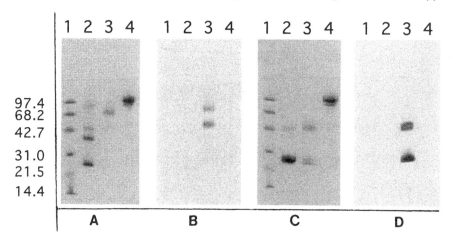

Figure 36 SDS polyacrylamide Phast gels of native and Nanogold labeled proteins, with development by Coomassie Blue or silver-enhancement. Lane 1 is a protein molecular weight standard (values listed on left are in kDa), lane 2 is native Fab', lane 3 is Nanogold-Fab', and lane 4 is F(ab')2. Gels A and C are developed with Coomassie Blue and gels B and D are developed with a silver enhancer (LI Silver, Nanoprobes). A and B are gels of samples that were not heated before running, and C and D are gels of samples heated to 100°C in 1.3% SDS for 5 min before running. Gels A and B were identical except for staining, as were gels C and D. The unheated samples show native and Nanogold labeled Fab' to run anomalously, showing bands >50 kDa, whereas F(ab')2 runs at ~100 kDa as expected. After heating (gels C and D), the Fab' runs as expected, showing bands at 50 kDa, and single light or heavy chains at 25 kDa. The Nanogold labeled Fab' bands are nearly indistinguishable from the native Fab' bands in this case (gel C, lanes 2 and 3). In all cases, the silver-enhancement specifically developed the Nanogold labeled proteins selectively (gels B and D), and unlabeled proteins did not develop (gels B and D, lanes 1, 2, and 4). In addition, Nanogold bands with silver-enhancement were intense in <5 min, whereas Coomassie staining took 1 h (followed by 1 h of destaining).

cluster appeared to retard the migration by that amount on the gel, so that labeled and unlabeled proteins could be distinguished. Another example was provided by Boeckh and Wittmann (1991), where another ribosomal protein was labeled and run on PAGE, and the labeled protein was retarded by the weight of the cluster. Staining was by Coomassie and conventional gel silver staining.

Further work in our laboratories has shown that Nanogold also may be used directly in gels; some samples however indicated no detectable retardation corresponding to the weight of the cluster. Perhaps it is small enough or embedded well enough to only marginally alter the protein migration. A further study showed that if the sample (including protein and SDS) was not heated before running, some gold labeled protein bands were retarded. However, under these conditions (no heating), some native proteins also run anomalously. If, however, they were boiled for 5 min before running (with SDS), native proteins ran as expected, and the Nanogold labeled bands then also ran identically to unlabeled protein (Figure 36).

An interesting finding was that silver-enhancement of gold cluster-labeled proteins run on gels yielded intense bands after only a few minutes of development, compared to the usual 1 to 4 h required for regular gel staining. This intense staining was with a stoichiometry of one gold cluster per protein molecule. Unlabeled protein bands do not develop with even extended silver-enhancement times. Use of this property permits one to selectively stain and determine labeled subunits, for example. A parallel gel using Coomassie Blue would indicate total protein distribution (e.g., Figure 36).

There is one restriction to using gold clusters on gels: The gold clusters are sensitive to reducing agents, such as β-mercaptoethanol (BME) or dithiothreitol (DTT); they decompose over a period of time in these reagents. Therefore, these gold clusters are not very useful for reducing gels, but may be used successfully with native or (nonreducing) SDS gels.

The specific staining of Nanogold labeled subunits on gels was used by Wilkens and Capaldi (1992) to identify the subunit labeled in a multicomponent system, F1 ATPase. They also did immunological identification of the bands to confirm their result. In their application, they saw a slight retardation of the Nanogold-labeled subunit compared to the native unlabeled subunit; this difference roughly corresponded to the weight of the Nanogold (15 kDa).

The common silver stain specifically designed for staining protein or nucleic acid gels was found to more effectively (densely) stain gold cluster labeled bands; the normal staining procedure and approximate time may be used. The use of this stain shows the gold labeled bands at first, and then the nonlabeled bands develop. This pattern is in contrast to the usual silver enhancers for gold which only develop the gold labeled bands.

5.13 Summary

A recent advance in immunogold technology has been the use of molecular gold instead of colloidal gold. A number of advantages are realized by this approach, such as stable covalent, site-specific attachment, small probe size and absence of aggregates for improved penetration. Silver-enhancement has led to improved and unique results for electron and light microscopy, as well as their use with blots and gels.

Acknowledgments

The authors thank Kyra Carbone, Inan Feng, and Beth Lin for excellent technical assistance, Martha Simon and Frank Kito for operation of the STEM, and Joseph Wall for fruitful discussions. We also thank Drs. Sarah Bacon, Susan J.-H. Tao-Cheng, Dale Vandré, Richard Burry, and Yair Gazitt for contributing beautiful immunocytochemical micrographs. This work was supported by the Office of Health and Energy Research of the U.S. Department of Energy.

References

Bartlett, P. A., Bauer, B., and Singer, S. J. 1978. Synthesis of water-soluble undecagold cluster compounds of potential importance in electron microscopic and other studies of biological systems. *J. Am. Chem. Soc.* 100: 5085.

Baschong, W. and Wrigley, N. G. 1990. Small colloidal gold conjugated to Fab fragments or to immunoglobulin G as high-resolution labels for electron microscopy: a technical overview. *J. Electron. Microsc. Tech.* 14: 313.

Boeckh, T. and Wittmann, H.-G. 1991. Synthesis of a radioactive labeled undecagold cluster for application in X-ray structure analysis of ribosomes. *Biochim. Biophys. Acta* 1075: 50.

Boisset, N., Pochon, F., Chwetzoff, S., Barray, M., Delain, E., and Lamy, J. 1992. Electron microscopy of α_2-macroglobulin with a thiol ester bound ligand. *J. Struct. Biol.* 108: 221.

Burry, R. W., Vandré, D. D., and Hayes, D. M. 1992. Silver enhancement of gold antibody probes in pre-embedding electron microscopic immunocytochemistry. *J. Histochem. Cytochem.* 40: 1849.

Gu, J., D'Andrea, M., Yu, C.-Z., and McGrath, L. B. 1993. Quantitative evaluation of indirect immunogold-silver electron microscopy. *J. Histotechnol.* 16: 19.

Hainfeld, J. F. 1987. A small gold-conjugated antibody label: Improved resolution for electron microscopy. *Science* 236: 450.

Hainfeld, J. F. 1990. STEM analysis of Janssen Auroprobe One. In: *Proc. XIIth Int. Cong. Elec. Microsc.* p. 954. Bailey, G. W. (Ed.), San Francisco Press, San Francisco.

Hainfeld, J. F., Sprinzl, M., Mandiyan, V., Tumminia, S. J., and Boublik, M. 1991. Localization of a specific nucleotide in yeast tRNA by scanning transmission electron microscopy using an undecagold cluster. *J. Struct. Biol.* 107, 1–5.

Hainfeld, J. F. 1992. Site specific cluster labels. *Ultramicroscopy* 46: 135.

Hainfeld, J. F. and Furuya, F. R. 1992. A 1.4-nm gold cluster covalently attached to antibodies improves immunolabeling. *J. Histochem. Cytochem.* 40: 177.

Hainfeld, J. F., Furuya, F. R., Carbone, K., Simon, M., Lin, B., Braig, K., Horwich, A. L., Safer, D., Blechschmidt, B., Sprinzl, M., Ofengand, J., and Boublik, M. 1993. High resolution gold labeling. In: *Proc. 51st Ann. Meet. Micros. Soc. Amer.* p. 330. Bailey, G. W. and Rieder, C. L. (Eds.), San Francisco Press, San Francisco.

Hayat, M. A. (Ed.) 1989. *Colloidal Gold: Principles, Methods, and Applications.* Vol. 1. pp. 177, 225, 270, 212, 218, 307, 503. Academic Press, San Diego.

Holgate, C. S., Jackson, P., Cowen, P. N., and Bird, C. C. 1983. Immunogold-silver staining: New method of immunostaining with enhanced sensitivity. *J. Histochem. Cytochem.* 31: 938.

Horisberger, M. and Clerc, M. F. 1985. Labeling of colloidal gold with protein A. A quantitative study. *Histochemistry* 82: 219.

Hsu, Y.-H. 1984. Immunogold for detection of antigen on nitrocellulose paper. *Anal. Biochem.* 142: 221.

McPartlin, M., Mason, R., and Malatesta, I. 1969. Novel cluster complexes of gold(0)-gold(1). *J. Chem. Soc. Chem. Commun.* 334.

Milligan, R. A., Whittaker, M., and Safer, D. 1990. Molecular structure of F-actin and location of surface binding sites. *Nature* 348: 217.

Moeremans, M., Daneels, G., Van Dijck, A., Langanger, G., De Mey, J. 1984. Sensitive visualization of antigen-antibody reactions in dot and blot immune overlay assays with immunogold and immunogold/silver staining. *J. Immunol. Methods* 74: 353.

Pollard-Knight, D., Simmonds, A. C., Schaap, A. P., Akhavan, H., and Brady, M. A. W. 1990. Nonradioactive DNA detection on Southern blots by enzymatically triggered chemiluminescence. *Anal. Biochem.* 185: 353.

Reardon, J. E. and Frey, P. A. 1984. Synthesis of undecagold cluster molecules as biochemical labeling reagents. 1. Monoacyl and mono[N-(succinimidooxy)succinyl] undecagold clusters. *Biochemistry* 23: 3849.

Safer, D., Bolinger, L., and Leigh, J. S. 1986. Undecagold clusters for site-specific labeling of biological macromolecules: simplified preparation and model applications. *J. Inorg. Biochem.* 26: 77.

Stierhof, Y.-D., Humbel, B. M., Hermann, R., Otten, M. T., and Schwarz, H. 1992. Direct visualization and silver enhancement of ultra-small antibody-bound gold particles on immunolabeled ultrathin resin sections. *Scan. Microsc.* 6: 1009.

Skutelsky, E. and Roth, J. 1986. Cationic colloidal gold-a new probe for the detection of anionic cell surface sites by electron microscopy. *J. Histochem. Cytochem.* 34: 693.

Vandré, D. D. and Burry, R. W. 1992. Immunoelectron microscopic localization of phosphoproteins associated with the mitotic spindle. *J. Histochem. Cytochem.* 40: 1837.

Wall, J. S., Hainfeld, J. F., Bartlett, P. A., and Singer, S. J. 1982. Observation of an undecagold cluster compound in the scanning transmission electron microscope. *Ultramicroscopy* 8: 397.

Weinstein, S., Jahn, W., Hansen, H., Wittman, H. G., and Yonath, A. 1989. Novel procedures for derivatization of ribosomes for crystallographic studies. *J. Biol. Chem.* 264: 19138.

Weiser, H. B. 1933. *Inorganic Colloid Chemistry.* Vol. 1. pp. 21–57. Wiley, New York.

Wilkens, S. and Capaldi, R. A. 1992. Monomaleimidogold labeling of the γ subunit of the *Escherichia coli* F_1 ATPase examined by cryoelectron microscopy. *Arch. Biochem. Biophys.* 299: 105.

Wu, B., Mahony, J. B., and Chernesky, M. A. 1990. A new immune complex dot assay for detection of rotavirus antigen in faeces. *J. Virol. Meth.* 29: 157.

Yang, H., Reardon, J. E., and Frey, P. A. 1984. Synthesis of undecagold cluster molecules as biochemical labeling reagents. II. Bromoacetyl and maleimido undecagold clusters. *Biochemistry* 23: 3857.

Chapter 6

Use of TEM, SEM, and STEM in Imaging 1-nm Colloidal Gold Particles

York-Dieter Stierhof, René Hermann, Bruno M. Humbel, and Heinz Schwarz

Contents

6.1   1-nm Gold Colloids

One nanometer colloidal gold preparations first used by Leunissen and colleagues (for review see Leunissen and Van De Plas, 1993) belong to the smallest particulate markers presently used in immunoelectron microscopy (IEM). One nanometer gold markers (IgG, Fab, Streptavidin, and BSA conjugates; a reactive protein A conjugate is not available) are offered by several companies, e.g., AuroProbe One (formerly Janssen, now Amersham,

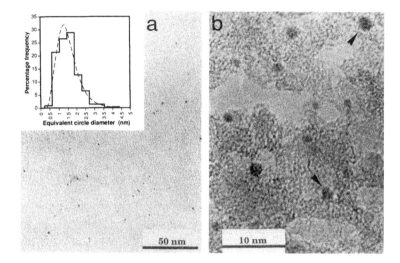

Figure 1   (a) TEM micrograph of a standard gold sol obtained as described by Duff et al. (1993a) on carbon film. Inset: Particle size distribution. The fit to a log normal curve is shown ($<d>_g$ = 1.42 mm, σ_g = 1.47). (b) High resolution TEM image (JOEL JEM-200CX, C_s = 0.52, C_c = 1.05 mm, 200 kV). Arrowheads point to particles showing lattice fringes. Reprinted with permission from Duff et al. (1993a). Copyright 1994 American Chemical Society.

U.K.), GP-ULTRA SMALL gold (Aurion, Wageningen, NL; Leunissen and Van De Plas, 1993), and ultrasmall (1 nm) gold (formerly BioCell, now British BioCell International, Cardiff, UK). None of these companies give a detailed characterization of their gold preparations, but all observations suggest that they contain heterogeneously sized gold particles in the range of ~1 to 3 nm, neglecting bound proteins (Hainfeld, 1990; Stierhof et al., 1991a, 1992; see below). For comparison, the recently introduced gold compounds Nanogold and undecagold (Nanoprobes, USA; Hainfeld and Furuya, 1992; Hainfeld, 1989; see Chapter 5) are in the same size range: Nanogold is 2.7 nm, undecagold 2.0 nm in diameter including their organic shells, but without coupled IgG or Fab fragment; their electron dense core is only 1 to 1.4 nm and 0.8 nm in diameter, respectively. IgGs can bind more than one 1-nm gold particle (Hainfeld, 1990; Leunissen and Van De Plas, 1993), whereas Fab fragments bind only one (Hermann et al., 1991).

Recently, Duff et al. (1993a,b) described the preparation of colloidal gold in the range of 1 to 2 nm by reduction with alkaline tetrakis(hydroxymethyl)phosphonium chloride of hydrogen tetrachloroaurate(III) (Figure 1). The authors also demonstrated the difficulties in characterizing the gold dispersion, e.g., in regard to size distribution, stability of particles, and composition. Particles 1.5 nm and larger often showed lattice fringes, consistent with metallic gold, in the TEM. Smaller clusters may be amorphous, or alternatively dynamic processes could occur on a time scale shorter than the photographic exposure. The coefficient of variation was about 37% for 0.9- and 1.5-nm particle preparations (Figure 1) with a mean cluster nuclearity (number of gold atoms) of 34 and 170, respectively (Duff et al., 1993a). In contrast to larger gold colloids, unconjugated 1-nm particles are not stable and coalesce with time, both in solution (Leunissen and Van De Plas, 1993) and in the electron beam (Duff et al., 1993b). However, 1-nm gold particles coupled to protein are stable over years (a gift of J. Leunissen [Janssen] from 1988 is still in use). We also include in this chapter small gold colloids obtained after reduction with thiocyanate of $HAuCl_4$ (1.5 to 3 nm, Baschong et al., 1985; Baschong and Wrigley, 1990). In contrast to Nanogold (Hainfeld and Furuya, 1992), 1-nm gold colloids are stable at temperatures above 50°C, which may be important for pre-embedding labeling studies without silver-enhancement.

One-nanometer gold markers were introduced in IEM (including *in situ* hybridization studies) for several reasons. It is well known from numerous microscopic studies that the reduction of gold size is inversely related to the label density and that smaller gold markers penetrate permeabilized specimens and to a lesser extent, thawed ultrathin Tokuyasu cryosections easier (Horisberger, 1981; Slot and Geuze, 1983; Tokuyasu, 1984; Sautter, 1986; Walther and Müller, 1986; Yokota, 1988; Leunissen and De Mey, 1989; Stierhof and Schwarz, 1989; Stierhof et al., 1991b). This fact is most probably due to reduced electrostatic repulsion of smaller gold colloids (due to their reduced negative surface charge) and due to less steric hindrance. All of this should result in a better detection limit for antigens in biological samples. Furthermore, 1-nm gold markers should allow a higher spatial resolution.

In this chapter we describe the application of 1-nm gold markers in transmission electron microscopy (TEM), scanning electron microscopy (SEM), and scanning transmission electron microscopy (STEM), and discuss advantages and limitations. Another section deals with improving the contrast of 1-nm gold particles by silver-enhancement, especially when bound to immunolabeled ultrathin resin and cryosections. We evaluate different silver-enhancement solutions in regard to quality and discuss problems associated with this technique.

6.2 TEM Application

There are only a few examples in TEM that use unenhanced 1-nm gold as marker molecules. This is due to their low contrast (an intrinsic property) not allowing intense heavy metal staining of the specimen in resin sections nor in negatively stained preparations. This has been shown in pre-embedding (Stierhof, unpublished results) and on-section labeling experiments for AuroProbe One (Amersham; Stierhof et al., 1991a), GP-ULTRA SMALL gold (Aurion; Stierhof et al., 1992), and ultrasmall (1 nm) gold (British BioCell International; Dulhanty et al., 1993). It can be clearly seen that 1-nm gold probes consist of heterogeneously sized particles. With conventional TEMs, larger particles (~3 nm) can be easily detected, whereas smaller ones are difficult to observe (Figure 2a; Stierhof et al., 1991a, 1992; Dulhanty et al., 1993). The same problem exists with negatively stained preparations, e.g., macromolecules as shown by Baschong and Wrigley (1990) for bacteriophage T4 polyheads and viral hemagglutinin molecules immunolabeled with "thiocyanate" gold. The authors successfully used aqueous sodium silicotungstate as stain.

Although not yet shown, 1-nm gold markers should be suitable as a marker for cryoelectron microscopy, e.g., of ice-embedded specimens, since Nanogold worked successfully (Boisset et al., 1992).

6.3 SEM Application

Biological structures are usually imaged in the SEM by secondary electrons (SE). These electrons, which are emitted from the specimen with energies ≤50 eV, are generated within the sample by the passage of high-energy electrons. They mainly provide a topographic image of the surface of the observed biological object. Another type of signal (besides the topographic SE-image) must be used for the unambiguous localization of small metal markers. Hoyer and co-workers (1979) introduced this idea in SEM, combining the topographic SE signal with the metal specific X-ray signal. So far, the X-ray signal has not been suitable for imaging gold particle distributions and very small gold particles for several reasons (e.g., one by one identification, long counting times, low spatial resolution). Since the availability of powerful detectors for backscattered electrons (BSE) (Autrata et al., 1986, 1992), this type of signal is now routinely used for imaging heavy metal markers. BSE are electrons of the primary electron beam which, having been elastically or inelastically

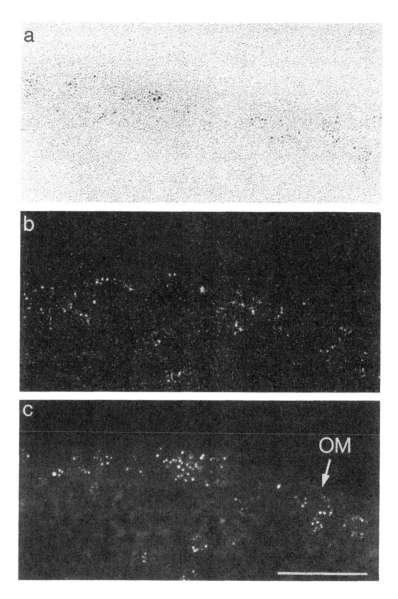

Figure 2   Imaging of 1-nm gold particles (Aurion) bound to immunolabeled ultrathin resin (HM20) sections of *E. coli* cells. The outer membrane protein OmpA is labeled, sections are unstained. (a) TEM image (Philips 201), accelerating voltage 60 kV, objective aperture 30 μm, primary magnification ×70,000. (b) SEM (BSE) image (Hitachi S-900), in-lens field emission SEM, annular YAG single crystal BSE detector, accelerating voltage 20 kV, objective aperture 20 μm, spot size <1 nm, primary magnification ×200,000. (c) STEM image (Philips CM20 FEG/STEM), Schottky field emission gun, spot size 1.5 nm, probe current ~500 pA, accelerating voltage 200 kV, HAADF detector, primary magnification ×205,000. OM: outer membrane. Bar = 100 nm. With permission from Stierhof et al. (1992).

scattered in the specimen, leave the specimen surface again (with energies close to that of the primary electron beam). BSE display an atomic number-dependent-image (their yield increases with increasing Z-number), outlining the heavy metal markers, which are much heavier than the biological material. The SE and BSE signals are easily separated by different detector systems due to the large energy difference between SE and BSE, and can be recorded simultaneously. To obtain sufficient structure information gold labeled samples have to be coated with, e.g., carbon or chromium (see below). The size of colloidal gold particles that can be used as markers depends on the beam diameter and accelerating voltage of the SEM and on the efficiency and sensitivity of the BSE detector employed.

Colloidal gold particles >15 nm can also be located in the SE image alone (Horisberger and Rosset, 1977; Sieber-Blum et al., 1981; Trejdosiewicz et al., 1981; DeHarven et al., 1984; Pawley and Albrecht, 1988; Osumi et al., 1992). BSE imaging is necessary, however, for accurate localization of smaller colloidal gold (5 to 15 nm: Walther et al., 1983; Walther and Müller, 1986; 4 to 18 nm: Albrecht et al., 1989). While colloidal gold >15 nm can still be imaged in conventional SEMs, field-emission SEMs (FESEMs), which provide a much smaller electron beam diameter with sufficient brightness at higher magnifications, are necessary for the detection of smaller colloidal gold by backscattered electrons.

However, only in-lens field-emission SEMs, which provide electron beam diameters of less than 1 nm at 30 kV accelerating voltage, are suitable instruments for imaging 1 nm colloidal gold particles. In these instruments, the sample can be placed within the lens at a position where the electron beam can be focused, thus reducing beam aberrations. Currently available FESEMs include the Hitachi S-5000, JEOL S-890 or ISI DS-130-FE. The highest possible accelerating voltage (i.e., 30 kV) should be used since the electron beam diameter increases with reduced accelerating voltage (4 nm at 1kV in the Hitachi S-900, the precursor model of the S-5000; Nagatani et al., 1987), decreasing the resolution in both the SE and the BSE images. Beam damage and contamination are serious problems when working at the high primary magnifications of more than 100,000 times necessary for localizing the 1-nm gold label (e.g., 25 to 35 electrons per $Å^2$ at 30 kV and ×100,000). In practice a photograph of the area of interest has to be taken with the first scan over this area.

A standard Everhart Thornley SE-detector satisfies the requirements to get a high resolution SE-image. In in-lens FESEMs, the detector is placed just before the objective lens (Koike et al., 1971). This results in high collection efficiency without taking up the already limited space in the objective lens gap. Care must be taken, however, in choosing an appropriate, highly sensitive BSE detector, in order to observe 1-nm colloidal gold. An optically improved BSE detector of the YAG type (equipped with an annular Yttrium Aluminium garnet single crystal; Autrata et al., 1986, 1992), placed just above the specimen in the objective lens gap, fulfills this requirement. The Hitachi S-900 FESEM, which is equipped with such a detector, has been successfully used by different groups to examine 1-nm colloidal gold particles adsorbed onto carbon grids (Erlandsen et al., 1990; Müller and Hermann, 1990; Pawley, 1992). Such instrumentation may now allow the resolution of individual target molecules, provided that the molecule by which the gold marker is bound is sufficiently small. A labeling precision within a 5 nm range, suitable for submolecular mapping studies, may be obtained by direct coupling of Fab, or Fab' fragments to 1-nm colloidal gold markers.

In SEM, however, there are very few immunological applications of unenhanced 1-nm colloidal gold markers. One of the most important reasons is that even the best SEM instruments and the smallest possible marker systems are unable to extract high resolution information from distorted samples: The same care that has been put into optimization of the marker and detection system has also to be taken in the optimization of the specimen preparation techniques. Procedures based on cryoimmobilization are able to preserve the structural integrity at molecular dimensions. Chemical fixation followed by dehydration and critical point drying of the sample, a routine SEM preparation protocol, cause structural artefacts in the specimens (Lee, 1984; Walther et al., 1984). These can be partially avoided by rapid freezing and freeze-drying of the samples, either directly or after freeze-substitution (see e.g., Steinbrecht and Müller [1987], for a review of these techniques). Shrinking and partial collapse is further reduced when the specimens are kept cooled for subsequent coating and SEM examination (Hermann and Müller, 1991a). Metal coating techniques and SEM-observation at low temperatures are therefore necessary prerequisites for reliable preservation of structural information at molecular dimensions.

SEM specimens are usually coated before observation to enhance their stability, electron conductivity, and the topographic localization of the SE signal. Thin, fine grained and continuously thick heavy metal coats, e.g., platinum or tungsten, can be successfully employed for a merely structural description of biological surfaces. However, these metal films would obscure the underlying signal from 1-nm colloidal gold markers. Metals are usually replaced by pure carbon coatings in immunological SEM studies employing larger colloidal gold. Carbon-coated biological samples yield low lateral resolution in the SEM due to their large electron interaction volume (Joy, 1984). A way out of this dilemma is to use thin light metal films, e.g., chromium. Cooled specimens can be chromium coated by planar magnetron sputtering or by electron beam shadowing, e.g., using double-axis rotary shadowing (Hermann and Müller, 1991b). Chromium coatings permit accurate localization of the SE signal at the specimen surface (Peters, 1986) and at the same time the unambiguous detection of 1-nm colloidal gold in the BSE mode.

An immunological application, combining the high topographic resolution achievable with modern SEM-instrumentation and specimen preparation techniques with very small marker systems, is shown in Figure 3: (a) SE and (b) BSE images of a T-even bacteriophage (TuII*-46; Schwarz et al., 1983) labeled with 1-nm colloidal gold-Fab fragments against tail fiber proteins (Hermann et al., 1991). The specimens were frozen by plunging them into liquid ethane, freeze-dried, and subsequently double-axis rotary shadowed with chromium. The T-even phage is clearly visible in the SE image; the BSE image, depicting the 1-nm colloidal gold particles, permits the unambiguous localization of the colloidal gold-Fab fragments. For comparison, negatively stained bacteriophages, unlabeled and labeled with unmarked Fabs, are shown in Figure 4.

One-nanometer colloidal gold-Fab fragments can be detected in SEM on complex structures where TEM negative staining or SEM studies with unmarked Fab fragments or IgGs would fail because they cannot be discerned from structural details or contaminants of similar size, especially on highly structured surfaces. Examples are Fab binding sites on the phage head (Figure 3b) and labeling studies on the surfaces of bulk specimens, e.g., red blood cells (Müller and Hermann, 1990). Furthermore, 1-nm gold label can be imaged on ultrathin sections using backscattered electrons (Figure 2b), whereas the corresponding SE image of the carbon-coated section reveals only the surface relief.

6.4   STEM Application

In principle, two types of interaction between beam electrons and small gold particles can be used for imaging by darkfield STEM. Bragg-scattered electrons which have interacted with the regular spaced planes of atoms in the crystal, and (elastically) forward-scattered Rutherford electrons, which contain atomic number information (Z^2 dependence). Bragg-scattered electrons are less suitable, because their intensity depends strongly on the orientation toward the incident electron beam of the gold atom planes in the crystal (Otten et al., 1992). Moreover, the crystalline organization of very small gold particles in 1-nm gold preparations is doubtful (Duff et al., 1993a). Forward-scattered Rutherford electrons have been shown to be suitable for imaging 1-nm gold labels.

The detectability of small gold particles in a STEM depends on the number of gold atoms per cluster and the mass thickness of the particle as well as on the beam diameter and the electron beam current. In order to obtain the required spatial resolution to separate single particles, the beam diameter has to be less than twice the particle size. On the other hand, the beam current which in general decreases with smaller beam diameters has to be sufficiently high to be able to record the amount of signal necessary for discrimination

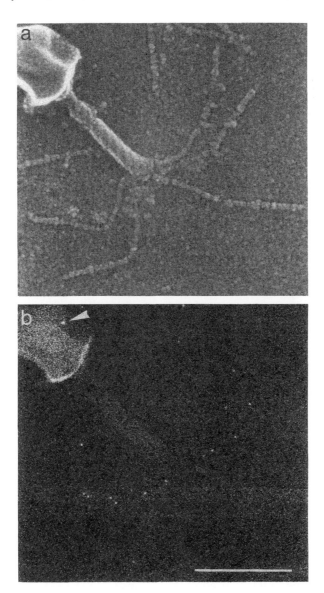

Figure 3 SEM micrographs of rapidly frozen, freeze-dried, and chromium coated TuII*-46 bacteriophage labeled with 1-nm gold (Aurion)-Fab fragments against tail fiber proteins. (a) SE image. (b) BSE image. The arrow points to an antigenic site on the phage head. Primary magnification ×300,000, accelerating voltage 30 kV, for details see Hermann et al. (1991). Bar = 100 nm. With permission from Hermann et al. (1991).

between gold particles and background noise. These conditions are fulfilled, e.g., in the Philips CM20 field emission gun (FEG)/STEM equipped with a Schottky field emitter (Mul et al., 1991; Otten et al., 1992; Stierhof et al., 1992). One nanometer gold particles, e.g., bound to ultrathin resin sections, can be imaged using a beam diameter of 1.5 nm and a beam current of about 500 pA (Otten et al., 1992; Stierhof et al., 1992).

Another problem is the selection of forward-scattered Rutherford electrons and the exclusion of undesired Bragg-scattered electrons which dominate the conventional STEM darkfield image at low scattering angles. This is possible by recording high-angle scattered Rutherford electrons, because Bragg-scattering is reduced at higher scattering angles. For this purpose, in the Philips CM20 FEG/STEM the high-angle annular darkfield (HAADF)

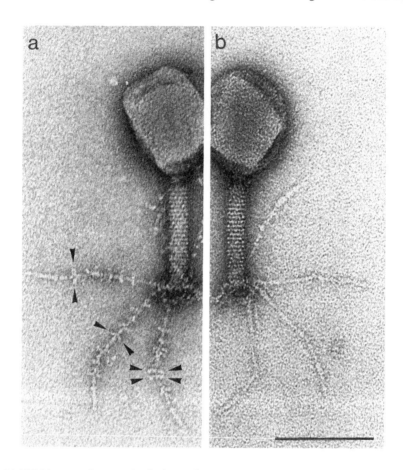

Figure 4   (a) TEM image of a negatively (uranyl acetate) stained bacteriophage TuII*-46 labeled as in Figure 3. Arrowheads point to some pairs of bound Fab fragments. (b) Unlabeled bacteriophage. Bar = 100 nm.

detector is mounted at a relatively short distance to the object (100 mm), accepting a conus of electrons scattered with angles between 2 and 6°. The signal produced by Rutherford-scattered electrons is similar to that obtained with BSEs, but at least a factor of 100× higher at the high accelerating voltages used in the Philips CM20 STEM (200 kV).

Figures 2c (detail) and 5 (overview) show ultrathin Lowicryl HM20 sections of E. coli cells, on-section immunolabeled with 1-nm gold (Aurion) for the outer membrane protein OmpA. The completely unstained section shows ultrastructural details like membranes, ribosomes, and cytoplasmic protein clumps (inclusion bodies). The gold particles located on the outer membrane, the periplasmic space and the inclusion bodies can be easily distinguished from the structure contrast. Heavy metal staining has to be avoided, as it would obscure the gold label (Otten et al., 1992). This imaging mode also clearly visualizes the variable colloid size of the used gold probe. Interestingly, 1-nm gold particles can even be imaged at relatively low primary magnifications (×50,000), allowing an overall presentation of a labeled bacterium. This technique should be also useful for pre-embedding IEM using unenhanced 1-nm gold markers.

6.5   Advantages and Limitations of 1-nm Gold Markers

After five years of application of 1-nm gold colloids in IEM a more detailed evaluation is possible.

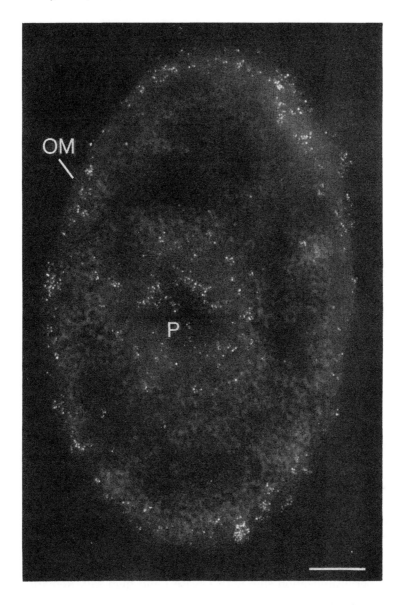

Figure 5 STEM overall presentation of an ultrathin *E. coli* resin (HM20) section labeled for OmpA using 1-nm gold markers (Aurion). In this overproducing strain OmpA can be detected on the outer membrane (OM), periplasm, and cytoplasmic protein clumps (P) (Freudl et al., 1986). Gold particles were visualized using HAADF STEM (see Figure 2). Primary magnification ×105,000. Bar = 100 nm. With permission from Stierhof et al. (1992).

6.5.1 *Specimen Penetration*

Studies on the light and electron microscopic level have shown that 1-nm gold markers penetrate permeabilized (extracted) cells or tissue (sections) more easily than larger colloids (LM: e.g., Leunissen and Van De Plas, 1993; EM: e.g., De Graaf et al., 1991). Therefore, 1-nm gold markers have become an important alternative to enzyme based marker systems (with their well-known problems like diffusion of the reaction product, lowered spatial resolution and low contrast).

Interesting differences between 1-nm gold and FITC conjugated antibodies were demonstrated by Sibon and co-workers (Sibon et al., 1995). In cells permeabilized with Triton

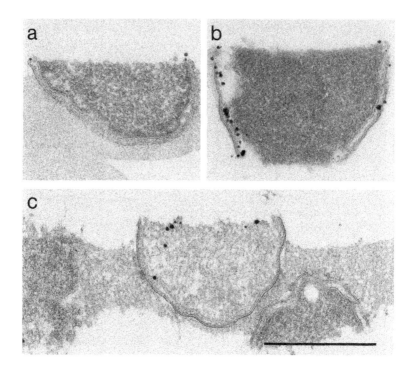

Figure 6 Thawed frozen section permeability to 1-nm gold markers (Aurion), silver enhanced using acidic silver lactate. Semithick (~500 nm) *E. coli* cryosections mounted on Epon blocks were immunolabeled and thereafter embedded in Epon. Ultrathin resin sections were cut perpendicular to the original cryosection surface. (a) In intact cells the OmpA label is restricted to the thawed frozen section surface. (b) In plasmolized cells the 1-nm gold label is able to enter the artificially dilated periplasm (in both cases, an OmpA deletion mutant protein accumulating in the periplasm is labeled). (c) After labeling of the DNA associated H-NS protein the marker molecules can be seen mainly at the thawed frozen section surface, and sometimes within the section. Bar = 500 nm.

X-100, hybridization with a digoxigenin labeled 28S DNA probe resulted in strong labeling of the nucleoli after incubation with a FITC labeled antidigoxigenin antibody. In contrast, 1-nm gold (Aurion) coupled antibodies also entered the nucleus, but were unable to penetrate the nucleoli. The nucleoli became only accessible for the 1-nm gold marker after an additional permeabilization step with DNase I and high salt. Remarkably, there are reports describing penetration of 1-nm gold markers into nonpermeabilized glutaralde-hyde fixed PtK2 cells (Leunissen and Van De Plas, 1993) and formaldehyde/glutaralde-hyde fixed and borohydride treated nerve cells (Van Lookeren Campagne et al., 1992). It is not clear whether in this example the membrane permeability is influenced by the borohydride treatment.

The situation is also complex using thawed cryosections of fixed, but nonpermeabilized specimens prepared after the method of Tokuyasu (Tokuyasu, 1984; Griffiths, 1993). Here, the (local) matrix density at the site of interest decides to which extend 1-m gold markers are able to enter the cryosection. On chicken erythrocyte sections, 1-nm gold label in the cytoplasm and in hetero- and euchromatin of the nucleus is restricted to the cryosection surface (Stierhof and Schwarz, 1989; Stierhof et al., 1991b). Some penetration could be observed with tubulin label on sections of the unicellular parasite *Leishmania* (Stierhof and Schwarz, 1989, Stierhof et al., 1991b) and of Chinese hamster ovary cells (Leunissen and Van De Plas, 1992). As an example, 1-nm gold labeling of different compartments in *E. coli* is shown in Figure 6 using transversely sectioned Epon embedded semithin cryosections (for the method see Stierhof and Schwarz, 1989). Labeling of the outer membrane protein

OmpA in *E. coli* cells is restricted to the original cryosection surface (Figure 6a), whereas in plasmolyzed cells, the periplasmic space becomes accessible to 1-nm gold markers (Figure 6b; in both cases an OmpA deletion mutant protein accumulating in the periplasm is labeled). Furthermore, cytoplasmic inclusion bodies consisting of overproduced OmpA are only accessible to the marker at the cryosection surface. In contrast, the loosely packed DNA containing region (a typical artifact of chemical fixation at ambient temperature) sometimes allows labeling of the DNA binding protein H-NS (Dersch et al., 1993) within the cryosection (Figure 6c). In general, penetration is poor, and is mainly restricted to areas of lower matrix density or to damaged regions (e.g., crevasses, holes). Most probably the accessibility of antigens located very close to the cryosection surface is improved. There may be other factors beside the (original) matrix density which hamper labeling of structures located within a cryosection, as discussed by Stierhof and Schwarz (1989) and Griffiths (1993), e.g., extensive cross-linking of the surrounding matrix by the fixative or masking epitopes. Because 1-nm gold markers do not enter thawed cryosections in significant amounts, a strategy was developed to extract the biological material in a controlled way before chemical fixation and cryosectioning by treating with streptolysin O, a pore forming toxin (Krijnse-Locker et al., 1994). Streptolysin O treatment leads to loss of cytoplasmic material thus improving the accessibility of membrane bound antigens located within the cryosection. There is no evidence that 1-nm gold markers enter resin sections (for larger gold colloids, see Stierhof and Schwarz, 1989; Stierhof et al., 1991b and references therein).

6.5.2   Label Density

As mentioned above, it is a well-known fact that the smaller the IgG or Fab coupled gold colloid is, the higher the label density becomes. This situation has been also shown for resin sections labeled with 1-nm gold markers (Stierhof et al., 1991a; Dulhanty et al., 1993), and thawed Tokuyasu cryosections (Nielsen and Bastholm, 1990; Humbel and Biegelmann, 1992) or in pre-embedding cell surface labeling experiments (Farrell and Pardridge, 1991) and is also true for *in situ* hybridization studies (Sibon et al., 1995).

6.5.3   Influence of Gold Coupling to Antigen-Antibody (Fab) Binding Efficiency

An interesting question is whether 1-nm gold coupled IgGs or Fab fragments result in the same antigen-binding properties (efficiency) as unconjugated IgGs or Fab fragments. Two examples give strong evidence that this is not the case. Baschong and Wrigley (1990) showed in negative staining experiments that purified viral hemagglutinin homotrimers, which theoretically are able to bind three Fab fragments, bind often two, sometimes three unconjugated Fab fragments, but no more than two "thiocyanate" gold (1 to 2 nm) coupled ones. The authors also observed conformational changes of Fab arms in IgG molecules after complexing with gold colloids. Hermann and colleagues (1991) convincingly demonstrated a similar effect for labeling tail fibers of the T-even bacteriophage TuII*-46 (Schwarz et al., 1983). The monovalent Fab fragments prepared from a rabbit serum bind pairwise (opposite to each other) to the tail fibers (Figure 4a,b; Schwarz and Henning, 1984; Riede et al., 1987). However, when coupled to 1-nm gold colloids (Aurion), these Fab fragments bind only separately, never pairwise, as was seen in a SEM using BSE imaging (Figure 3b; Hermann et al., 1991). In these cases one single gold colloid was conjugated to the "end" of the Fab fragment (Baschong and Wrigley, 1990; Hermann et al., 1991). Both examples strongly suggest that even very small gold colloids can induce changes in the interaction of the Fab-molecule and the adjacent epitope thus preventing the binding of the adjacent Fab fragment. Therefore, unconjugated Fab fragments and probably IgGs show a higher label density when compared to 1-nm gold coupled ones, and in consequence, the detection limit for antigens is lower than for antibodies or fragments coupled to colloidal gold. This also means that in principle a ratio of one epitope to one primary antibody (fragment) conjugated to 1-nm gold cannot be expected.

6.5.4 Spatial Resolution

The highest possible resolution with gold markers can be achieved, when 1-nm gold colloids are directly bound to Fab fragments. Thus the maximum distance from gold to the epitope in the phage labeling experiment is about 5 nm (the gold colloid obviously binds preferentially opposite to the antigen binding region; Baschong and Wrigley, 1990; Hermann et al., 1991).

The high labeling resolution obtainable with phages or macromolecules as shown above cannot be expected for on-section or thawed cryosection labeling experiments, nor for immunolabeled and thereafter embedded specimens. In these cases the resolution is mainly determined by the electron contrast of the labeled structure and the resin section thickness. Fifty to eighty nanometer-thick sections show considerable superimposition effects of structural details and the exact localization of the label in the range of a few nanometers is, to say the least, difficult or impossible. Statistical methods (Kemler and Schwarz, 1989), tilting of the section, and electron microscope tomography applied to resin sections (Mehlin et al., 1992; Starink et al., 1995) can improve the spatial resolution.

6.6 Silver-Enhancement of 1-nm Gold Particles Bound to Immunolabeled Ultrathin Resin and Cryosections

Due to the small size and low contrast of 1-nm gold particles and due to high-resolution microscopes normally being unavailable, improving the marker contrast is a prerequisite for easy identification of 1-nm gold particles in most biological applications. A well-known possibility is the deposition of metallic silver on the gold surface after the immunolabeling step, the so-called silver-enhancement or autometallography (for references see Chapter 1). This technique has been shown to be suitable for applications in TEM (e.g., Scopsi, 1989) as well as in SEM (e.g., Namork, 1991). The chemical process is described in detail in Chapters 1 and 2; another intensification process, so-called gold toning, is described in Chapter 13. Silver-enhancement is not without its problems. This is reflected by the fact that numerous enhancing procedures have been developed for colloidal gold intensification (for review see Scopsi, 1989; Larsson, 1989). The difficulties associated with this technique include control in regard to reproducibility, homogeneous growth, efficiency (meaning a high ratio of enhanced to unenhanced or poorly enhanced particles), and deposition of nonspecific silver precipitates. In this chapter, we focus mainly on the application of silver-enhancement to on-section immunolabeled ultrathin resin sections and thawed cryosections. Ultrathin resin or cryosections serve as a suitable test system for the quality of the silver enhancer to be evaluated. The surface bound gold particles exhibit a similar accessibility to the reagents, thus largely eliminating problems associated with varying diffusion rates of the silver-enhancement components, e.g., into gold labeled (permeabilized) tissue. This may result in different reaction conditions in thick specimens and, in consequence, in different growth of gold particles (see Chapters 1 and 2).

In general, silver-enhancement procedures differ greatly in quality especially when applied to 1-nm gold particles. To profit maximally from the advantages of 1-nm gold markers, one should use a method which is able to enhance all bound particles in a reproducible and even manner. This is true also for light microscopic applications. An inefficient silver-enhancement drastically reduces the label density and may contribute to the disappointing results reported by De Valk et al. (1991) and Dulhanty et al. (1993).

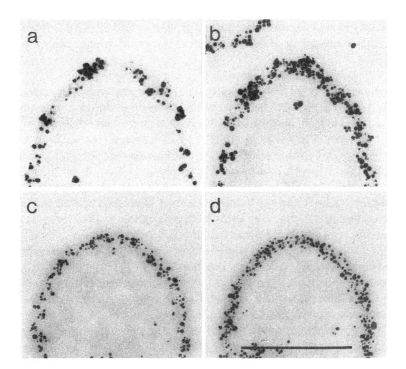

Figure 7 Comparison of 1-nm gold (Aurion) or Nanogold (Nanoprobes) labels silver enhanced with different intensifying solutions applied to OmpA labeled *E. coli* resin sections. (a) 1-nm gold, R-Gent (Aurion), 20°C, 6 min. (b) 1-nm gold, R-Gent containing 30% gum arabic, 42°C, 30 min. (c) 1-nm gold, acidic silver lactate (Danscher, 1981), 20°C, 17 min. (d) Nanogold, acidic silver lactate (Danscher, 1981), 20°C, 25 min. Bar = 500 nm.

6.6.1 1-nm Gold Enhancement

We have compared four commercial and four published enhancers in detail: R-Gent (Aurion), IntenSE M (Amersham), HQ SILVER™ (Nanoprobes), SEKL15 (British BioCell), an acidic and a neutral developer containing silver lactate (Danscher, 1981; Lah et al., 1990), a silver acetate-containing developer (Hacker et al., 1988), and a Metol/Ilford L4 containing solution (Bienz et al., 1986). The results can be summarized as follows (see Stierhof et al., 1991a, 1992):

1. R-Gent, IntenSE M, and Metol/Ilford L4 solutions are less suitable for intensification of 1-nm gold particles (Figure 7a): The efficiency appears to be low, the particle growth is irregular, and enhanced particles tend to be angular rather than round. Also, SEKL 15 exhibits a reduced enhancement efficiency.
2. The quality of Metol/Ilford L4 and especially of R-Gent and IntenSE M can be considerably improved by adding high amounts (20 to 33%) of the so-called protective colloid gum arabic (Figure 7b, SEKL15 was not tested but an improvement is also expected). Liesegang (1911) and Danscher (1981) demonstrated that the protective colloid gum arabic improves the performance of the enhancer solution. In the case of 1-nm gold it could be clearly shown that in the presence of gum arabic the efficiency of enhancement is considerably higher, and the enhanced particles are more equal in size and appear round (Stierhof et al., 1991a, 1992). However, silver deposition is slowed down. The role of protective colloids like gum arabic is discussed in more detail in Chapter 2.
3. Enhancers already containing the protective colloid gum arabic, the neutral and

acidic silver lactate solutions, and most probably, HQ SILVER™, result in a considerably higher efficiency of enhancement and more homogeneously sized particles (Figure 7c).

4. The silver acetate enhancer, resulting in a relatively high efficiency without the protective colloid (e.g., showing a higher enhancement efficiency as the original R-Gent and IntenSE M kits), can also be improved by adding gum arabic.

5. It appears that at least most of the 1-nm gold particles can be developed with "high-quality" enhancers (Stierhof et al., 1992). Nevertheless, the evaluation of enhancement efficiency is problematic. First, very small unenhanced gold particles could be overlooked; and larger, unenhanced gold particles have to be differentiated from weakly enhanced particles. Second, very small silver precipitates in the range of 1-nm gold (see below) have to be distinguished from gold containing particles. This discrimination is possible, e.g., using EDX microanalysis (Felsmann et al., 1992), but requires the separate examination of all small particles.

6. It was not possible to get particles of uniform size after enhancement of 1-nm gold preparations using the enhancers mentioned above. There may be several reasons for this: (1) the original gold probes consist of heterogeneously sized particles. To test this, we also enhanced Nanogold-IgG bound to resin sections. Although Nanogold was shown to be a very homogeneously sized gold compound with a core diameter of 1 to 1.4 nm (Hainfeld and Furuya, 1992; see also Chapter 5), silver-enhancement according to Danscher (1981) results in particles similar to the 1-nm gold preparations (Figure 7d). This means that variable particle sizes cannot be the only reason for uneven growth; (2) the surface properties of 1-nm gold particles probably vary (e.g., exposure of different planes of the crystal lattice and differing protein layer thickness on the gold surface), thus influencing the initiation of silver deposition and further growth; and (3) gold particles in close vicinity may influence the silver deposition on their neighbors. Hacker and colleagues have also compared different developers in this volume.

6.6.2   Nanogold Enhancement

We also evaluated Nanogold intensification for three different enhancer solutions. The acidic as well as the neutral silver lactate and gum arabic containing solutions and the commercial HQ SILVER™ enhancer gave excellent results (e.g., acidic silver lactate: Figure 7d). In contrast, the SEKL 15 kit, which results in a relatively high enhancement efficiency when applied to 1-nm gold particles, developed very few Nanogold particles. Burry and colleagues (1992) achieved good results for enhancement of 1-nm gold (Aurion) and Nanogold in pre-embedding experiments using a neutral gum arabic and silver lactate containing developer with N-propyl-gallate instead of hydroquinone as reducing agent.

6.6.3   Light Microscopic Application

For several reasons silver-enhancement of gold labeled ultrathin resin sections is also useful for light microscopic experiments (Stierhof et al., 1992; for thick sections, see Scopsi [1989] and references therein, and Chapter 2). First, such sections provide a good overview of a large area of labeled cells. Second, they allow high-resolution labeling experiments on the light microscopic level: Due to the fact that gold markers do not penetrate the section, most, if not all, of the gold label can be brought into the focal plane. In consequence, blurring due to label outside the plane of focus which is generally a problem of thick specimens in light microscopy, is minimized. Together with the thinness of the resin section, it results in a high spatial resolution, which in the z direction is even higher than that achieved with a confocal laser scanning microscope.

Silver-enhanced gold on ultrathin resin sections can be best imaged using phase contrast or epi-polarization illumination (Stierhof et al., 1992; for thick sections and whole mount preparations, see Scopsi [1989] and references therein and Chapter 11). It needs to be noted that the gold/silver label can interfere with fluorescence labels in double labeling experiments: The gold/silver particles can also be shown by epi-illumination with filter systems usually used to detect fluorochrome markers. Therefore, care must be taken in double labeling experiments to unequivocally discriminate between both types of markers (Stierhof et al., 1992; for additional problems with combined immunofluorescence and gold/silver labeling see Goodman et al., 1991).

Especially with some fast working enhancers, e.g., the original IntenSE M and the Metol/Ilford L4 solution, we observed an influence of the local gold particle density on the silver growth rate, resulting in bigger silver particles in areas of low gold label density and smaller ones in areas of high gold label density (Stierhof et al., 1991a). This effect is especially important for light microscopy because it can suggest gold label densities which do not exist.

6.7   Practical Aspects of Silver-Enhancement

1. Illumination by the electron beam causes instability of the silver layer on on-section labeled resin sections and leads to loss of metallic silver during subsequent storage in the presence of air (Figure 8a–c). The electron dose and storage time dependent effect can be also demonstrated for larger gold colloids covered with silver. It can be prevented by keeping the silver enhanced and illuminated specimen under vacuum. Furthermore, on thawed cryosections redistribution/dislocation of silver (embedded in methyl cellulose) occurs in areas which were illuminated by the electron beam and thereafter stored in the presence of air. The silver seems to diffuse in small particles away from the original position at the gold colloid (Figure 8d,e). Areas not exposed to the electron beam are not altered (Figure 8f).

2. Staining of sections with heavy metals like uranyl acetate and lead citrate does not interfere with subsequent silver-enhancement. Uranyl acetate and lead citrate are washed out during the silver-enhancement procedure; only areas already exposed to the electron beam retain the heavy metal stain. For problems with osmium tetroxide fixation after silver-enhancement (e.g., pre-embedding IEM) which normally results in loss of silver from silver-enhanced gold particles, see Burry et al. (1992) and Chapter 14. In contrast to osmium tetroxide fixation, fixation/staining with aqueous uranyl acetate before dehydration is possible and recommended.

3. Enhancement after exposure to the electron beam of unenhanced sections is possible even after staining with uranyl acetate and lead citrate. However, the area already illuminated by the electron beam cannot be enhanced. The same is true for repeated enhancement of previously silver intensified and heavy metal stained sections. In both cases the effect appears to be dose dependent (Stierhof et al., 1992). Possibly exposure to the electron beam leads to masking or modifying the gold or silver surface thus preventing the deposition of new silver layers.

4. Nonspecific silver precipitates generally appear to be negligible when the enhancement procedure is thoroughly carried out. Field emission darkfield STEM reveals a number of small particles which exhibit low contrast and are randomly dispersed on the section surface of some sectioned *E. coli* cells. EDX microanalysis of those particles strongly suggest that they consist only of silver (Felsmann et al., 1992). They cannot be detected on unenhanced sections, but on unlabeled, silver enhanced ones and are therefore a silver-enhancement artifact (Stierhof et al., 1992). Due to their low contrast, these small silver precipitates are not of practical importance. The possibility of "specific" silver background due to endogeneous metals like sulfides

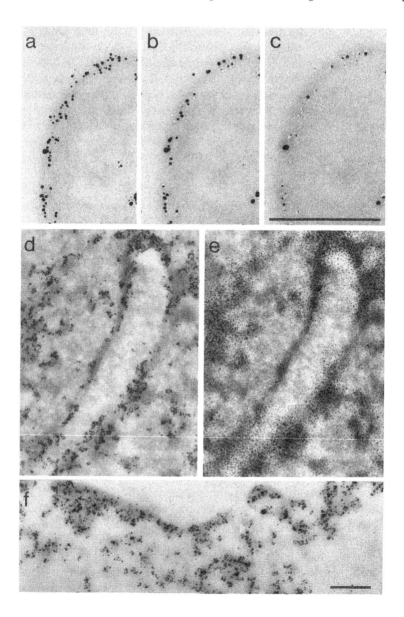

Figure 8   Loss of silver (a–c: Lowicryl section, identical area of an *E. coli* cell labeled for OmpA) and dislocation/redistribution of silver (d–f: thawed frozen section, identical area (d,e) of a HeLa cell nucleus labeled for DNA) after silver-enhancement (Danscher, 1981). (a) First illumination. (b) Second photograph taken 18 d after enhancement. (c) Third photograph taken 84 d after enhancement. (d) First illumination. (e) Second photograph taken three years after enhancement. (f) Same section as in e, but an area which was not illuminated before. Bar = 500 nm.

or selenides within the biological sample has to be kept in mind. Veznedaroglu and Miller (1992) could prevent undesired enhancement of zinc by pretreatment with a zinc specific chelator.

6.8   Enhancement Protocols

Sections have to be thoroughly washed with bidistilled water to remove ions (especially chloride ions) which could interfere with the silver-enhancement process. Enhancement is carried out by incubating the grids on a drop of ~100 µl of the enhancer solution. Although most of the enhancers can even be used in daylight, we recommend covering the grids or working under red safe light conditions, as this delays self-nucleation of silver ions. The reaction is stopped by transferring the grids to bidistilled water or to a photographic fixative (Larsson, 1989). Before staining with uranyl acetate and lead citrate, components like gum arabic have to be completely removed by thoroughly washing the grids. It has to be noted that higher temperatures and moving the grids during incubation on the enhancer solution speed up the deposition of silver on the gold surface (Humbel and Biegelmann, 1992).

6.8.1   Acidic Silver Lactate (Danscher, 1981)

0.6 ml gum arabic (33% in bidistilled water) plus 0.1 ml citrate buffer (2.55 g citric acid plus 2.35 g trisodium citrate dihydrate, add bidistilled water to make 10 ml, pH 3.8) plus 0.15 ml hydroquinone (0.85 g in 15 ml bidistilled water) plus 0.15 ml silver lactate (0.11 g in 15 ml bidistilled water). Gum arabic, citrate buffer, and hydroquinone can be premixed and stored in a freezer. Silver lactate has to be stored separately in a freezer. Incubation temperature 20 to 22°C, incubation time for 1-nm gold ~20 min, for Nanogold ~25 min.

6.8.2   Neutral Silver Lactate (Lah et al., 1990)

See above, replace citrate buffer by 0.2 M Hepes buffer, pH 6.8. Incubation temperature 20 to 22°C, incubation time ~3 min for 1-nm gold, for Nanogold 4 to 5 min.

6.8.3   Silver Acetate (Hacker et al., 1988)

Mix equal amounts of 0.5% hydroquinone in 0.5 M citrate buffer (see above), and 0.2% silver acetate. Final concentration of 16.5% gum arabic was prepared from a stock solution of 33% gum arabic. Incubation temperature 20 to 22°C; incubation time ~90 min for 1-nm gold (but only ~20 min without gum arabic).

6.8.4   HQ SILVER™ (Nanoprobes)

Equal amounts of the three components initiator, moderator, and activator were mixed before use (see instructions of Nanoprobes). Incubation temperature 20 to 22°C, incubation time ~3 min for 1-nm gold and ~5.5 min for Nanogold.

6.8.5   IntenSE M (Amersham), R-Gent (Aurion), SEKL15 (British BioCell)

Equal amounts of developer and enhancer were mixed before use (see instructions of Amersham, Aurion, and British BioCell). Final concentrations of 20 or 33% gum arabic in IntenSE M and R-Gent were prepared from stock solutions of 50% gum arabic. Incubation temperature 42°C, incubation time 20 min (20% gum arabic) or 30 min (33% gum arabic) (but only ~6 min without gum arabic at 20 to 22°C).

6.8.6   Grid Material (Stierhof et al., 1992)

Nickel grids are useful for all enhancers mentioned above. If sections have been already mounted on copper grids, the neutral silver lactate enhancer is recommended. Copper grids are not suitable for IntenSE M, R-Gent, and for the acidic silver lactate and the silver acetate solution. Gold grids are useful for all enhancers with the exception of R-Gent. The commercial HQ SILVER™ enhancer and the SEKL15 kit were only tested with nickel grids.

6.9 Summary

Preparations of 1-nm colloidal gold markers presently available consist of heterogeneously sized particles in the range of 1 to 3 nm in diameter. The main advantage in using them for IEM is higher sensitivity (higher label density) when compared to larger gold markers. One-nanometer gold markers penetrate especially permeabilized biological specimens more easily, due to less steric hindrance and reduced electrostatic repulsion of the negatively charged gold surface. However, uncoupled or fluorochrome coupled antibodies are even more sensitive. The main disadvantage of 1-nm gold particles is their low contrast when used without enhancement which requires high resolution SEMs and STEMs equipped with sensitive BSE or HAADF detectors, respectively. To date there have been only a few applications of this technology addressing biological problems.

Silver-enhancement allows the use of 1-nm gold markers even in conventional electron microscopy and light microscopy. As shown by on-section labeling, enhancers containing the protective colloid gum arabic (e.g., Danscher, 1981) are characterized by high efficiency, even silver deposition, and low nonspecific silver precipitation. Even with "high quality" enhancers 1-nm gold colloids and Nanogold cannot be enhanced to a homogeneously sized particle population. Repeated enhancement of the same section even after heavy metal staining is possible. Unfortunately, the silver layer around the gold particles becomes unstable after direct exposure to the electron beam and subsequent storage in the presence of air, which results in loss or redistribution/dislocation of silver in the previously illuminated area.

Acknowledgments

We thank R. Koebnik and E. Bremer for bacterial strains and anti-H-NS antibodies, R. Kelsh and M. Fuchs for comments on the manuscript, and G. Müller for photographic work.

References

Albrecht, R. M., Prudent, J. R., Simmons, S. R., Pawley, J. B., and Choate, J. J. 1989. Observation of colloidal gold labeled platelet microtubules: high voltage electron microscopy and low voltage-high resolution scanning electron microscopy. *Scan. Microsc.* 3: 273–278.

Autrata, R., Hermann, R., and Müller, M. 1992. An efficient BSE single crystal detector for SEM. *Scanning* 14: 127–135.

Autrata, R., Walther, P., Kriz, S., and Müller, M. 1986. A BSE scintillation detector in the (S)TEM. *Scanning* 8: 3–8.

Baschong, W., Lucocq, J. M., and Roth, J. 1985. Thiocyanate gold: Small (2–3 nm) colloidal gold for affinity cytochemical labeling in electron microscopy. *Histochemistry* 83: 409–411.

Baschong, W. and Wrigley, N. G. 1990. Small colloidal gold conjugated to Fab fragments or to immunoglobulin G as high-resolution labels for electron microscopy: A technical overview. *J. Electron Microsc. Tech.* 14: 313–323.

Bienz, K., Egger, D., and Pasamontes, L. 1986. Electron microscopic immunocytochemistry: Silver enhancement of colloidal gold marker allows double labeling with the same primary antibody. *J. Histochem. Cytochem.* 34: 1337–1342.

Boisset, N., Grassucci, R., Penczek, P., Delain, E., Pochon, F., Frank, J., and Lamy, J. N. 1992. Three-dimensional reconstruction of a complex of human α_2-macroglobulin with monomaleimido Nanogold ($Au_{1.4nm}$) embedded in ice. *J. Struct. Biol.* 109: 39–45.

Burry, R. W., Vandré, D. D., and Hayes, D. M. 1992. Silver enhancement of gold antibody probes in pre-embedding electron microscopic immunocytochemistry. *J. Histochem. Cytochem.* 40: 1849–1856.

Danscher, G. 1981. Histochemical demonstration of heavy metals: A revised version of the sulphide silver method suitable for both light and electron microscopy. *Histochemistry* 71: 1–16.

De Graaf, A., Van Bergen en Henegouwen, P. M. P., Heijne, A. M. L., Van Driel, R., and Verkleij, A. J. 1991. Ultrastructural localization of nuclear matrix proteins in HeLa cells using silver enhanced ultra small gold probes. *J. Histochem. Cytochem.* 39: 1035–1045.

DeHarven, E., Leung, R., and Christensen, H. 1984. A novel approach for scanning electron microscopy of colloidal gold-labeled cell surfaces. *J. Cell Biol.* 99: 53–57.

Dersch, P., Schmidt, K., and Bremer, E. 1993. Synthesis of the *Escherichia coli* K-12 nucleoid-associated DNA-binding protein H-NS is subjected to growth-phase control and autoregulation. *Mol. Microbiol.* 8: 875–889.

De Valk, V., Renmans, W., Segers, E., Leunissen, J., and De Waele, M. 1991. Light microscopical detection of leukocyte cell surface antigens with a one-nanometer gold probe. *Histochemistry* 95: 483–490.

Duff, D. G., Baiker, A., and Edwards, P. P. 1993a. A new hydrosol of gold clusters. 1. Formation and particle size variation. *Langmuir* 9: 2301–2309.

Duff, D. G., Baiker, A., Gameson, I., and Edwards, P. P. 1993b. A new hydrosol of gold clusters. 2. A comparison of some different measurement techniques. *Langmuir* 9: 2310–2317.

Dulhanty, A. F., Junankar, P. R., and Stanhope, C. 1993. Immunogold labeling of calcium ATPase in sarcoplasmic reticulum of skeletal muscle: Use of 1-nm, 5-nm, and 10-nm gold. *J. Histochem. Cytochem.* 41: 1459–1466.

Farrell, C. L. and Pardridge, W. M. 1991. Ultrastructural localization of blood-brain barrier-specific antibodies using immunogold-silver enhancement techniques. *J. Neurosci. Meth.* 37: 103–110.

Felsmann, M., Stierhof, Y.-D., Humbel, B. M., and Otten, M. T. 1992. Detection of ultra-small gold labels with a FEG TEM/STEM. *Beitr. Elektronenmikroskop. Direktabb. Oberfl.* 25: 15–18.

Freudl, R., Schwarz, H., Stierhof, Y.-D., Gamon, K., Hindennach, I., and Henning, U. 1986. An outer membrane protein (OmpA) of *E. coli* undergoes a conformational change during export. *J. Biol. Chem.* 261: 11355–11361.

Goodman, S. L., Park, K., and Albrecht, R. M. 1991. A correlative approach to colloidal gold labeling with video-enhanced light microscopy, low-voltage electron microscopy, and high-voltage electron microscopy. In: *Colloidal Gold: Principles, Methods, and Applications.* Vol. 3. pp. 370–409. Hayat, M. A. (Ed.), Academic Press, San Diego.

Griffiths, G. 1993. *Fine Structure Immunocytochemistry.* Springer, Berlin.

Hacker, G. W., Grimelius, L., Danscher, G., Bernatzky, G., Muss, W., Adam, H., and Thurner, J. 1988. Silver acetate autometallography: An alternative enhancement technique for immunogold-silver staining (IGSS) and silver amplification of gold, silver, mercury, and zinc in tissues. *J. Histotechnol.* 11: 213–221.

Hainfeld, J. M. 1990. STEM analysis of Janssen AuroProbe One. In: *Proc. XIIth Cong. Electron Microsc.* Vol. 3. p. 954. Peachey, L. D. and Williams, D. B. (Eds.), San Francisco Press, San Francisco.

Hainfeld, J. F. 1989. Undecagold-antibody method. In: *Colloidal Gold: Principles, Methods, and Applications.* Vol. 2. pp. 413–429. Hayat, M. A. (Ed.), Academic Press, San Diego.

Hainfeld, J. F. and Furuya, F. R. 1992. A 1.4-nm cluster covalently attached to antibodies improves immunolabeling. *J. Histochem. Cytochem.* 40: 177–184.

Hermann, R., Schwarz, H., and Müller, M. 1991. High precision immunoscanning electron microscopy using Fab fragments coupled to ultra-small colloidal gold. *J. Struct. Biol.* 107: 38–47.

Hermann, R. and Müller, M. 1991a. Prerequisites of high resolution scanning electron microscopy. *Scan. Microsc.* 5: 653–664.

Hermann, R. and Müller, M. 1991b. High resolution biological scanning electron microscopy: A comparative study of low temperature metal coating techniques. *J. Electron Microsc. Tech.* 18: 440–449.

Horisberger, M. 1981. Colloidal gold: A cytochemical marker for light and fluorescent microscopy and for transmission and scanning electron microscopy. *Scan. Electron Microsc.* II: 9–13.

Horisberger, M. and Rosset, J. 1977. Colloidal gold, a useful marker for transmission and scanning electron microscopy. *J. Histochem. Cytochem.* 25: 295–305.

Hoyer, L. C., Lee, J. C., and Bucana, C. 1979. Scanning immunoelectron microscopy for the identification and mapping of two or more antigens on cell surfaces. *Scan. Electron Microsc.* III: 629–636.

Humbel, B. M. and Biegelmann, E. 1992. A preparation protocol for postembedding immunoelectron microscopy of *Dictyostelium discoideum* cells with monoclonal antibodies. *Scan. Microsc.* 6: 817–825.

Joy, D. C. 1984. Beam interactions, contrast and resolution in the SEM. *J. Microsc.* 136: 241–258.

Kemler, R. and Schwarz, H. 1989. Ultrastructural localization of the cell adhesion molecule uvomorulin using site-directed antibodies. In: *Cell to Cell Signals in Mammalian Development*. NATO ASI Series. Vol. H26. pp. 145–152. De laat et al. (Eds.), Springer-Verlag, Berlin.

Koike, H., Ueno, K., and Suzuki, M. 1971. Scanning device combined with conventional microscope. In: *Proc. 29th Ann. Meeting Electron Microsc. Soc. Am.* pp. 28–29. Claitor's, Baton Rouge.

Krijnse-Locker, J., Ericsson, M., Rottier, P. J. M., and Griffiths, G. 1994. Characterization of the budding compartment of mouse hepatitis virus: Evidence that transport from the RER to the Golgi complex requires only one vesicular transport step. *J. Cell Biol.* 124: 55–70.

Lah, J. J., Hayes, D. M., and Burry, R. W. 1990. A neutral pH silver development method for the visualization of 1-nm gold particles in pre-embedding electron microscopic immunocytochemistry. *J. Histochem. Cytochem.* 38: 503–508.

Larsson, L.-I. 1989. Immunocytochemical detection systems. In: *Immunocytochemistry: Theory and Practice*. pp. 77–145. Larsson, L.-I. (Ed.), CRC Press, Boca Raton, FL.

Lee, R. M. K. W. 1984. A critical appraisal of the effects of fixation, dehydration and embedding on cell volume. In: *The Science of Biological Specimen Preparation*. pp. 61–70. Revel, J.-P., Barnard, T., and Haggis, G. H. (Eds.), SEM Inc., AMF O'Hare, Chicago.

Leunissen, J. L. M. and De Mey, J. R. 1989. Preparation of colloidal gold probes. In: *Immuno-Gold Labeling in Cell Biology*. pp. 3–16. Verkleij, A. J. and Leunissen, J. L. M. (Eds.), CRC Press, Boca Raton, FL.

Leunissen, J. L. M. and Van De Plas, P. 1993. Ultrasmall gold probes and cryoultramicroscopy. In: *Immuno-Gold Electron Microscopy on Virus Diagnosis and Research*. pp. 327–348. Hyatt, A. D. and Eaton, B. T. (Eds.), CRC Press, Boca Raton, FL.

Liesegang, R. E. 1911. Die Kolloidchemie der histologischen Silberfärbungen. *Kolloid Beihefte* 3: 1–46.

Mehlin, H., Daneholt, B., and Skoglund, U. 1992. Translocation of a specific premessenger ribonucleoprotein particle through the nuclear pore studied by electron microscope tomography. *Cell* 68: 605–613.

Müller, M. and Hermann, R. 1990. Towards high resolution SEM of biological objects. In: *Proc. XIIth Int. Congr. Electron Microsc.* Vol. 3. pp. 4–5. Peachy, L. D. and Williams, D. B. (Eds.), San Francisco Press, San Francisco.

Mul, P. M., Bormans, B. J. H., and Otten, M. T. 1991. Design of the CM20 FEG. *Philips Electron Optics Bulletin* 130: 53–62.

Nagatani, T., Saito, S., Sato, M., and Yamada, M. 1987. Development of an ultrahigh resolution scanning electron microscope by means of a field emission source and in-lens system. *Scan. Microsc.* 1: 901–909.

Namork, E. 1991. Double labeling of antigenic sites on cell surfaces imaged with backscattered electrons. In: *Colloidal Gold: Principles, Methods, and Applications*. Vol. 3. pp. 188–205. Hayat, M. A. (Ed.), Academic Press, San Diego.

Nielsen, M. H. and Bastholm, L. 1990. Improved immunolabeling of ultrathin cryosections using antibody conjugated with 1-nm gold particles. In: *Proc. XIIth Int. Congr. Electron Microsc.* Vol. 3. pp. 928–929. Peachy, L. D. and Williams, D. B. (Eds.), San Francisco Press, San Francisco.

Osumi, M., Yamada, N., Kobori, H., and Yaguchi, H. 1992. Observation of colloidal gold particles on the surface of yeast protoplasts with UHR-LVSEM. *J. Electron Microsc.* 41: 392–396.

Otten, M. T., Stenzel, D. J., Cousens, D. R., Humbel, B. M., Leunissen, J. L. M., Stierhof, Y.-D., and Busing, W. M. 1992. High-angle annular darkfield STEM imaging of immunogold labels. *Scanning* 14: 282–289.

Pawley, J. and Albrecht, R. 1988. Imaging colloidal gold labels in LVSEM. *Scanning* 10: 184–189.

Pawley, J. B. 1992. LVSEM for high resolution topographic and density contrast imaging. *Adv. Electronics Electron Phys.* 83: 203–274.

Peters, K.-R. 1986. Metal deposition by high-energy sputtering for high magnification electron microscopy. In: *Advanced Techniques in Biological Electron Microscopy.* Vol. 3. pp. 101–166. Koehler, J. K. (Ed.), Springer-Verlag, Berlin.

Riede, I., Schwarz, H., and Jähnig, F. 1987. Predicted structure of tail-fiber proteins of T-even type phages. *FEBS Lett.* 215: 145–150.

Sautter, C. 1986. Immunocytochemical labeling of enzymes in low temperature embedded plant tissue: The precurser of glyoxysomal malate dehydrogenase is located in the cytosol of watermelon cotyledon cells. In: *The Science of Biological Specimen Preparation for Microscopy and Microanalysis 1985.* pp. 215–227. SEM Inc., AMF O'Hare, Chicago.

Schwarz, H. and Henning, U. 1984. Comparative immuno-electron microscopy of T-even bacteriophages. In: *Proc. 8th Europ. Cong. Electr. Microsc.* Vol. 2. pp. 1565–1569. Csanady, A., Röhlich, P., and Szabo, D. (Eds.).

Schwarz, H., Riede, I., Sonntag, I., and Henning, U. 1983. Degrees of relatedness of T-even type *E. coli* phages using different or the same receptors and topology of serologically cross-reacting sites. *EMBO J.* 2: 375–380.

Scopsi, L. 1989. Silver-enhanced colloidal gold. In: *Colloidal Gold: Principles, Methods, and Applications.* Vol. 1. pp. 251–295. Hayat, M. A. (Ed.), Academic Press, San Diego.

Sibon, O. C. M., Cremers, F. F. M., Humbel, B. M., Boonstra, J., Verkleij, A. J. 1995. Localization of nuclear RNA by pre- and post-embedding in situ hybridization using different gold probes. *Histochem. J.* 27: 35–45.

Sieber-Blum, M., Sieber, F., and Yamada, K. M. 1981. Cellular fibronectin promotes adrenergic differentiation of quail neural crest cells in vitro. *Exp. Cell Res.* 133: 285–295.

Slot, J. and Geuze, H. J. 1983. The use of protein A-colloidal gold (pAg) complexes as immunolabels in ultrathin sections. In: *Immunohistochemistry.* pp. 323–346. Cuello, A. C. (Ed.), IBRO Handbook Series, Wiley, Chichester.

Starink, J. J. P., Humbel, B. M., Verkleij, A. J. 1995. Three-dimensional localization of immunogold markers using two tilted electron microscope recordings. *Biophys. J.* 68: 1–10.

Steinbrecht, R. A. and Müller, M. 1987. Freeze-substitution and freeze-drying. In: *Cryotechniques in Biological Electron Microscopy.* pp. 149–171. Steinbrecht, R. A. and Zierold, K. (Eds.), Springer-Verlag, Berlin.

Stierhof, Y.-D., Humbel, B. M., and Schwarz, H. 1991a. Suitability of different silver enhancement methods applied to 1 nm colloidal gold particles: An immunoelectron microscopic study. *J. Electron Microsc. Tech.* 17: 336–343.

Stierhof, Y.-D., Schwarz, H., Dürrenberger, M., Villiger, W., and Kellenberger, E. 1991b. Yield of immunolabel compared to resin sections and thawed cryosections. In: *Colloidal Gold: Principles, Methods, and Applications.* Vol. 3. pp. 87–115. Hayat, M. A. (Ed.), Academic Press, San Diego.

Stierhof, Y.-D., Humbel, B. M., Hermann, R., Otten, M.T., and Schwarz, H. 1992. Direct visualization and silver enhancement of ultra-small antibody-bound gold particles on immunolabeled ultrathin resin sections. *Scan. Microsc.* 6: 1009–1022.

Stierhof, Y.-D. and Schwarz, H. 1989. Labeling properties of sucrose-infiltrated cryosections. *Scan. Microsc.* Suppl. 3: 35–46.

Tokuyasu, K.T. 1984. Immunocryoultramicrotomy. In: *Immunolabeling for Electron Microscopy.* pp. 71–82. Polak, J. M. and Varndell, M. (Eds.), Elsevier, Amsterdam.

Trejdosiewicz, L. K., Smolira, M. A., Hodges, G. M., Goodman, S. L., and Livingston, D. C. 1981. Cell surface distribution of fibronectin in cultures of fibroblasts and bladder derived epithelium: SEM-immunogold localization compared to immunoperoxidase and immunofluorescence. *J. Microsc.* 123: 227–236.

Van Lookeren Campagne, M., Dotti, C. G., Jap Tjoen San, E. R. A., Verkleij, A. J., Gispen, W. H., and Oestreicher, A. B. 1992. B-50/GAP43 localization in polarized hippocampal neurons *in vitro:* An ultrastructural quantitative study. *Neuroscience* 50: 35–52.

Veznedaroglu, E. and Miller, T. A. 1992. Elimination of artifactual labeling of hippocampal mossy fibers seen following pre-embedding immunogold-silver technique by pretreatment with zinc chelator. *Microsc. Res. Tech.* 23: 100–101.

Walther, P., Ariano, B. H., Kriz, S. R., and Müller, M. 1983. High resolution SEM detection of protein-A gold (15 nm) marked surface antigens using backscattered electrons. *Beitr. Elektronenmikroskop. Direktabb. Oberfl.* 16: 539–545.

Walther, P. and Müller, M. 1986. Detection of small (5–15 nm) gold-labeled surface antigens using backscattered electrons. In: *The Science of Biological Specimen Preparation 1985.* pp. 195–201. SEM Inc., AMF O'Hare, Chicago.

Walther, P., Müller, M., and Schweingruber, M. E. 1984. The ultrastructure of the cell surface and plasma membrane of exponential and stationary phase cells of *Schizosaccharaomyces pombe,* grown in different media. *Arch. Microbiol.* 137: 128–134.

Yokota, S. 1988. Effect of particle size on labeling density for catalase in protein A-gold immunocytochemistry. *J. Histochem. Cytochem.* 36: 107–109.

Chapter 7

Quantitative Evaluation of Immunogold-Silver Staining

Jiang Gu, Michele Forte, Nancyleigh Carson, Gerhard W. Hacker,
Michael D'Andrea, and Robyn Rufner

Contents

7.1 Introduction

Immunogold-silver staining (IGSS) is one of the most sensitive techniques in immunocytochemistry for both light microscopy (LM) and electron microscopy (EM). It provides a
distinct black granular reaction at the LM level and round, highly electron dense labeling
at the EM level. It is suitable for many different applications. Despite its high antigen
detecting sensitivity, this technique has not been widely used because certain precautions
are required for its performance. To better understand the property of immunogold-silver
amplification, it is important to have an appreciation of the characteristics and rate by
which the gold-silver particles grow in size. We conducted a quantitative evaluation at
both LM and EM levels, using three silver salts, i.e., silver acetate, silver lactate, and silver
nitrate. Five sizes of gold particles were evaluated for durations ranging from 1 to 17 min
under controlled conditions. The sizes of the electron dense granules or the density of the
stain were measured quantitatively for each variable and the data analyzed comparatively.
Optimal conditions were determined.

7.2 Materials and Methods

Atrial natriuretic peptide (ANP) in the rat atria was chosen as a study model because of its
abundance, uniformity, and well-defined distribution in the secretory granules in the
cytoplasm of atrial myocardiocytes (Gu et al., 1989). Right atria of 12 adult rats (Wistar
Kyoto, 200 to 300 g in body weight) were removed under general anesthesia and dissected
into 0.5 mm^3 pieces. After the specimens were fixed in a mixture of 2% glutaraldehyde and
1% formaldehyde for 3 h at 4°C, they were washed in a sodium cacodylate buffer (0.06 M,

pH 7.3) and embedded in Epon. Thin sections (60 nm thick) were cut and mounted on uncoated 300 mesh nickel grids.

The indirect immunogold staining, which consisted of employing a polyclonal antiserum against the 28 amino acid peptide hormone alpha-ANP, was performed (Gu and Gonzalez-Lavin, 1988; Gu and D'Andrea, 1989; Gu et al., 1989). The characteristics of the primary antibody (Peninsula Labs, Belmont, CA) were discussed previously (Gu et al., 1989). After the grids were incubated with nonimmune goat serum in 0.1 M Tris buffer for 20 min at room temperature, primary antiserum (1:2000, in 0.1 M Tris buffer) was applied and incubated for 20 h at 4°C in a humid chamber. The secondary goat-antirabbit antibody was labeled with 5 different sizes of colloidal gold (Amersham, Arlington Heights, IL), according to DeRoe et al. (1987). Dilutions of 1:100, 1:80, 1:30, 1:20, and 1:5, were used for 1-, 5-, 15-, 30-, and 40-nm colloidal gold particles, respectively (Roth, 1982). A previous study established that these dilutions gave optimal results for these gold labeled antibodies and antigens (Gu and D'Andrea, 1989). The various gold particle sizes were exposed to three silver amplification (autometallography) solutions (silver acetate, silver lactate, and silver nitrate) at increasing incremental time intervals from 1 to 17 min at 20°C. The compositions of the silver amplification solutions were 6 mM for each silver salt, and 33 mM for hydroquinone in a citrate buffer (pH 3.5 to 3.8) (Danscher, 1981; Hacker et al., 1988).

Six samples were tested for each particle size and duration producing a mean of statistical significance. Grids were jet washed with distilled water and counterstained with uranyl acetate (5% in 50% alcohol for 10 min) and lead citrate (0.4% for 10 min) without Lugol's iodine pretreatment. Grids were examined with a transmission electron microscope operating at 60 kV. Controls consisted of: (1) normal rabbit serum as the first layer instead of the primary antiserum, (2) preabsorbed primary antiserum that was incubated with specific antigen before immunostaining at a concentration of 10 nmol/ml, and (3) replacement of the secondary antibody with one that was not labeled with colloidal gold.

The diameters of the specifically labeled gold-silver particles on electron micrographs were measured manually using a computerized morphometric image analyzer (Universal Image I, Universal Image Corp, West Chester, PA). The longest axis of each particle was measured when particle irregularity was present. Each sample provided more than 25 secretory granule measurements, noting that results from different grids were evaluated comparatively. For LM immunogold staining, the indirect immunogold staining procedure (Gu et al., 1981) was performed followed by silver amplification with the same three autometallographic solutions and time range as for EM. The results were also evaluated using a computerized image analyzer.

Electron microscope double immunostaining was attempted with the IGSS method (Holgate et al., 1983; Hacker et al., 1985; Lackie et al., 1985) Two antigens, Alpha-ANP and its anti-parallel dimer, Beta-ANP, were chosen as the model both of which had been previously demonstrated in cytoplasmic granules in atrial myocardiocytes (Gu and D'Andrea, 1989). Immunostainings of Alpha-ANP and Beta-ANP were carried out on two sides of the same grid, one on each side, consecutively. Secondary antibodies for both antigens consisted of 5 nm gold labeled goat-antirabbit IgG. After gently placing the grid on top of a drop of autometallographic solution with Beta-ANP facing down for 3 min, the grid was washed in distilled water and counterstained. The gold particles on the other side of the grid for Alpha-ANP were not enlarged. Controls for double immunostaining were the same as those for single immunostaining. Immunostaining of each antigen was also performed independently as positive controls along with double immunostaining with 5-nm gold for both antigens, without silver amplification. Additional double immunostaining was performed with 5- and 15-nm gold labelings for the two antigens, respectively (Vandell et al., 1982; Bendayan, 1982; Gu and D'Andrea, 1989). The results were analyzed comparatively.

7.3   Results

All of the immunogold-silver methods tested gave positive immunostaining of ANP in the myocardiocytes of the rat atrium. All of the positive and negative controls worked properly. At the LM level, the dark black staining was confined to the cytoplasm of the myocytes with higher concentrations around the parinuclear region. The connective tissue (i.e., smooth muscle cells and endothelial cells) was negative. At the EM level, the positivity was confined to the cytoplasmic granules.

Of the three silver salts, silver acetate provided the best results in staining quality and reproducibility both for LM and EM. Representative micrographs from IGSS for the five sizes of colloidal gold particles with silver acetate amplification are presented in Figures 1 to 5. In summary, for electron microscopic IGSS, regardless of gold particle size, the particles enlarged at an accelerated rate. The larger the particle became, the more rapidly it increased in size (Figures 6 to 10). At 3 min of autometallography, all of the gold labeling particles were enlarged with silver acetate and no gold particles remained at their original size. For each size of gold particle, there appeared to be a limit to silver amplification, above which the particle shape or size became too large or distorted to be of any practical value. The upper limits for silver acetate amplification of 1-, 5-, 15-, 30-, and 40-nm gold, were 10, 8, 6, 5, and 4 min, respectively. With the prolongation of time, the autometallography tended to form large aggregations and gradually fused together to form smaller numbers of large irregularly shaped particles. This gradually masked the ultrastructure and lost value for antigen detection.

Silver lactate and silver nitrate did not give as strong or clear IGSS as 6 mM silver acetate. Without silver amplification, the 1-nm gold particles were not visible, even at very high magnification (Figure 1a) to (200,000x). A silver treatment of 3 to 7 min gave the best diameter range of 1-nm gold to 12 to 26 nm in diameter. When the silver treatment time went beyond 10 min, the variations of the gold-silver particles enlarged significantly in size, thereby obscuring the specificity of the reaction. Although the 1-nm immunogold-silver labeling method has a very high detecting sensitivity, more labeling particles were detected on average sized cytoplasmic secretory granules with 5-nm gold. Modifying the concentrations of the 1-nm gold-labeled antibodies did not affect this phenomenon.

The results obtained at the LM level correlated well with ultrastructural observations. The optimal incubation for LM IGSS averaged 2 min longer than for EM. Therefore, IGSS is one of the few immunostaining procedures that can be used effectively at both the LM and EM levels to identify the same tissue structure at very different magnifications. Nevertheless, this potential has yet to be fully explored.

When performed under optimal conditions, silver amplification does not create unspecified labeling. This was evidenced by completely negative results obtained with silver treatment alone (Figure 12). Silver amplification did not affect subsequent counterstaining by uranyl acetate and lead citrate. The counterstaining, however, cannot be performed before silver amplification since the autometallographic solution bleached the counterstain. In addition, exposure of thin sections to high electron beam prevented the irradiated gold particles from absorbing silver grains and becoming enlarged.

The two different antigens were demonstrated in double immunostaining procedures. More antigens were detected by the silver amplification method. The two antigens were differentiated by the double immunogold-silver method, which enlarged the isolated particles to 12 to 15 nm in diameter for Beta-ANP; those for Alpha-ANP remained at 5 nm (Figure 13). The new double immunostaining method did not diminish the antigen detection sensitivity of the staining method when used individually. Also, it did not create any nonspecific labeling. It should be noted that the gold-silver particles resulting from silver amplification were not as uniform as gold particles alone.

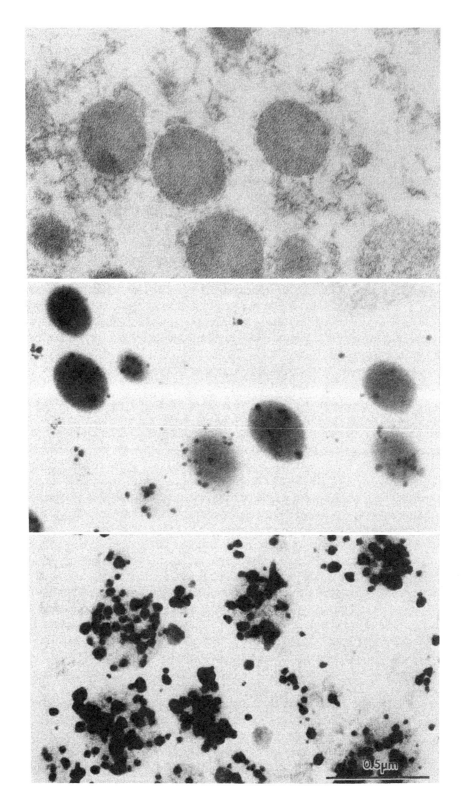

Figure 1   1-nm colloidal gold diluted 1:100. (a) Control, no silver, the gold particles are not visible. (b) Optimal, exposure at 5 min silver treatment. (c) Over, exposure at 10 min silver treatment.

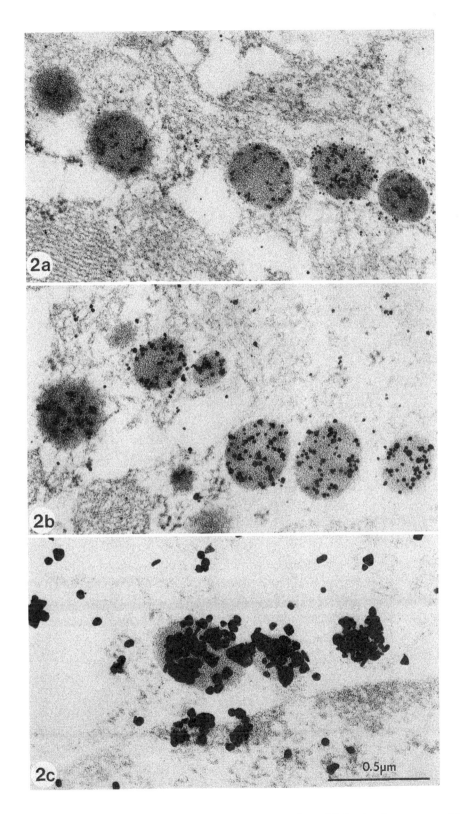

Figure 2   5-nm colloidal gold diluted 1:80. (a) Control, no silver. (b) Optimal, exposure at 3 min silver treatment. (c) Over, exposure at 5 min silver treatment.

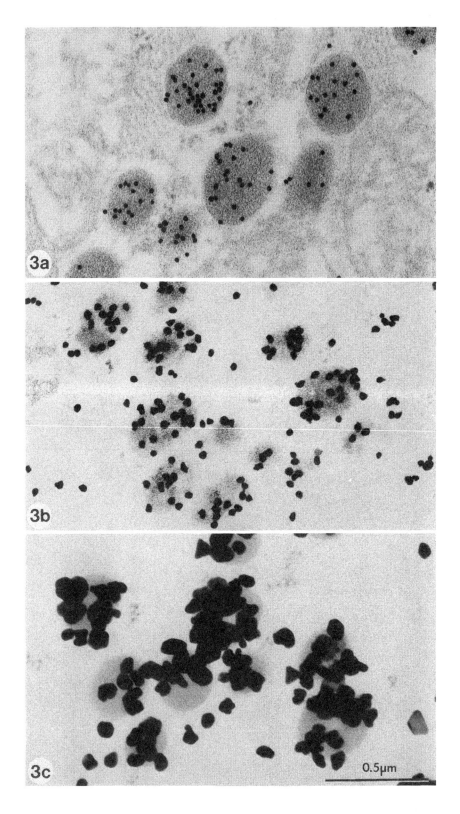

Figure 3 15-nm colloidal gold diluted 1:30. (a) Control, no silver. (b) Optimal, exposure at 3 min silver treatment. (c) Over, exposure at 12 min silver treatment.

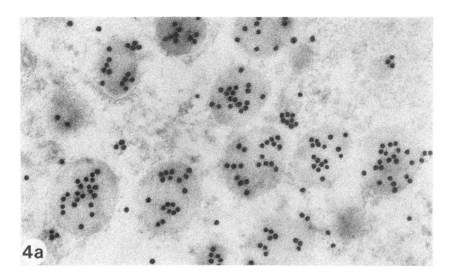

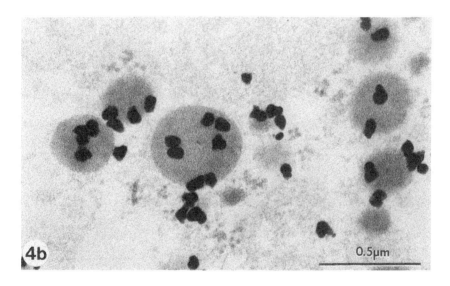

Figure 4 30-nm colloidal gold diluted 1:20. (a) Control, no silver. (b) Optimal, exposure at 3 min silver treatment.

7.4 Discussion

The IGSS technique remains among the most sensitive of antigen detection methods for morphologists (Holgate et al., 1983; Danscher and Norgaard, 1983; Roth, 1982; De Waele et al., 1986, 1988; Hacker et al., 1985 and in this volume; Springall et al., 1981; Lackie et al., 1985; Otsuki et al., 1990; Nielson and Bastholm, 1990). It is especially useful when a very high contrast between the reporting signal and the background is desired. When used correctly, the reporting signals are amplified and the labeling gold particles enlarge without creating a background. This gives the flexibility of controlling the desired particle size for the EM or the staining darkness for LM. At the same time, it permits examination of immunostaining results at lower magnifications and detection of smaller amounts of

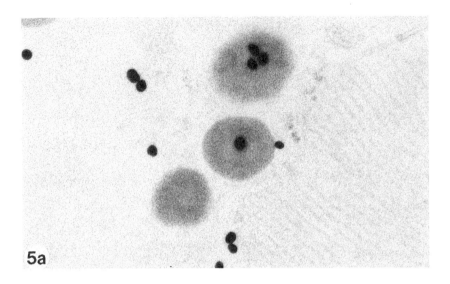

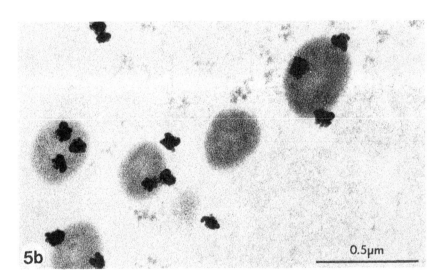

Figure 5 40-nm colloidal gold diluted 1:5. (a) Control, no silver. (b) Optimal exposure at 3 min silver treatment.

antigen with higher dilutions of antibodies. Thus, a better appreciation of antigen distribution in relation to tissue morphology may be obtained.

The growth rates of gold-silver particles were demonstrated quantitatively. There appeared to be a nonlinear relationship between particle size and silver amplification. The growth appeared exponential within 8 to 10 min. Beyond this time, variations in particle size and shape prevented an immunocytochemical value. It was also observed that small gold particles were directly related to increased detection sensitivity in this method due to increased tissue penetration and higher partial density to the same number of tissues antigens (Gu and D'Andrea, 1989). Therefore, autometallography not only enlarged gold particle size to make visualization easier but also retained the antigen detecting sensitivity of small gold particles. If larger gold particles were used directly, the antigen detecting

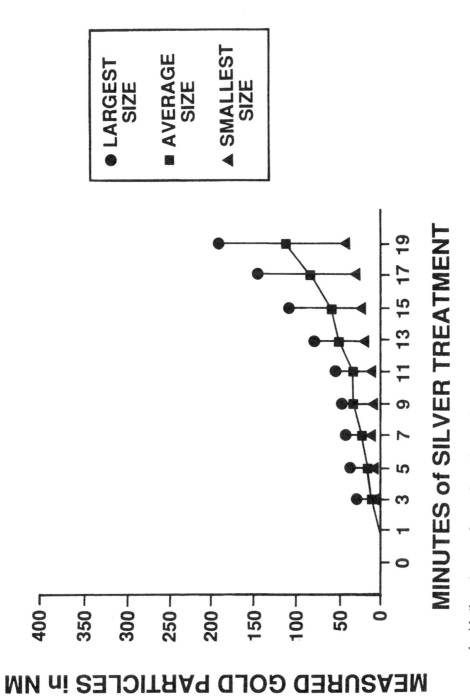

Figure 6 Summary of gold-silver size growth rates for each size of gold in 6 m*M* silver acetate solution at the tested time range: 1-nm *vs.* silver treatment enlarging variability.

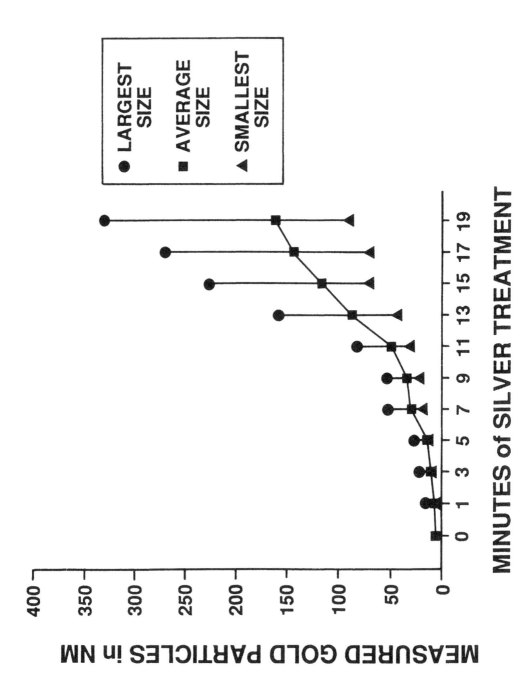

Figure 7   Summary of gold-silver size growth rates for each size of gold in 6 m*M* silver acetate solution at the tested time range: 5-nm vs. silver treatment enlarging variability.

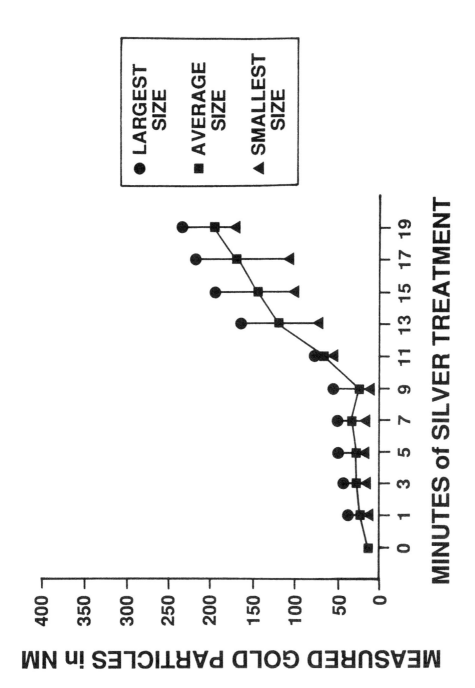

Figure 8 Summary of gold-silver size growth rates for each size of gold in 6 m*M* silver acetate solution at the tested time range: 15-nm vs. silver treatment enlarging variability.

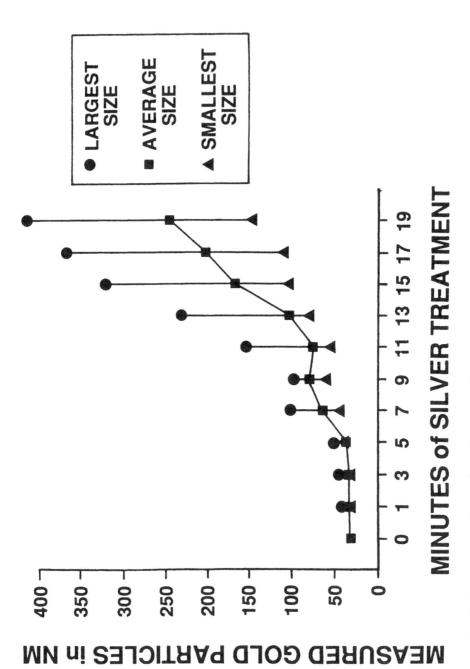

Figure 9   Summary of gold-silver size growth rates for each size of gold in 6 mM silver acetate solution at the tested time range: 30-nm vs. silver treatment enlarging variability.

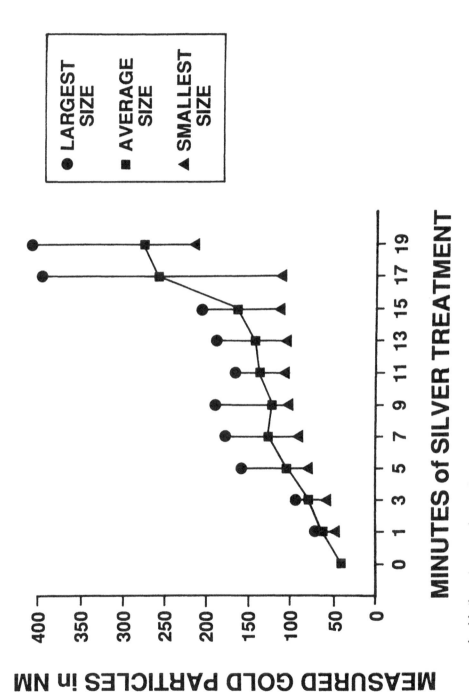

Figure 10   Summary of gold-silver size growth rates for each size of gold in 6 mM silver acetate solution at the tested time range: 40-nm vs. silver treatment enlarging variability.

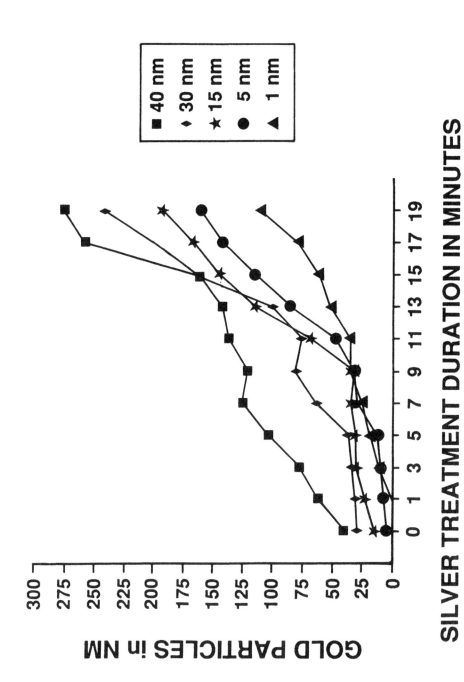

Figure 11   Summary of gold-silver size growth rates for each size of gold in 6 m*M* silver acetate solution at the tested time range: gold size vs. silver treatment duration.

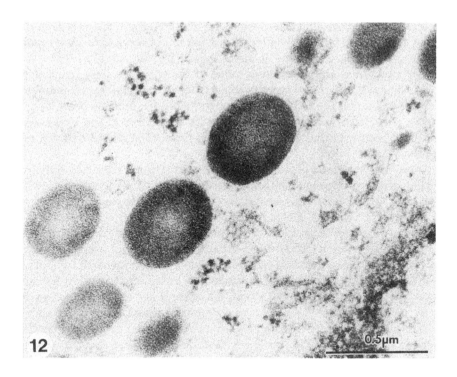

Figure 12 Control received 6 min silver treatment without gold. No background was created.

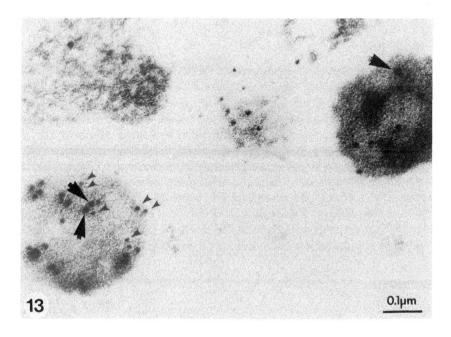

Figure 13 Preliminary result of double immunolabeling employing gold-silver amplification technique. On one side of the grid, 5-nm gold-labeled to Alpha-ANP (small arrowheads); on the other side of the grid 5-nm gold with 2 min silver treatment labeled to Beta-ANP (large arrowheads).

sensitivity would have been reduced. Nevertheless, silver-enhanced 1-nm gold did not provide denser antigen labeling than 5-nm gold. Possibly, the 1-nm gold-labeled antibodies were not all enlarged by the silver treatment. Similar observations were reported when demonstrating cell surface antigens at the LM level (De Valck et al., 1991).

The IGSS method is able to create a range of particle sizes when used with various silver-enhancement times. It also permits enlargement of particle size after the preparation has been reviewed under the EM without counterstaining. It must be noted that there is no further silver amplification of areas irradiated with an electron beam. However, this does not prevent gold-silver particles in other areas of the section from becoming enlarged to predetermined sizes.

The widely used method of double immunogold staining uses two different sizes of gold particles for two different antigens (Vandell et al., 1982; Bendayan, 1982; Gu and D'Andrea, 1989). The detection sensitivity for each is related to each gold particle size (Gu and D'Andrea, 1989; Gu et al., 1993), but not to each other. It is cumbersome to quantitatively compare two antigens because the number of gold labeled particles for each antigen has no proportional relationship to the comparative amounts of antigenic molecules detected. Silver amplification makes it possible to detect two antigens on two different sides of a grid with gold labeled particles of the same size, being able to further enlarge the gold particles on one side of the grid with silver. All of the 5-nm or larger gold particles were enlarged by silver-enhancement. This would provide a direct quantitative comparison of two antigens. Of course, the detecting sensitivities of the two primary antibodies should be the same in order to make this comparison. Keeping in mind also that silver-enlarged particles are not as uniform in size as the gold particles, one must be careful to control the duration. Also, silver-amplified particles may overlap gold particles on the other side of the grid. Overall, gold-silver double staining has both advantages and disadvantages when compared with the current method of using two different sizes of gold particles on the same grid. However, it does provide an alternative for direct quantitative comparison of two antigens.

7.5   Conclusion

The immunogold-silver technique possesses a number of merits over other methods: (1) For a given number of antigenic sites, it can increase detection sensitivity by enlarging gold particle sizes without losing labeling density or creating background; (2) Particle size enlargement can be controlled to suit individual needs. The rate and limitations of the amplification, using silver acetate in particular, were evaluated by our study; (3) It may provide a variety of gold-silver particle sizes for multiple labeling techniques; (4) Identical cells and antigens on the same tissue samples may be viewed with LM and EM by means of using the same IGSS method on the same or adjacent sections. The immunogold-silver stained sections for LM may be re-embedded, processed, cut, and counterstained for ultrastructural examinations. This makes it one of the few choices for evaluating the distribution of the same antigens at both the LM and the EM levels. The data obtained from our study provide a quantitative guideline for future studies.

7.6   Summary

The morphologic characteristics of metallic silver amplification of immunogold staining at both the LM and EM levels were evaluated quantitatively. Five sizes of gold particles (1, 5, 15, 30, and 40 nm) were evaluated for different amplification durations (1 to 17 min) using three different silver salts. Atrial natriuretic peptide in rat atria was used as a model. Silver acetate was found to be superior to either silver lactate or silver nitrate in enhancing gold particles. The size of the particles grew in an accelerated fashion within certain time intervals. Beyond this duration, the variation in particle sizes and shapes increased out of the range to be of any practical value. A semiquantitative correlation between the particle size and the duration of silver treatment was derived. The results obtained at both the LM and the EM level correlated well with each other and did not appear to affect the background of the staining in our experiment. The silver enhanced 1-nm gold was not as sensitive as the 5-nm gold. A double immunostaining procedure using IGSS was attempted with particles of the same size but different silver amplification durations on two sides of the EM grid. It was found to be feasible. Our study provides a quantitative basis for IGSS and can be used as a guide for various applications.

References

Bendayan, M. 1982. Double immunocytochemical labeling applying the protein A-gold technique. *J. Histochem. Cytochem.* 30: 81.

Danscher, G. and Norgaard, J. O. R. 1983. Light microscopic visualization of colloidal gold on resin-embedded tissue. *J. Histochem. Cytochem.* 31: 1394.

DeRoe, C., Courtoy, P. J., and Baudhuin, P. 1987. A model of protein-colloidal gold interactions. *J. Histochem. Cytochem.* 35: 1191.

De Valck, V., Renmans, W., and Segers, E. 1991. Light microscopical detection of leukocyte cell surface antigens with a one-nanometer gold probe. *Histochemistry* 95: 483.

DeWaele, M., DeMey, J., Renmans, W., Laleur, C., Reynaert, P. H., and Van Camp, B. 1986. An immunogold-silver staining method for the detection of cell surface antigens in light microscopy. *J. Histochem. Cytochem.* 34: 935.

DeWaele, M., Renmans, W., Segers, E., Jochmans, K., and Van Camp, B. 1988. Sensitive detection of immunogold-silver staining with darkfield and epipolarization microscopy. *J. Histochem. Cytochem.* 36: 679.

Gu, J. and Gonzalez-Lavin, L. 1988. Light and electron microscopic localization of atrial natriuretic peptide in the hypertrophic ventricles of spontaneously hypertensive rats. *J. Histochem. Cytochem.* 36: 1239.

Gu, J. and D'Andrea, M. 1989. Comparison of detection sensitivities of different sized gold particles with electron microscopic immunogold staining using atrial natriuretic peptide in rat atria as a model. *Am. J. Anat.* 185: 264.

Gu, J., D'Andrea, M., and Seethapathy, M. 1989. Atrial natriuretic peptide and its mRNA in overload and overload-released ventricles of rat. *Endocrinology* 125: 2066.

Gu, J., D'Andrea, M., Yu, C., Forte, M., and McGrath, L. B. 1993. Quantitative evaluations of indirect immunogold-silver electron microscopy. *J. Histotechnol.* 16: 19.

Hacker, G. W., Grimelius, L., Danscher, G., Bernatzky, G., Muss, W., Adam, H., and Thurner, J. 1988. Silver acetate autometallography: an alternative enhancement technique for immunogold-silver (IGSS) and silver amplification of gold, silver, mercury and zinc in tissues. *J. Histotechnol.* 11.

Hacker, G. W., Springall, D. R., van Noorden, S., Bishop, A. E., Grimelius, L., and Polak, J. M. 1985. The immunogold-silver staining method, a powerful tool in histopathology. *Virchows Arch. (Pathol. Anat.)* 406: 449.

Holgate, C., Jackson, P., Cowen, P., and Birdd, C. C. 1983. Immunogold-silver staining: New method of immunostaining with enhanced sensitivity. *J. Histochem. Cytochem.* 31: 938.

Knight, D. P. 1977. Cytological staining methods in electron microscopy. In: *Staining Methods for Sectioned Material.* Vol. 5. pp. 25–68. Lewis, P. R. and Knight, D. P. (Eds.), Elsevier Biomedical Press, Amsterdam.

Lackie, P. M., Hennessey, R. J., Hacker, G. W., and Polak, J. M. 1985. Investigation of immunogold-silver staining by electron microscopy. *Histochemistry* 83: 545.

Moeremans, M., Daneels, G., VanDijck, A., Langanger, G., and De Mey, J. 1984. Sensitive visualization of antigen-antibody reaction in dot and blot immune overlay assays with immunogold and immunogold/silver staining. *J. Immunol. Methods* 74: 353.

Nielson, M. H. and Bastholm, L. 1990. Improved immunolabeling of ultrathin cryosections using antibody conjugated with 1nm gold particles. In: *Biological Sciences Proceedings, XIIth International Congress for Electron Microscopy.* Vol. 3. p. 928. Peachey, L. D. and Williams, B. D. (Eds.), San Francisco Press.

Otsuki, Y., Maxwell, L. E., Magari, S., and Kubo, H. 1990. Immunogold-silver staining method for light and electron microscopic detection of lymphocyte cell surface antigens with monoclonal antibodies. *J. Histochem. Cytochem.* 38: 1215.

Roth, J. 1982. Applications of immunocolloids in light microscopy. Preparation of protein A-silver and protein A-gold complexes and their application for localization of single and multiple antigens in paraffin sections. *J. Histochem. Cytochem.* 30: 691.

Springall, D. R., Hacker, S. E., Grimelius, L., and Polack, J. M. 1981. The potential of immunogold-silver staining method for paraffin sections. *Histochemistry* 81: 603.

Vandell, I. M., Tapia, F. T., Probert, L., and Buchan, A. M. 1982. Immunogold staining procedure for the localization of regulatory peptides. *Peptides* 3: 259.

Chapter 8

Rapid Cooling and Freeze-Substitution for Immunogold-Silver Staining

Paul Monaghan

Contents

8.1 Introduction

In any ultrastructural analysis of antigen distribution using antibody markers, there will always be the question: How does the localization pattern observed relate to the *in vivo* situation? Every stage in the processing from live sample to immunolabeled section has the potential to modify the observed antigen-antibody binding pattern. Each step should introduce minimal antigenic and structural change if the resulting immunolabeling is to be interpreted in a meaningful manner. There is unfortunately no sample processing method (with perhaps the exception of pre-embedding methods for accessible antigens) which can be considered to offer the optimum combination of tissue architecture and antigen sensitivity in all circumstances (reviewed by Monaghan and Robertson, 1993).

Routine methods of chemical fixation and room temperature dehydration have a number of theoretical drawbacks when used for processing samples for immunocytochemistry. These include the relatively slow rate of fixation by aldehyde fixatives preventing the stabilization of rapidly occurring cellular events and the possible redistribution of cellular constituents. A further consideration is the effect of chemical fixatives upon the antigen-antibody reaction in the processed sample. The importance of this latter point will be

0-8493-2449-1/95/$0.00+$.50
© 1995 by CRC Press, Inc.

difficult to predict at the initial stages of investigations, and will depend on the precise antigen-antibodies concerned. For example, in a series of rat monoclonal antibodies raised against the oncogene c-erbB-2, 3 out of 10 clones produced antibodies which react to some extent with aldehyde fixed samples (Dean et al., 1993). In contrast, it has been much more difficult to produce antibodies recognizing keratins, for example, which react with aldehyde fixed samples.

Following aldehyde fixation (osmium tetroxide treatment of samples is generally deleterious to antibody labeling), room temperature dehydration has the disadvantage of removing a considerable proportion of lipids from samples (Weibull and Christiansen, 1986). The effects of lipid extraction can be minimized by dehydration at reduced temperature. This approach gives the progressive lowering of temperature or PLT method of processing its name. Low temperature dehydration is then followed by embedding in low temperature resin (Kellenberger et al., 1980; Carlemalm et al., 1982; Hobot, 1989; Robertson et al., 1992). The preparation of thawed cryosections by the method of Tokuyasu (1980, 1986) removes the dehydration step from sample preparation altogether, and is covered in Chapter 9 of this volume.

Replacement of chemical fixation by rapid freezing, or perhaps more correctly, rapid cooling (as discussed by Echlin, 1992, the aim is not to form ice crystals as indicated by the term freezing, thus rapid cooling being a more accurate description than rapid freezing) avoids these problems. The technique has much improved "time resolution" in that the living cells are stabilized in milliseconds rather than tens of seconds, with a recent estimate of freezing rate being 0.5 to 1.5 ms (Baatsen, 1993). In addition, the potential damage to antigens caused by cross-linking fixatives can be avoided.

Having cooled the sample, the ice must be removed from the specimen. This task can be accomplished either by freeze-drying, or by freeze-substitution in solvents such as methanol, ethanol, or acetone. Following removal of the water from the sample it can be embedded, sectioned, and hopefully, immunolabeled.

The advantages of rapid cooling followed by freeze-substitution have been briefly mentioned. These have been exploited in many morphological investigations (reviewed by Echlin, 1992; Hippe-Sanwald, 1993). For these studies the substitution medium contains fixatives glutaraldehyde, osmium tetroxide, uranyl acetate, acrolein, or mixtures of them. Samples thus substituted can be warmed to room temperature and embedded in epoxy resin.

When considering the use of freeze-substitution for immunolabeling studies, a number of points require consideration in the context of the specimen and antibodies in use. First, the choice of freezing method may be determined by the equipment available. If several methods are available, which is most suitable for the sample? Having frozen the sample, what substitution solvent will be used at what temperature and for how long and will it contain fixatives? If so, which one(s)? Is it proposed to embed and immunolabel the sample in epoxy resin, LR White, LR Gold, or one of the Lowicryl resins?

There can be no hard and fast rules to answer these questions, as the literature on the use of rapid freezing and freeze-substitution for immunolabeling is relatively sparse, and no consensus can be readily found. For every generalization — such as osmium tetroxide fixation being deleterious to immunolabeling (Roth et al., 1981; Bendayan, 1984) — there will be an exception. In this case, treatment with sodium metaperiodate can restore immunolabeling of some osmium-treated samples (Bendayan and Zollinger, 1983). Similarly, epoxy resins have a poor reputation for immunolabeling sensitivity, but nonetheless good results have been obtained with such resins. Clearly, success with methods generally considered to be less than optimum will depend upon the antigen lability, the characteristics of the antibodies, and even the concentration of antigen within the sample. Thus, strong immunolabeling can be readily obtained over hormone storage granules in epoxy resin embedded samples (Monaghan and Roberts, 1985).

Because no one single protocol can be considered "the" method, each stage of specimen preparation will be considered in some detail, and possible alternatives can then be discussed. In this way the advantages and drawbacks of the method can be considered in the context of alternative processing methods.

8.2 Cooling Methods

These are the crucial steps in the preparation of samples for freeze-substitution. No amount of care in the subsequent stages can overcome the damage to samples resulting from suboptimal cooling. When water is cooled, unless care is taken, it forms ice crystals. When these form within a biological specimen, they disrupt the cellular structure and cause translocation of solutes. The formation of ice crystals during cooling is a complex process and variations in the cooling technique (including extremely high pressures) can result in the formation of up to nine different forms of ice. Of these, three types of ice are of concern: hexagonal, cubic, and vitreous or amorphous ice. Hexagonal ice is most commonly found in frozen samples, and is the most stable form. Cubic ice is a metastable form of ice, and is most likely to be found in this context when amorphous ice is warmed above −150°C. Cubic ice is therefore important when considering the subsequent processing of a sample which has been successfully vitrified. Amorphous (or vitreous) ice is the ideal form for microscopy because it has no crystal structure to disrupt the sample. The formation of ice within biological samples has recently been extensively reviewed by Bachmann and Meyer (1987) and Echlin (1992).

The objective of all cooling methods for electron microscopy is to minimize damage to the sample caused by freezing. This process can be achieved only by cooling at sufficient speed to prevent the formation of large ice crystals. At its best, rapid cooling can cause the water to form amorphous ice — a glass-like state where no ice crystals have formed. In practice, although vitrification can be achieved, as long as the ice crystals formed within the sample remain of the size of a few nanometers, these will not interfere with many studies. Vitrification will only occur at high cooling rates, and this requires good thermal contact between the sample and a suitable cryogen. Unfortunately, the sample itself has a poor thermal conductivity and this is the limiting factor in the cooling process. In practice only approximately 10 to 15 μm of the sample in contact with the cryogen will achieve a satisfactory cooling rate and further into the sample, ice crystal damage will destroy the structure of the cells. Two methods most commonly employed for cooling biological samples for immunocytochemical studies are impact or slam cooling and plunge cooling. Further cooling methods which are less commonly used for immunocytochemical studies are jet cooling and spray cooling. The only method available for improving the depth of vitrification is high-pressure cooling (Moor and Hoechli 1970; Moor 1987; Studer et al., 1989), and it has considerable potential for immunocytochemical studies (Steinbrecht, 1993; McDonald and Morphew, 1993).

8.2.1 Impact Cooling

Impact cooling uses a solid cryogenic surface of high thermal conductivity and thermal capacity and the sample to be frozen is pressed against it in a controlled and reproducible manner. Its advantage is that relatively large samples can be successfully cooled. Often the solid surface is very pure copper which is cooled by liquid nitrogen or liquid helium, and commercial apparatus is obtainable from several manufacturers.

While there is some theoretical advantage to cooling with liquid helium, in practice the slight improvement in freezing over that obtained with liquid nitrogen is probably outweighed by the increased cost and limited availability of liquid helium (Sitte et al., 1987; Bald, 1987). The design of an impact cooling device must ensure that the copper block is not contaminated by condensation prior to freezing. A thin layer of ice on the cooling surface would greatly reduce the rate of cooling of the sample. It is also important that the sample, once "slammed" against the copper block, is held in contact with the block and does not bounce thus interrupting the cooling. Crushing of the sample is reduced by mounting the sample to be frozen on a small piece of plastic foam. Considerable discrepancy exists between devices in the speed of impact and the pressure which is applied to hold the sample in contact with the copper block. In practice, the optimum conditions vary from sample to sample.

Both solid tissues and cell suspensions can be cooled by this method, but care in specimen preparation is needed. With solid tissues, the surface needs to be as flat as possible to maximize the area in contact with the cryogenic surface, but as only the first 10 to 15 μm will be well frozen, mechanical and drying damage need to be minimized during the preparation immediately prior to cooling. Suspensions of cells or microorganisms can also be frozen by impact, but in order to prevent loss of the sample during the subsequent processing some form of encapsulation is required, and a number of alternatives have been proposed including agar, gelatin, etc.

8.2.2   Plunge Cooling

Plunge freezing immerses the sample to be cooled into a liquid cryogen, with liquid nitrogen cooled liquid propane or liquid ethane being the most common cryogens. They both share the advantages of being readily available and having good cooling characteristics. However, they also share the disadvantage of being highly flammable; great care must be taken in their disposal (Sitte, 1987).

Samples must be kept as small as possible, and may be held between two streamlined metal foils or held on the tip of a sample holder which should also be as small as possible. A number of publications have considered the optimum speed and distance of plunging (reviewed by Robards and Sleytr, 1985; Ryan et al., 1990). Briefly, the sample must be plunged at 2 to 4 m/sec^{-1} and must continue moving through the cryogen either by providing forced convection of the cryogen or a long plunge length.

8.2.3   Jet Cooling

An alternative to plunge cooling is to direct a jet of liquid cryogen (liquid nitrogen cooled propane or ethane) onto the sample which is protected by thin metal foils. Subsequent processing would be similar to samples which have been plunge cooled between thin metal foils.

8.2.4   Spray Cooling

Spray cooling directs a jet of small droplets of a suspension sample into a holder containing liquid cryogen, and as long as the droplets are kept around 20 μm or less, excellent cooling may be obtained. Handling of the sample during the subsequent substitution and embedding is less simple, however, than with the other methods.

8.2.5   High-Pressure Cooling

Increasing the depth of freezing to greater than 10 to 15 μm would obviously be a major advantage. While suspension samples can be frozen with a number of cells or microorganisms in the well-frozen region, with solid heterogeneous tissues it is more difficult to arrange to have the region of interest in the outer few microns of the sample. What is obviously needed in order to improve the depth of a well-frozen sample is to inhibit the formation of ice crystals during the cooling process. Applying pressure to the sample immediately prior to the cooling can significantly lower the rate of cooling needed to vitrify the sample. The method developed by Moor and co-workers (1970) and more recently by Müller (Müller and Moor, 1984; Studer, 1989) has the potential to vitrify up to several hundred μm depending on the sample. The success of high-pressure cooling depends to some extent on the sample, but vitrification of 200 μm samples of botanical specimens has been demonstrated with this method (Michel et al., 1991). The application of high pressure (2100 bar) depresses the minimum supercooling temperature and increases the viscosity of the water in the sample. Water expands as it freezes, and therefore the application of high pressure during the cooling process inhibits ice crystal formation.

The major disadvantage of this approach is that the equipment needed to apply these high pressures unfortunately is expensive. Also, the effects on the specimen of the brief application of 2100 bar are not clear, and some artifacts may be induced by this cooling method. Of interest also is the presence of cracks within blocks of high-pressure frozen samples which may result from volume changes occurring as the ice form changes when the pressure is removed from the specimen (Dahl and Staehelin, 1989). Nonetheless, with its ability to significantly increase the volume of sample that can be satisfactorily cooled, the potential of this method for immunocytochemical studies is considerable (Figure 1).

8.3   Freeze-Substitution

The ice in the frozen sample must be removed by incubation in an organic solvent in a method pioneered for electron microscopy by Fernandez-Moran (1957). The choice of substitution medium is determined by the freezing point of the solvent bearing in mind the temperature at which substitution is planned, the substitution time, and the percentage of water that the solvent can accommodate. The two most commonly used solvents are methanol and acetone, although a number of other possibilities exist including diethyl ether, ethanol, and propane. Methanol has the advantage that it is tolerant of up to 10% water, whereas acetone can only tolerate 1 to 2% water at −80°C (Humbel et al., 1983; Humbel and Müller, 1984). Dried acetone must therefore be used, possibly in conjunction with molecular sieve during the substitution process. Methanol has the additional advantage that substitution is rapid (hours rather than days for acetone) but it will almost certainly remove more of the lipid component of the sample as compared with acetone (Weibull and Christiansson, 1986). The question of whether there are immunocytochemical advantages to the use of different substitution solvents can only be determined for each individual antigen-antibody combination. The only way to completely avoid the use of substitution solvents is to freeze-dry the frozen samples (Chiovetti et al., 1987; Linner et al., 1986) and embed them at low temperature in Lowicryl resins, although this approach does not rule out any solvent effects of the resin prior to polymerization.

Substitution temperature is determined by two factors. The first factor is that the lower the temperature, the slower the substitution; the second factor is the limit set by the freezing point of the substitution medium. For example, methanol freezes at −93.9°C and acetone at −95.4°C. A detailed study of the effects of temperature on substitution times has been reported by Humbel (1983). Substitution is usually undertaken at −80 to −90°C. Having taken great care to vitrify at least a portion of your sample, this is theoretically a risky process, because at this temperature, vitrified ice should recrystallize and damage the sample. In practice although much warmer than the safe level of around −120°C, recrystallization does not always occur, and ice artifact-free micrographs have been published by numerous authors. The risk of ice recrystallization has been examined by Steinbrecht (1985), who has indicated that recrystallization is a relatively slow process. Short (10 min) exposures of temperatures as warm as −40°C caused no appreciable damage, but by 45 min at this temperature, ice crystal damage was apparent (Steinbrecht, 1985).

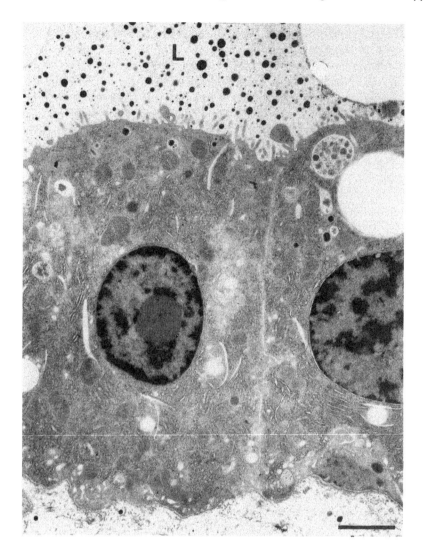

Figure 1 Electron micrograph of lactating mouse mammary gland. The gland was high-pressure cooled, freeze-substituted in acetone containing 1% osmium tetroxide for 72 h at –90°C and embedded in epoxy resin. The sections were contrasted with uranyl acetate and lead citrate. The cellular morphology closely resembles that seen in routinely processed material. The duct lumen (L) is distended with milk and lined by the luminal epithelial cells. Magnification ×7500; scale = 2 μm.

Substitution protocols vary considerably, and it is difficult to give one single protocol. As already discussed, methanol substitution is more rapid than acetone substitution. Considering first acetone substitution, while excellent results have been reported with a substitution protocol of 8 h at –90°C, 8 h at –60°C followed by 8 h at –30°C (Humbel et al., 1983) acetone substitution is recommended for 7 d by Edelmann (1989). Hobot (1989) recommends 64 to 88 h as a general substitution time. Methanol substitution for 36 h has given satisfactory results (Monaghan and Robertson 1990), but more recent experiments have reduced this time to 20 h (Figure 2). Methanol substitution has also been successfully applied using the 8 h at –90°C, 8 h at –60°C and 8 h at –30°C regime (Engfelt et al., 1986). Three days at –80°C is reported by Bittermann (1992) for a variety of substitution media and fixatives.

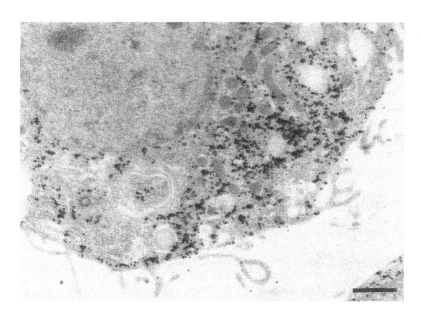

Figure 2 A cell from the human breast carcinoma cell line ZR-75-1, high-pressure cooled and freeze-substituted in methanol without additional fixatives for 20 h and embedded at –50°C in Lowicryl HM20. Thin sections were immunolabeled with mouse monoclonal antibody against the intermediate filament protein vimentin followed by 5 nm colloidal gold. The gold was silver-enhanced and the sections contrasted with uranyl actate and lead citrate. Mammalian cells processed in this manner exhibit lower contrast than when osmium is included in the substitution medium. Magnification ×12,100; scale = 1 μm.

The question is: To add fixatives to the substitution medium or not? The effects of fixatives in the substitution medium at substitution temperatures are difficult to determine, and there is some evidence that it is not until the sample is warmed that fixation takes place beyond about –50°C (Humbel et al., 1983). Fixation could be considered to be a generally bad thing for immunocytochemistry, whether for light or electron microscopy. In practice, some antibodies react equally in unfixed and fixed samples, but this cannot be relied on. While there are reports that fail to detect loss of immunoreactivity in rapid cooled samples which have been substituted in the presence of glutaraldehyde (Hobot et al., 1987; Hunziker and Herrmann, 1987), other reports have compared immunolabeling of samples which had been rapid cooled and freeze-substituted with and without fixatives. In samples substituted in the presence of fixatives, the immunolabeling was always reduced in the fixed samples (Nicholas et al., 1993; Bitterman et al., 1992; Schwarz and Humbel, 1989). Nonetheless, if the antibodies in question react in fixed samples, the improvement in stability obtained may outweigh the loss of sensitivity. A further advantage of the addition of fixatives to the substitution medium is the possible immobilization of antigens which may undergo some migration during substitution and resin infiltration (Schwarz and Humbel, 1989; Steinbrecht, 1993). Fixatives should be used when the samples are to be warmed to room temperature for epoxy resin embedding.

Glutaraldehyde will be more readily used with methanol as it can cope with the water introduced with the glutaraldehyde solution. Osmium tetroxide can be readily prepared in acetone and substitution in this solvent containing 1 or 2% osmium tetroxide is commonly employed. Acrolein, uranyl acetate, and various combinations of these fixatives have also been successfully used. Results may depend on the particular sample, however.

8.4   Resin Embedding

If fixatives have not been included in the substitution protocol, the samples will probably be embedded in Lowicryl resin. These resins were developed for low temperature embedding, and comprise two hydrophilic (K4M, K11M) and two hydrophobic (HM20, HM23) resins (Kellenberger et al., 1980; Carlemalm, 1982). In addition to their low viscosity, they are also particularly appropriate for immunolabeling studies because they have low reactivity toward biological molecules. This has been postulated to result in a relatively uneven surface of thin Lowicryl sections which could give improved access of reagents to antigenic sites within the tissue (Kellenberger, 1987). In contrast, Epoxy resins are, in general, inferior for immunolabeling studies (Shida and Ohga, 1990) and have much less pronounced surface irregularity. Infiltration times as recommended by the manufacturer give satisfactory infiltration under a range of conditions. A typical Lowicryl infiltration protocol is

25% resin : 75% solvent	60 min
50% resin : 50% solvent	60 min
75% resin : 25% solvent	60 min
100% resin	overnight
100% resin	60 min

Ultraviolet light polymerization times vary with the resin used, the polymerization temperature, the UV lamp output, and the geometry of the lamp to resin distance. For HM20 polymerized at –50°C, in a Leica AFS, 48 h at –50°C followed by at least 24 h at room temperature is required for polymerization to be satisfactory.

UV polymerization is an exothermic reaction which has received discussion over the damage that it may inflict upon the sample (Ashford et al., 1986; Weibull et al., 1986). However, provided the samples are held in contact with a large heat sink, the temperature rise can be less than one or two degrees.

In the early applications of immunolabeling of Lowicryl sections, it was believed that the choice of the hydrophilic resins Lowicryl K4M and K11M would have superior labeling properties when compared with the hydrophobic HM20 and HM23. In practice there seems to be little to choose between the resins in terms of immunolabeling sensitivity (Durrenberger et al., 1991). The choice may well be made more on the minimum temperature that the resin can be used at, rather than its hydrophobic or hydrophilic nature. Minimum temperatures for the Lowicryl resins are K4M –35°C, HM20 –50°C, K11M –60°C, and HM23 –80°C. Less than perfectly infiltrated or incompletely polymerized Lowicryl blocks are difficult to section, but if care is taken in the substitution infiltration and polymerization steps a clear, hard block will be produced. This sample may be a little brittle to trim and the initial thin sections of a recently trimmed block may be of poor quality. Careful facing of the block prior to thin sectioning will allow the block to be easily sectioned. Incomplete polymerization, particularly of HM20, is indicated by coloration of the block which may be either very pale pink or even pale yellow. Further polymerization will usually produce a clear block.

If fixatives have been used in the substitution medium, then there are more possibilities for choice of embedding medium. UV polymerized LR White or LR Gold acrylic resins can be used if osmium has not been part of the fixation protocol, and heat cured LR resins as well as epoxy resins can be used with any substitution mixture. Whether the heat polymerization of epoxy (or acrylic) resins will affect antibody binding will have to be determined for each situation.

8.5   Immunolabeling

The procedures for immunolabeling rapid cooled/freeze substituted samples are similar to those for other resin embedding methods. They consist of a nonspecific binding blocking stage, incubation with the primary antibody, washing, incubation with the second antibody or protein A gold conjugate, further washes, silver-enhancement, and finally contrasting. Numerous minor variations of this basic theme have been published.

8.5.1 Blocking Nonspecific Binding

The hydrophobic resins Lowicryl HM20 and HM23 gained a reputation for giving slightly increased background labeling in immunocytochemical studies, but care with blocking procedures can minimize this. Durrenberger et al. (1991) recommend 0.1 to 0.2% gelatin (as have many other authors) in all incubation solutions, but also suggest that the diluted antiserum is incubated for 1 h with polymerized resin in powder form prior to labeling. The resin is removed by centrifugation prior to use. Many further variations of methods for blocking nonspecific binding have been published. The following protocol has proved successful at keeping background labeling to low levels in both HM20 and K4M embedded samples. (Monaghan and Robertson, 1993).

- Incubate for 30 min in blocking buffer: Phosphate-buffered saline (PBS) plus 0.8% bovine serum albumin (BSA) plus 0.1% gelatin plus 5% fetal calf serum (FCS) (pH 7.4).
- Make up primary antibody in blocking buffer and incubate for 1 to 2 h. With many antibodies, improved levels of labeling will be achieved by incubation in the primary antibody overnight.
- Wash sections (3 to 5 min) in washing buffer: PBS plus 0.8% BSA plus 0.1% gelatin (pH 7.4), and incubate for 90 min in secondary antibody/protein A 1- or 5-nm gold conjugate diluted in blocking buffer.

The choice of size of colloidal gold is relatively simple. While 10 nm gold is perhaps the optimum size for detection in the electron microscope, it offers reduced sensitivity when compared with 5- and 1-nm markers. A good starting point, and indeed our standard protocol for both PLT and freeze-substituted samples in HM20 or K4M, is to use 5-nm colloidal gold which is silver enhanced to around 10 nm. Thus this gives the sensitivity of using 5-nm gold with the ease of detection of the larger sizes of gold markers. In experiments immunolabeling smooth muscle actin in human breast myoepithelial cells embedded by PLT in HM20, there was no detectable improvement in labeling when changing from 5- to 1-nm gold (Robertson, unpublished data).

8.5.2 Silver-Enhancement

One of the great advantages of using silver-enhancement methods for colloidal gold labeling is the opportunity they present for undertaking preliminary studies of resin-embedded material at the light microscope level. It is easier to label a number of large 0.5- or 1-μm sections of resin-embedded material with the antibody(ies) of interest and determine the optimum dilutions, and labeling patterns by light microscopy than going straight for thin section labeling. Thick sections for light microscopy should be cut with care (a light microscopy grade diamond, e.g., Diatome Histodiamond is particularly valuable at this stage), and collected onto subbed slides. These greatly enhance the adhesion of sections to the slides through the incubations. They can be prepared by making up 200 ml of subbing solution consisting of 1% gelatin plus 1% formalin in distilled water in the following manner. Dissolve 2 g gelatin in 194 ml distilled water, and warm to dissolve. Add 6 ml of 36% formalin and immerse microscope slides in this subbing solution. Drain and dry in 60°C oven. Lowicryl embedded material can be stained readily with a dilute solution (0.1%) of toluidine blue.

Once this preliminary work has been completed, thin sections can be labeled with suitably reduced silver-enhancement times. A particularly sensitive method for viewing light microscope sections which have been labeled with silver-enhanced colloidal gold is epi-polarized microscopy. The cost of modifying a light microscope for epi-polarization is not excessive, and this form of imaging is particularly good at detecting low levels of labeling. This sensitivity is advantageous for determining the distribution of rare antigens as well as being unforgiving in showing the presence of even small amounts of background labeling.

The use of thick sections for light microscopy can also take advantage of one of the more frustrating aspects of rapid cooled/freeze-substituted material. It can often be extremely difficult to determine the quality of cooling in any particular block by light microscopy. Areas of truly dreadful freezing damage can be detected by the presence of ice crystal damage particularly in red blood cells (if present), and a rather granular appearance in cell nuclei. It is rarely possible to detect less severe ice crystal damage at this resolution, and so the selection of a well-frozen piece of tissue has to be done at the ultrastructural level. For light microscopy, however, much of a block will show excellent morphology and the overall distribution of an antigen can be determined readily from such material.

Thin sections can either be collected on formvar (or similar) coated grids or uncoated grids. If the block is not perfectly embedded, then the incubations needed for immunolabeling may weaken the sections which develop numerous holes. In this hopefully rare case, the sections must be collected on coated grids. The disadvantage of this process is that only one side of the section will be accessible for immunolabeling thereby reducing the sensitivity by one half in one step.

8.6 Summary

The broad objective of immunolabeling studies is to add a degree of functional information to ultrastructural studies. There are numerous advantages of rapid cooling over aldehyde fixation as the initial stabilization step. As was clear from the preceding discussion, rapid cooling followed by freeze-substitution holds the potential for being one of the most sensitive methods for ultrastructural immunolabeling as well as providing excellent retention of cellular morphology. This approach is, however, still evolving and there are many opportunities for variations in cooling methods, substitution protocols, resin embedding, and immunolabeling protocols. Whether the additional complications of the technique over, for example, the PLT method will provide worthwhile improvements in labeling intensity will depend on the specific problem under investigation, but for many investigations this approach offers minimal antigenic change with maximal retention of the *in vivo* sample characteristics.

Acknowledgment

The Institute of Cancer Research is supported by funds from the Cancer Research Campaign and the Medical Research Council.

References

Ashford, A. E., Allaway, W. G., Gubler, F., Lennon, A., and Sleegers, J. 1986 Temperature control in Lowicryl K4M and glycol methacrylate during polymerisation: Is there a low temperature embedding method? *J. Microsc.* 144: 107.

Baatsen, P. H. W. W. 1993. Empirically determined freezing time for quick freezing with a liquid nitrogen cooled copper block. *J. Microsc.* 172: 71.

Bald, W. 1987. *Quantitative Cryofixation.* Adam Hilger, Bristol.

Bendayan, M. and Zollinger, M. 1984. Ultrastructural localisation of antigenic sites of osmium fixed tissues applying the Protein A gold technique. *J. Histochem. Cytochem.* 31: 101.

Bendayan, M. 1984. Protein A-gold electron microscopic immunocytochemistry: methods, applications and limitations. *J. Electron Microsc. Tech.* 1: 243.

Bitterman, A. G., Knoll, G., Nemeth, A., and Plattner, H. 1992. Quantitative immuno-gold labeling and ultrastructural preservation after cryofixation (combined with different freeze-substitution and embedding protocols) and after chemical fixation and cryosectioning. *Histochemistry* 97: 421.

Carlemalm, E., Garavito, R. M., and Villiger, W. 1982. Resin development for electron microscopy and an analysis of embedding at low temperature. *J. Microsc.* 126: 123.

Chiovetti, R., McGuffee, J., Little, S. A., Wheeler-Clark, E., and Brass-Dale, J. 1987. Combined quick freezing freeze-drying and embedding tissue at low temperature and in low viscosity resins. *J. Electron Microsc. Tech.* 5: 1.

Dahl, R. and Staehelin, L. R. 1989. High pressure freezing for the preservation of biological structure: theory and practice. *J. Electron Microsc. Tech.* 13: 165.

Dean, C., Styles, J., Valeri, M., Modjtahedi, H., Bakir, A., Babich, J. W., and Eccles, S. 1993. Growth factor receptors as targets for antibody therapy. In: *Mutant Oncogenes. Targets for Therapy.* p. 27. Lemoine, N. and Epenetos, A. (Eds.), Chapman and Hall, London.

Durrenberger, M., Villiger, W., Arnold, B., Humbel, B. M., and Schwarz, H. 1991. Polar or apolar Lowicryl resin for immunolabeling? In: *Colloidal Gold: Principles Methods and Applications.* Vol. 3. p. 73. Hayat, M. A. (Ed.), Academic Press, San Diego.

Echlin, P. 1992. *Low Temperature Microscopy and Analysis.* Plenum Press, New York.

Edelmann, L. 1991. Freeze-substitution and the preservation of diffusable ions. *J. Microsc.* 161: 217.

Fernandez-Moran H. 1957. Electron microscopy of nervous tissue. In: *Metabolism of the Nervous System.* p. 1. Richter, D. (Ed.), Pergamon Press, Oxford.

Hippe-Sanwald, S. 1993. Impact of freeze-substitution on biological electron microscopy. *Microsc. Res. Tech.* 24: 400.

Hobot, J. A. 1989. Lowicryls and low temperature embedding for colloidal gold methods. In: *Colloidal Gold: Principles Methods and Applications.* Vol. 2. p. 75. Hayat, M. A. (Ed.), Academic Press, San Diego.

Hobot, J. A., Bjornsti, M. A., and Kellenberger, E. 1987. Use of on-section immunolabeling and cryosubstitution for studies of bacterial DNA distribution. *J. Bacteriol.* 169: 2055.

Humbel, B., Marti, T., and Müller, M. 1983. Improved structural preservation by combining freeze-substitution and low temperature embedding. *Beitr. Elektronmikroskop. Direktabb. Oberfl.* 6: 585.

Hunziker, E. B. and Herrmann, W. 1987. In situ localisation of cartilage extracellular matrix components by immunoelectron microscopy after cryotechnical tissue processing. *J. Histochem. Cytochem.* 35: 647.

Kellenberger, E., Carlemalm, E., Villiger, W., Roth, J., and Garavito, R. M. 1980. Low denaturation embedding for electron microscopy of thin sections. Chemische Werke Lowi GmbH, P.O. Box 1660, D-8264 Waldkraiburg, Germany.

Kellenberger, E., Dürrenberger, M., Villiger, W., Carlemalm, E., and Wurtz, M. 1987. The efficiency of immunolabel on Lowicryl sections compared to theoretical predictions. *J. Histochem. Cytochem.* 35: 959.

Linner, J. G., Livesey, S. A., Harrison, D., and Steiner, A. L. 1986. A new technique for removal of amorphous phase tissue water without ice crystal damage: a preparative method for ultrastructural analysis and immunoelectron microscopy. *J. Histochem. Cytochem.* 34: 1123.

McDonald, K. and Morphew, M. K. 1993. Improved preservation of ultrastructure in difficult-to-fix organisms by high pressure freezing and freeze substitution. I. *Drosophila melanogaster* and *Strongylocentrotus purpuratus* embryos. *Microsc. Res. Tech.* 24: 465.

Michel, M., Hillmann, T., and Müller, M. 1991. Cryosectioning of plant material frozen at high pressure. *J. Microsc.* 163: 3.

Monaghan, P. and Roberts, D. J. B. 1985. Immunocytochemical evidence for neuroendocrine differentiation in human breast carcinomas. *J. Pathol.* 147: 281.

Monaghan, P. and Robertson, D. 1990. Freeze-substitution without aldehyde or osmium fixatives: ultrastructure and implications for immunocytochemistry. *J. Microsc.* 158: 355.

Monaghan, P. and Robertson, D. 1993. Immunolabeling techniques for electron microscopy. Post-embedding techniques. In: *Immunocytochemistry, A Practical Approach.* p. 43. Beesley, J. E. (Ed.), Oxford University Press, Oxford.

Moor, H. and Hoechli, M. 1970. The influence of high pressure freezing on living cells. In: *Proc. 7th Int. Cong. EM. Grenoble.* Vol. 1. p. 449. Societe Francais de Microscopie Electronique.

Moor, H. 1987. Theory and practice of high pressure freezing. In: *Cryotechniques in Biological Electron Microscopy.* p. 175. Steinbrecht, R. A. and Zierold, K. (Eds.), Springer-Verlag, Berlin.

Müller, M. and Moor, H. 1984. Cryofixation of thick specimens by high pressure freezing. In: *The Science of Biological Specimen Preparation.* p. 131. Revel, J.-P., Barnard, T., and Haggis, G. (Eds.), SEM, Inc., AMF O'Hare, Chicago.

Nicholas, M.-T. and Bassot, J. M. 1993. Freeze-substitution after fast-freeze fixation in preparation for immunocytochemistry. *Microsc. Res. Tech.* 24: 474.

Robards, A. W. and Sleytr, U. B. 1985. Low temperature methods in biological electron microscopy. In: *Practical Methods in Electron Microscopy.* Vol. 10. Glauert, A. M. (Ed.), Elsevier, Amsterdam.

Robertson, D., Monaghan, P., Clarke, C., and Atherton, A. J. 1992. An appraisal of low temperature embedding by progressive lowering of temperature into Lowicryl HM20 for immunocytochemical studies. *J. Microsc.* 168: 85.

Roth, J., Bendayan, M., Carlemalm, E., Villiger, W., and Garavito, R. M. 1981. Enhancement of structural preservation and immunocytochemical staining in low temperature embedded pancreatic tissue. *J. Histochem. Cytochem.* 29: 663.

Schwarz, H. and Humbel, B. M. 1989. Influence of fixatives and embedding media on immunolabeling of freeze-substituted cells. *Scan. Microsc. Suppl.* 3: 57.

Shida, H. and Ohga, R. 1990. Effect of resin use in the post-embedding procedure on immunoelectron microscopy of membranous antigens with special reference to sensitivity. *J. Histochem. Cytochem.* 38: 1687.

Sitte, H., Edelmann, L., and Neumann, K. 1987. Cryofixation without pre-treatment at ambient pressure. In: *Cryotechniques in Biological Electron Microscopy.* p. 88. Steinbrecht, R. A. and Zierold, K. (Eds.), Springer-Verlag, Berlin.

Studer, D., Michel, M., and Müller, M. 1989. High pressure freezing comes of age. *Scan. Microsc. Suppl.* 3: 253.

Steinbrecht, R. A. 1985. Recrystallisation and ice crystal growth in a biological specimen, as shown by a simple freeze-substitution method. *J. Microsc.* 140: 41.

Steinbrecht, R. A. 1993. Freeze-substitution for morphological and immunocytochemical studies in insects. *Microsc. Res. Technique* 24: 488.

Tokuyasu, K. T. 1980. Immunocytochemistry on ultrathin frozen sections. *Histochem J.* 12: 381.

Tokuyasu, K. T. 1986. Immunocryoultramicrotomy. *J. Microsc.* 143: 139.

Weibull, C. 1986. Temperature rise in Lowicryl resins during polymerisation by ultraviolet light. *J. Ultrastruct. Mol. Strict. Res.* 97: 207.

Weibull, C. and Christiansson, A. 1986. Extraction of proteins and membrane lipids during low temperature embedding of biological material for electron microscopy. *J. Microsc.* 142: 79.

Chapter 9

Thin Cryosections and Immunogold-Silver Staining

Lone Bastholm and Folmer Elling

Contents

9.1 Introduction

During the past 2 to 3 decades immunoelectron microscopy (IEM) has emerged as a powerful tool in the visualization of molecules at the ultrastructural level and as a bridge between cell biology and molecular biology, between morphology and test tube studies. The overall challenge in IEM is to obtain both optimal morphology and optimal retention of complementarity between the antigen-antibody under study. For each antigen there is a balance between the preservation of its cellular localization and its antigenicity. The primary step in the ultrastructural localization of an antigen is thus to identify the appropriate fixation and other preparatory procedures.

The second step is embedding of the tissue samples. Because the introduction of cryoprotection in sucrose eliminates freezing artifacts (Tokuyasu, 1973), the water/sucrose in cells and tissues could replace resins as the embedding material. This procedure coupled with the development of easy-to-handle cryoultramicrotomes has made cryosectioning useful for IEM, since the morphology is adequately preserved and the antigenicity of proteins and peptides is retained better than in conventional resin-embedded material. Snap freezing and cryosectioning is also a far more rapid method than resin embedding and conventional thin sectioning.

The third step is the choice of electron dense markers. Early IEM employed ferritin and peroxidase as markers. The introduction by Faulk and Taylor (1971) of colloidal gold particles as markers gave rise to a new era due to their electron density and uniform size. The very small particles with diameters of 1 nm or smaller further increased the detection efficiency (Van Bergen en Henegouwen and Van Lookeren Campagne, 1989). These small particles are virtually unidentifiable in biological specimens. This problem can be circumvented by using silver-enhancement with a physical developer introduced by Danscher and Nörgaard (1983), facilitating examination of the specimens at conveniently low magnifications.

We describe some of the methodologies in immunocytochemistry on thin cryosections employing silver-enhanced colloidal gold particles as markers. We shall focus on advantages and disadvantages and identify the pitfalls as we have encountered them in our laboratory. We have for the past 10 years used the mouse pituitary gland and antimouse growth hormone (generously donated by National Institute of Arthritis, Diabetes and Kidney Diseases of the National Hormone and Pituitary Program, University of Maryland, School of Medicine, USA) as a model system in the determination of appropriate dilutions of new batches of gold-conjugated secondary antibodies and silver enhancers. This system has also been very useful in the introduction of IEM to beginners, and we use the system to illustrate the routine procedures and protocols in our laboratory.

9.2   Cryosections in Immunocytochemistry

9.2.1   Advantages

Tissues and cells cryoprotected with sucrose require minimal fixation prior to freezing. We have found that fixation with as low as 2% paraformaldehyde and 0.01% glutaraldehyde in 0.1 M phosphate buffer retains sufficient morphology; such gentle fixation renders most antigens detectable and this step is the only potential denaturation step for antigens. One of the major advantages of this combination of fixatives is that the small paraformaldehyde molecule causes limited cross-linking and thus little alteration in the physicochemical properties of proteins. The paraformaldehyde fixation is reversible, but addition of glutaraldehyde in a very low concentration is an effective countermeasure which causes acceptable cross-linking of proteins enhancing the morphology without damaging the antigenicity of most antigens. We have found that in some cases an antigen is undetectable even after fixation in 0.01% glutaraldehyde in 0.1 M phosphate buffer, and in some of these cases we have succeeded with paraformaldehyde at a higher concentration (8% in 0.1 M phosphate buffer) for a short period (10 min), but the morphology is somewhat inferior to that obtained in the paraformaldehyde/glutaraldehyde fixed materials.

The embedding procedure encompasses cryoprotection in 2.3 M sucrose for 20 to 30 min, removal of excess sucrose prior to placing the tissue block on a silver pin and snap freezing in liquid nitrogen; the frozen tissue or cells can be cut immediately with a cryoultramicrotome. This procedure is far more rapid than conventional embedding in Epon or Lowicryl and the antigens are not damaged by organic solvents and the high temperatures required for most resin embedding procedures. The procedure for cryosectioning is described in detail by Hagler (1993). No "foreign" embedding media will hinder access to antigens. Immunoelectron microscopy using cryosections provides high-detection efficiency with low background. It is possible to obtain considerably larger thin sections by cryomicrotomy compared with thin resin-embedded sections. The delineation of ultrastructure of cryoprotected cryosections has been claimed to be equal or superior to that of resin-embedded sections (Tokuyasu, 1984).

9.2.2 Disadvantages

Normally, the samples are frozen on silver pins. However, this requires storage capacity in liquid nitrogen containers, which are expensive to purchase and to maintain. Another disadvantage is the difficulty in identifying the area of interest found on the survey section for the trimming required for thin sectioning, because one cannot identify hallmarks on the surface of the tissue block. This process is rather easy on resin-embedded materials. Osmium tetroxide is not used in IEM because it is very reactive and frequently destroys antigenicity of proteins and peptides (Williams and Faulkner, 1993). The lack of osmium implies that cryosections do not have the high contrast of resin-embedded glutaraldehyde/OsO_4-fixed sections.

9.2.3 Practical Hints

The storing procedure may be modified to reduce the consumption of silver pins and space in the nitrogen canisters by storing cells and tissues for up to several weeks in 1% paraformaldehyde in 0.1 M phosphate buffer at 4°C. We have also frozen sucrose cryoprotected cells and tissues without mounting on silver pins and stored the specimens in Eppendorph tubes in nitrogen containers. For sectioning, the specimens are thawed in 2.3 M sucrose, cut into appropriate size, mounted on silver pins, and then frozen.

Thin cryosections must be kept moist until the immunolabeling is performed. It is often convenient to cut a batch of sections for subsequent immunolabeling. In our experience the labeling intensity is unaffected even after several days when the sections are kept on the nickel grids face down on droplets of 1% paraformaldehyde in 0.1 M phosphate buffer. Prior to immunolabeling, the paraformaldehyde is removed by washing with the buffer. It is also possible to store sections floated on 2.3 M sucrose placed in a Petri dish in the freezer at –18°C for up to 2 months (Hagler, personal communication). After thawing, the sucrose can be washed out in 0.1 M phosphate buffer (3 times for 10 min each) prior to immunolabeling. These modified storing procedures do not reduce labeling intensity.

9.3 Immunocytochemical Detection Systems

Successful detection of antigens at the electron microscopical level depends on the specimen preparation, the antibody, and the electron dense marker.

9.3.1 Direct Immunocytochemistry

In direct immunocytochemistry the antigen is detected by antibody conjugated with the gold marker. One of the advantages of the direct technique is that the localization of the antigen is more precise than with indirect immunocytochemistry where the gold particles are further away from the antigen. In staining for more than one antigen the advantages are high speed and simplicity because antibodies against different antigens may be applied simultaneously. Identification of more than one antigen requires antibodies conjugated with gold particles of different sizes; the antibodies may be applied simultaneously or sequentially. Despite the obvious advantages, the direct technique is not often used because of the inconvenience of having to conjugate each primary antibody. Most primary antibodies are available in only small quantities and significant amounts are lost during the conjugation procedure.

9.3.2   Indirect Immunocytochemical Methods

Indirect or sandwich techniques employ two or more layers with the specific antibody as the first layer and colloidal gold-conjugated anti IgG to the first species as the second layer. The indirect method has several advantages and the most obvious is that one IgG-gold batch can be used to identify all primary antibodies from one species. Furthermore, each primary antibody may bind three to five secondary antibodies, thus increasing the number of marker particles several fold. The indirect technique is also convenient in simultaneous double and triple immunolabeling (Nielsen and Bastholm, 1993).

9.3.3   Controls

A plethora of possibilities exist for unspecific staining in IEM due to the multitude of antigenic sites in cells and tissues, to antisera containing known and sometimes unknown cross-reacting antibodies, and to the fact that a variety of proteins and other molecules may bind to the antibody molecule at sites other than its antigen combining site. This diversity calls for appropriate controls. All control stainings must be performed identical with the original staining with respect to dilution of sera, length of incubation, and use of protein blocking agents. Control staining should ensure that the specific antibody is bound only to the antigen in question and that the secondary antibody is bound to the primary antibody and not to unspecific components in the section.

Indirect IEM control procedures include:

1. Preabsorption of the primary polyclonal antibody with the purified specific antigen.
2. Replacement of the primary monoclonal antibody with a nonsense monoclonal of the same subclass.
3. Omission of the primary antibody.

These controls should result in unlabeled specimens. Control staining is discussed in detail by Nielsen and Bastholm (1993).

9.3.4   Gold Particles

As indicated above, the introduction of gold particles (Faulk and Taylor, 1971) is a step forward for IEM for several reasons:

1. Colloidal gold particles are highly electron dense and when larger than 5 nm, they are easily recognized in biological specimens.
2. Colloidal gold particles can be manufactured with narrow size distribution.
3. Colloidal gold particles of different sizes can be used simultaneously in double labeling.
4. Colloidal gold particles can be silver enhanced, a feature especially useful for visualization and identification of very small particles for light and electron microscopy.

Colloidal gold is negatively charged and may be conjugated to most proteins under appropriate conditions of pH and concentration. The binding of macromolecules at their isoelectric point to colloidal gold is a noncovalent adsorption which is stable and the binding rarely affects the biological activity of the tagged molecule. Particles of decreasing diameters have been developed in an attempt to increase labeling density and detection efficiency. Gold particles are superior to other markers in IEM, and 1 nm particles are superior to larger particles (e.g., 10 nm) because the ultrasmall probes cause less steric hindrance and penetrate better than larger particles. Additionally, one IgG molecule will bind more than one 1 nm gold particle (Leunissen and van de Plas, 1993). We have thus found that 1-nm particles conjugated to secondary IgG clearly increased the labeling density in the detection of growth hormone and laminin in mouse pituitary gland in comparison with 5-nm gold particle-secondary IgG (Nielsen and Bastholm, 1990).

9.4   Silver-Enhancement of Colloidal Gold Particles

As indicated above, ultrasmall particles are virtually unrecognizable in biologic material and therefore they must be enhanced. Amplification of gold particles was simultaneously introduced by Danscher and Nørgaard (1983) and Holgate et al. (1983). The exact process by which silver ions adhere to the colloidal gold with the reducing hydroquinone is not known, but the metallic silver atoms are so closely attached to the gold particle that they represent a true concentric expansion around the gold and the growth will continue as long as the developer contains silver ions and reducing molecules. Gum arabic solution must be added to the silver ion/hydroquinone solution in order to avoid autotriggered nucleation of silver and also to reduce the speed of development to a manageable protocol. Silver-enhancement is discussed in detail by Scopsi (1989) and Danscher et al. (1993); also, see Chapters 1 and 2 in this volume. We have shown that, in contrast to Danscher and Nørgaard (1985), the acidic physical developer containing gum arabic can be employed as an intensifier of gold probes for thin sections without any corrosion of the nickel grids or unspecific silver precipitates (Bastholm et al., 1986; Scopsi et al., 1986). Unspecific silver nucleation on copper grids is shown in Figure 1a. It is important that each step of silver-enhancement be carried out in carefully cleaned glassware and with high purity chemicals and redistilled water from a glass distillator.

9.4.1   Commercial Developer

The silver-enhancement procedure (the Danscher method) with gum arabic ensures a uniform increase in size of gold particles. It is important to carry out the silver-enhancement at a constant temperature (e.g., at room temperature, 20 to 22°C) in the dark. Commercial developers do not contain gum arabic; they contain only silver ions and a reducing agent which are mixed prior to use. We have found that commercial developers can be used for 5-nm or larger particles, but not for 1-nm particles. We use commercial developers in preliminary detection of a given antigen for obtaining a survey impression of the localization. It is important to establish a procedure with a 2 to 4 min development time, because longer development (e.g., 10 min) will cause unspecific self nucleation of silver (Figures 1b, 2, and 3). The development can be done in daylight. Also it shows at magnification ≥10,000× that deletion of gum arabic in the developer results in uneven and nonspherical particles (Figure 4a) (Scopsi et al., 1986).

Recently, a commercial, purified gum arabic was introduced which can be easily dissolved in water (Leunissen and van de Plas, 1993). We mix this solution diluted to 30% in distilled water 1:1 with a commercial developer solution with a neutral pH (Aurion, Holland) and have found that this provides a uniform enlargement of 5-nm or larger gold particles (Figure 5a). One-nanometer particles, however, can only be enhanced with the Danscher developer (Figure 6a).

It is important to note that silver-enhanced ultrasmall gold particles are not nearly as uniform in size as silver-enhanced 5-nm or larger particles (Figure 6b). The reason for this could be (1) ultrasmall particles are not as uniform in size as the larger particles, and (2) one IgG molecule will bind two ultrasmall gold particles, which are located so close to each other that they, by subsequent silver-enhancement, will appear as one large electron dense

Figures 1–8 are electron micrographs of thin cryosections of growth hormone producing cells from mouse anterior pituitary gland fixed in 2% paraformaldehyde and 1% glutaraldehyde. Sections were collected at Formvar-coated nickel grids apart from Figure 1a, where Formvar-coated copper grid was used. The sections were indirectly stained with rabbit antigrowth hormone 10 to 25 μg/ml as the first layer. The second layer varied as indicated in the individual legends. In Figure 8, the cells were also stained for tubulin with monoclonal antitubulin 1:200, which was mixed with the antigrowth hormone. The sections were contrasted with uranyl acetate in methyl cellulose.

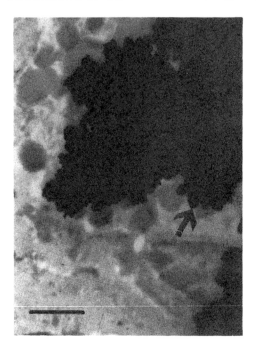

Figure 1a The sections were collected on a Formvar-coated copper grid. The secondary immunolabel was 5-nm gold-conjugated goat-antirabbit IgG 1:75. The gold particles were silver-enhanced according to Danscher for 12 min. Note the silver precipitates and corrosion (arrow) induced by the copper. Bar = 1 μm.

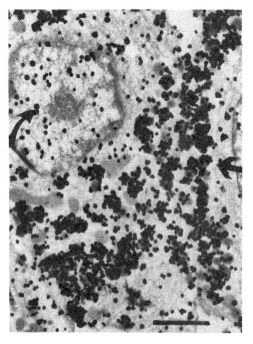

Figure 1b Immunostaining as in Figure 1a. The gold particles were silver-enhanced for 10 min with commercial developer (Aurion). Specific staining is localized to growth hormone containing granules (arrow), but also unspecific self-nucleation of silver is seen (curved arrow). Bar = 1 μm.

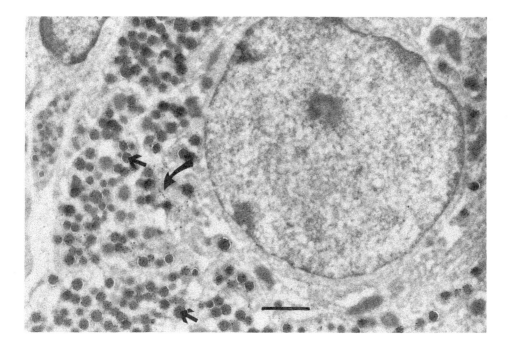

Figure 2   Immunostained as in Figure 1a and silver-enhanced for 2 min with commercial developer (Aurion). At low magnification growth hormone can be identified in granules (arrows) and in the Golgi stack (curved arrow). Bar = 1 μm.

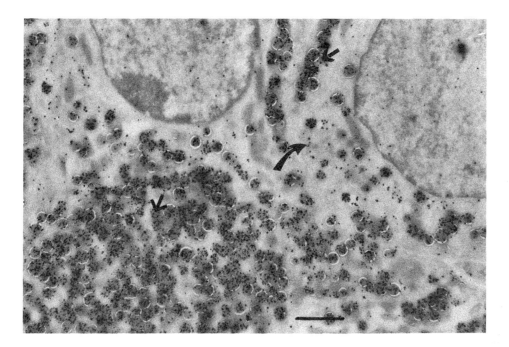

Figure 3   Immunostained as in Figure 1a and silver-enhanced for 4 min with commercial developer. Note that enhanced particles are not as uniformly spherical as in Figure 2. Growth hormone is seen on granules (arrows) and in the Golgi stack (curved arrow). Bar = 1 μm.

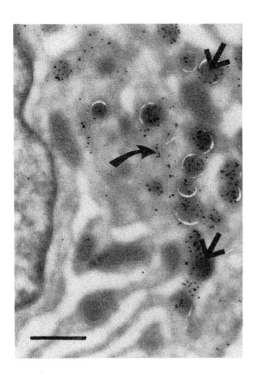

Figure 4a Procedures as in Figure 3. At higher magnification the silver-enhanced particles appear with uneven shapes due to lack of gum arabic. Growth hormone is seen on granules (arrows) and in the Golgi stack (curved arrow). Bar = 500 nm.

particle. Silver-enhanced ultrasmall particles are easily recognized at low magnifications (Figure 7). Silver-enhancement of double-labeled specimen can also be done as long as the development maintains expansion of the original particles with different diameters (Figure 8). The enhancement will improve detectability of the markers at low magnifications (Bastholm et al., 1987).

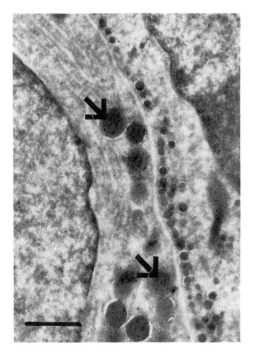

Figure 4b Immunostained as in Figure 3 and silver-enhanced for 15 min with commercial developer (Aurion) mixed 1:1 with 30% gum arabic (Aurion) resulting in small and even particles. For comparison see Figure 4a. Growth hormone is present on granules (arrows). Bar = 500 nm.

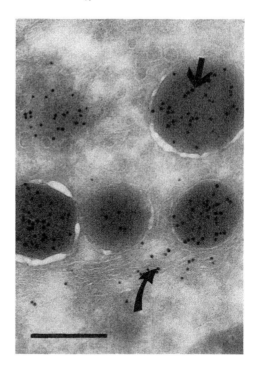

Figure 5a   Same as Figure 4B but at a higher magnification. Growth hormone is localized in the Golgi
stack (curved arrow). Bar = 200 nm.

9.4.2   *Controls*

It must be ensured that the silver-enhancement is specific, i.e., only colloidal gold particles
and not other substances in the section have induced the reduction of silver ions. It is often
possible to identify the gold particle inside the silver shelf, but we always silver enhance
a section untreated with gold labeled molecules.

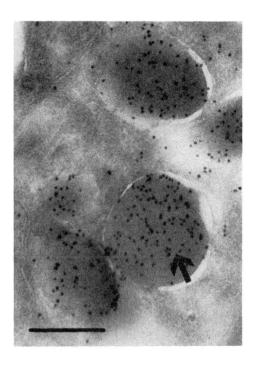

Figure 5b   Silver-enhanced for 12 min according to
Danscher. At this magnification there is no significant difference in shapes of the silver-enhanced
particles compared with the commercial developer
used in Figure 5a. Growth hormone is present on
granule (arrow). Bar = 200 nm.

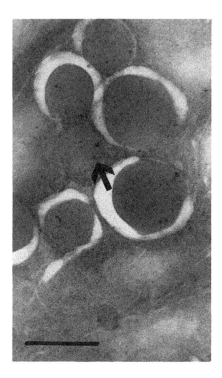

Figure 6a The secondary antibody (1-nm gold-conjugated goat-antirabbit IgG 1:40) was silver-enhanced for 40 min with commercial developer (Aurion) mixed 1:1 with gum arabic (Aurion). The labeling of growth hormone granules is sparce (arrow). Bar = 200 nm.

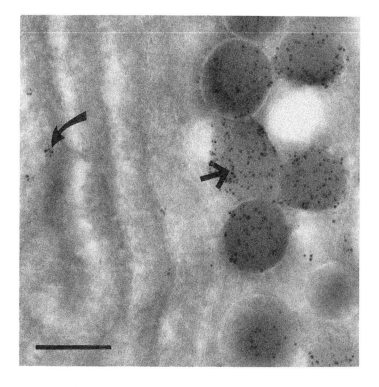

Figure 6b Immunostained as in Figure 6a and silver-enhanced for 15 min according to Danscher. Note that the enhanced particles are not as uniform in size as silver-enhanced 5-nm gold. Growth hormone is seen on granules (arrow) and in the endoplasmatic reticulum (curved arrow). Bar = 200 nm.

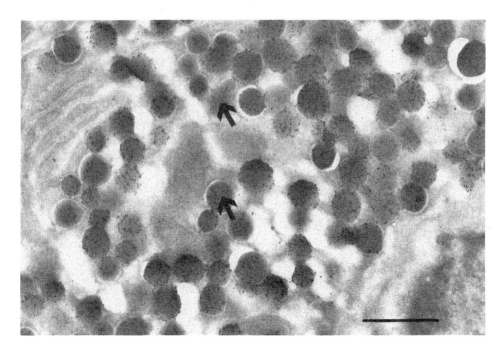

Figure 7 Immunostained and silver-enhanced as in Figure 6b. It is easy to identify the silver-enhanced particles even at a low magnification. Growth hormone granules (arrows). Bar = 500 nm.

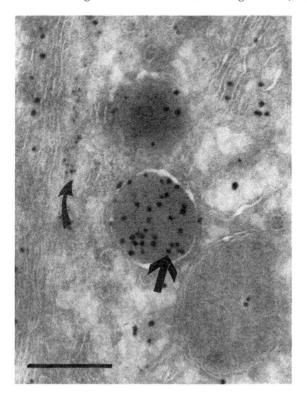

Figure 8 Labeling of growth hormone with 5-nm gold-conjugated secondary antibody and tubulin identified with secondary 1-nm gold-conjugated goat-antimouse IgG 1:40. Silver-enhancement for 15 min according to Danscher. The small particles identifying tubulin (curved arrow) can easily be distinguished from the larger particles demonstrating growth hormone (arrow). Bar = 200 nm.

9.5 Conclusions

Immunoelectron microscopy on thin cryosections employing silver-enhanced colloidal gold as electron dense marker is an attractive method for the localization of antigens because:

- it is a rapid method in comparison with traditional IEM on resin-embedded material
- the antigens retain recognizable epitopes due to the gentle fixation required
- the ultrasmall gold particles penetrate the tissue easily compared with larger particles, including ferritin
- the marker particle is located very close to the antigen even by a two layer technique
- double labeling can be performed
- detectability of the markers is improved at low magnifications.

9.6 Appendices

9.6.1 Incubation and Silver-Enhancement Procedures

During incubation and silver-enhancement the grids are kept floating on the droplets with the upper side kept dry, and the procedures are carried out at a shaking table, facilitating penetration and distribution of the reagents.

9.6.1.1 Immunoelectron Microscopy Protocol for a Two-Layer Indirect Method Using 5-nm Gold Particles as Marker

1. Phosphate-buffered saline (PBS), 3 times for 10 min each.
2. 0.05 M glycine in PBS for 20 min (in order to quench free aldehyde groups).
3. Two changes of PBS for 5 min each.
4. 0.05 M Tris-buffered saline (TBS)(pH 7.4) for 5 min.
5. 0.05 M TBS (pH 7.4) with 10% goat serum for 30 min.
6. Specific antibody 10 to 25 μg/ml in 0.05 M TBS (pH 7.4) with 0.25% bovine serum albumin (BSA) at 4°C overnight.
7. 0.05 M TBS (pH 7.4) with 0.1% BSA, five changes for 5 min each.
8. 0.02 M TBS (pH 8.2) with 0.1% BSA, five changes for 5 min each.
9. Goat-antimouse or goat-antirabbit IgG conjugated to 5 nm gold particles diluted in 0.02 M TBS (pH 8.2) with 0.25% BSA, 30 min at room temperature.
10. 0.02 M TBS (pH 8.2), five changes for 5 min each.
11. 0.1 M phosphate buffer (PB) (pH 7.4) for 5 min.
12. Fixation in 2% glutaraldehyde in 0.1 M phosphate buffer (pH 7.4) for 5 min.
13. Distilled water, four changes for 5 min each.
14. Silver-enhancement by Danscher method or by commercial silver enhancer kit, time depending on the desired size of gold particles.
15. Distilled water, five changes for 5 min each.
16. 2% methylcellulose with 0.4% uranyl acetate for 10 min for protection against air drying artifacts and for contrasting.

9.6.1.2 Immunoelectron Microscopy Protocol for a Two-Layer Indirect Method Using 1-nm Gold Particles as Marker

1–6. Steps are performed as described above.
7. 0.05 M TBS (pH 7.4) with 0.1% BSA for 5 min.
8. PBS (pH 7.4) with 0.8% BSA and 0.1% Cold Water Fish Skin Gelatin (CW-FSG), five changes for 8 min each.
9. Goat-antimouse or goat-antirabbit IgG conjugated to 1-nm gold particles diluted in PBS (pH 7.4) + 0.8% BSA and 0.1% CW-FSG at room temperature for 1 h.

10. PBS, six changes for 10 min each.

11–16. The remaining steps are performed as described in Appendix A, 11–16 except for silver-enhancement which always is performed according to the Danscher method.

9.6.1.3   *Modified Preparation of Physical Developer*
Solution A

- 3 ml 50% gum arabic stock diluted 1:1 with distilled water (for preparation of stock solution, see below)
- 3 ml distilled water
- 1.5 ml hydroquinone solution (0.55 g/15 ml distilled water)
- 1 ml citric acid monohydrate solution (2.55 g/10 ml distilled water)
- 1 ml trisodium citrate dihydrate (2.35 g/10 ml distilled water)

Mix all reagents in a plastic test tube, wrapped with aluminum foil and keep at –20°C.

Solution B

- 1 ml silver lactate solution (0.11 g/15 ml distilled water)

Keep in Eppendorf tubes wrapped in aluminum foil at –20°C. Before use, thaw solutions A and B and mix thoroughly. Develop in the dark.

Stock solution of gum arabic
Dissolve gum arabic (Merck) 1 kg + 2 L of distilled water under vigorous stirring for 3 to 4 h and leave at room temperature for 4 days. Filter through multiple layers of gauze. Stock solution can be stored in aliquots at –20°C.

Acknowledgments

The technical assistance of Ms. Annette Hauberg Fischer is greatly appreciated. Mr. Bent Børgesen has skillfully prepared the photographic works and Dr. Mogens Spang-Thomsen is thanked for valuable discussions of the manuscript.

References

Bastholm, L., Scopsi, L., and Nielsen, M. H. 1986. Silver-enhanced immunogold staining of semithin and ultrathin cryosections. *J. Electron Microsc. Tech.* 4: 175.

Bastholm, L., Nielsen, M. H., and Larsson, L. 1987. Simultaneous demonstration of two antigens in ultrathin cryosections by a novel application of an immunogold staining method using primary antibodies from the same species. *Histochemistry* 87: 229.

Danscher, G., Hacker, W. G., Grimelius, L., and Nørgaard, R. J. O. 1993. Autometallographic silver amplification of colloidal gold. *J. Histotechnol.* 16: 201.

Danscher, G. and Nørgaard, R. J. O. 1983. Light microscopic visualization of colloidal gold on resin-embedded tissue. *J. Histochem. Cytochem.* 12: 1394.

Danscher, G. and Nørgaard, R. J. O. 1985. Ultrastructural autometallography: a method for silver amplification of catalytic metals. *J. Histochem. Cytochem.* 33: 706.

Faulk, W. and Taylor, G. 1971. An immunocolloid method for the electron microscope. *Immunocytochemistry* 8: 1081.

Hagler, H. K. 1993. Cryo-sectioning. In: *Immunoelectron Microscopy in Virus Research and Diagnosis.* pp. 309–325. Hyatt, A. D. and Eaton, B. T. (Eds.), CRC Press, Boca Raton, FL.

Holgate, C., Jackson, P., Cown, P., and Bird, C. 1983. Immunogold silver-staining. New method of immunostaining with enhanced sensitivity. *J. Histochem. Cytochem.* 31: 938.

Leunissen, J. L. M., and van de Plas, P. 1993. Ultrasmall gold probes and cryoultramicrotomy. In: *Immunoelectron Microscopy in Virus Research and Diagnosis.* pp. 327–348. Hyatt, A. D. and Eaton, B. T. (Eds.), CRC Press, Boca Raton, FL.

Nielsen, M. H. and Bastholm, L. 1990. Improved immunolabeling of ultrathin cryosections using antibody conjugated with 1-nm gold particles. *Proc. XIIth Internat. Cong. Electron Microsc.* p. 928.

Nielsen, M. H. and Bastholm, L. 1993. Multiple labeling of thin sections. In: *Immuno-Gold Electron Microscopy in Virus Diagnosis and Research.* pp. 231–256. Hyatt, A. D. and Eaton, B. T. (Eds.), CRC Press, Boca Raton, FL.

Scopsi, L., Larsson, L., Bastholm, L., and Nielsen, M. H. 1986. Silver-enhanced colloidal gold probes as markers for scanning electron microscopy. *Histochemistry* 86: 35.

Scopsi, L. 1989. Silver-enhanced colloidal gold method. In: *Colloidal Gold: Principles, Methods and Applications.* pp. 251–259. Hayat, M. A. (Ed.), Academic Press, San Diego.

Tokuyasu, K. T. 1973. A technique for ultramicrotomy of cell suspension and tissue. *J. Cell Biol.* 57: 551.

Tokuyasu, K. T. 1984. Immuno-cryoultramicrotomy. In: *Immunolabeling for Electron Microscopy.* pp. 71–82. Polak, J. M. and Varndell, I. M. (Eds.), Elsevier, Amsterdam.

Van Bergen en Henegouwen, P. M. P. and Van Lookeran Campagne, M. 1989. Subcellular localization of phosphoprotein B-50 in isolated presynaptic terminals and in young adult rat brain using a silver-enhanced ultra-small gold probe. *AuroFile 2, Janssen Life Science,* Beerse, Belgium.

Williams, G. V. and Faulkner, P. 1993. Detection of viral and cellular structures by post-embedding immunocytochemistry. In: *Immunoelectron Microscopy in Virus Research and Diagnosis.* pp. 178–230. Hyatt, A. D. and Eaton, B. T. (Eds.), CRC Press, Boca Raton, FL.

Chapter 10

Microwave Fixation and Microwave Staining Methods for Microscopy

Gary R. Login and Ann M. Dvorak

Contents

10.1 Introduction

Cell biologists, pathologists, and histotechnologists have used microwave technology in the laboratory since 1970 and have shown that it has many methodological advantages over traditional fixation and staining methods. For example, microwave energy accelerates reaction rates by orders of magnitude, improves the detection of a variety of biological molecules, and provides excellent morphology of biological specimens for light and electron microscopy (Login and Dvorak, 1994b). In addition, the use of microwave methods has made possible some key contributions to cell biology such as the visualization of open images of synaptic vesicles in rat brain (Mizuhira et al., 1990), the study of the dynamics of vitamin D receptors in dermal fibroblasts (Barsony et al., 1990), and the ultrastructural localization of chymase in rat peritoneal mast cells (Login et al., 1987a).

In this chapter we describe a variety of microwave fixation and staining methods that are useful for immunogold-silver staining of biological samples. Because excellent specimen fixation and handling is important to all histochemical and immunohistochemical procedures, we describe newly generated microwave methods in current use as applied to fixation for morphologic analysis. In addition, we report antigens that are better preserved by microwave fixation methods compared to immersion fixation in aldehyde or in alcohol or to freeze fixation. Recognizing that reproducible microwave fixation and staining is made difficult by the variable performance of large cavity microwave ovens (e.g., household microwave ovens), we provide systematic approaches for calibrating microwave ovens. We further discuss the application of our step-by-step methods for immunogold labeling of histamine and immunogold-silver staining of histamine and serotonin in rat peritoneal mast cells fixed by a fast microwave fixation method. The versatility of microwave procedures is also reviewed with respect to tissue section drying, microwave-assisted antigen retrieval, microwave-accelerated labeling with primary and secondary immunogold and ligand gold reagents, as well as microwave-assisted silver staining reactions.

10.2 Overview of How Microwave Irradiation is Used to Facilitate Fixation

Microwave methods offer the potential for enhanced localization of a variety of antigens in biological samples. Conventional tissue fixation modalities such as chemical or freeze fixation generally provide either excellent morphology or excellent preservation of antigens. For example, chemical fixatives cross-link immunoreactive sites of biological molecules (Bullock, 1984) and freeze fixation may not prevent diffusion of biological molecules during the immunostaining procedure (Abbate et al., 1993). Microwave irradiation used in combination with chemical or freeze fixation provides additional control over the fixation process. In this section, we review a variety of microwave fixation methods that offer enhanced fixation speed (Login et al., 1986, 1991; Login and Dvorak, 1994b), minimized diffusion of small biological molecules (Barsony and Marx, 1991; Login et al., 1992b) and limited chemical cross-linking (Login et al., 1987b; Login and Dvorak, 1994b).

Microwave fixation describes a variety of methods where microwave irradiation is used in the fixation process. Certain limitations and possible advantages are associated with different microwave fixation methods. For example, the use of aldehydes during irradiation may influence the preservation of antigenicity of cellular macromolecules (Login et al., 1987a,b; Boon et al., 1988; Leong et al., 1988). Therefore, we have found it useful to classify microwave fixation methods according to the following criteria (Login and Dvorak, 1994b): (1) chemical environment around the specimen during irradiation, (2) duration of microwave exposure, and (3) the temporal relationship between the use of microwave irradiation and the use of other chemical or physical fixation modalities. Using this classification system, five approaches to microwave fixation are defined (Login and Dvorak, 1994b).

10.2.1 Microwave Stabilization

Microwave stabilization refers to the preservation of structure by microwave irradiation, without the superimposed effects of chemical fixatives. Microwave stabilization for fixation was introduced by Mayers (1970) for biopsied tissue and by Stavinoha et al. (1970) for rodent brains *in situ*. Table 1 lists several antigens that are better preserved in specimens fixed by microwave stabilization as compared to immersion fixation in aldehyde. In addition, microwave stabilization results in the same quality and quantity of preservation of DNA as do chemical fixation methods (Jackson, 1991).

Microwave stabilization has important limitations. The ultrastructural morphology of brain is especially difficult to preserve in situ by microwave stabilization (Login and Dvorak, 1994a). Microwave stabilization is considered to be inadequate for most ultrastructural studies (Zimmerman and Raney, 1972; Bernard, 1974; Ikarashi et al., 1986).

10.2.2 Fast and Ultrafast Primary Microwave-Chemical Fixation

Fast and ultrafast, microwave-chemical fixation are used for primary fixation of biological specimens. In practice, specimens are irradiated by microwave energy in a chemical environment for seconds to minutes (fast) (Login, 1978; Chew et al., 1983; Hopwood et al., 1984; Leong et al., 1985; Login and Dvorak, 1985, 1988, 1994b; Jensen and Harris, 1989) or for milliseconds (ultrafast) (Login et al., 1986, 1989, 1991). In this approach, specimens are not exposed to chemical fixatives before or after microwave irradiation. Table 1 lists several antigens that are better preserved for immunohistochemical studies by fast, primary microwave-chemical fixation as compared to immersion fixation in aldehyde, immersion fixation in alcohol or to snap freezing.

10.2.3 Microwave Irradiation Followed by Chemical Fixation

Chemical fixatives are used in association with microwave irradiation to improve the uniformity of preservation following fast, primary microwave-chemical fixation. Specimens are immersed in a chemical environment (e.g., aldehyde mixtures) for minutes to hours *after* microwave irradiation (Leong et al., 1985; Mizuhira et al., 1990). Several antigens that are reportedly better preserved by this microwave approach than by conventional immersion fixation are listed in Table 1. For example, Kang et al. (1991) showed a twofold higher gold labeling density of cell surface-lectin binding sites in rust fungus-infected wheat leaves fixed by microwave irradiation followed by chemical fixation as compared to conventional aldehyde immersion fixation.

Table 1 Superior Detection of Biological Molecules by Histochemical, Immunohistochemical, Immunocytochemical, and Affinity Techniques in Specimens Fixed by Various Microwave Methods Compared to Fixation by Immersion in Aldehyde or in Alcohol or by Freezing

Microwave stabilization vs. immersion fixation in aldehyde

Alpha-1-Chymotrypsin (Leong et al. 1988)
cAMP (Barsony et al. 1990)
Carcinoembryonic antigen (Leong et al. 1988)
cGMP (Barsony et al. 1990)
Chromogranin (Leong et al. 1988)
Collagen III (Moran et al. 1988)
Collagen IV (Moran et al. 1988)
Desmin (Leong 1988)
Factor VIII (Leong et al. 1988)
Fibronectin (Moran et al. 1988)
Laminin (Moran et al. 1988)
Leu-1,14 (Leong et al. 1988)
Leukocyte common antigen (Leong 1988; Leong et al. 1988)
LN 1,2,3 (Leong 1988; Leong et al. 1988)
Neuron specific enolase (Leong et al. 1988)
PreS2 epitope hepatitis B surface antigen (Hsu et al. 1991)
Transglutaminase (Patterson and Bulard 1980)
Tubulin (Barsony and McKoy 1992)
UCHL1 (Leong et al. 1988)
Vimentin (Leong 1988; Leong et al. 1988)
Vitamin D receptor (Barsony and Marx 1990, 1991; Barsony and McKoy 1992)

Microwave stabilization vs. freeze fixation

cAMP (Barsony et al. 1990)
cGMP (Barsony et al. 1990)
Collagen III (Moran et al. 1988)
Collagen IV (Moran et al. 1988)
Fibronectin (Moran et al. 1988)
Laminin (Moran et al. 1988)

Fast, primary microwave-chemical fixation vs. immersion fixation in aldehyde

Amylase (Yamashina et al. 1990; Login et al. 1992b)
Chymotrypsinogen (Yamashina et al. 1990)
Collagen IV (Azumi et al. 1990; Yamashina et al. 1990)
Desmin (Azumi et al. 1990)
Factor VIII (Azumi et al. 1990)
Gamma-glutamyl transpeptidase (Yasuda et al. 1989)
Hyaluronan (Eggli et al. 1992)
Ia (Login and Dvorak 1994a)
Keratin AE1/3 (Login et al. 1987b)
Keratin AE-1 (Azumi et al. 1990)
Laminin (Yamashina et al. 1990)
Pathogenesis-related proteins from tobacco leaves (Benhamou et al. 1991)
Leu-1 (Login and Dvorak 1994b)
LN-2 (Azumi et al. 1990)

Fast, primary microwave-chemical fixation vs. immersion fixation in alcohol

LN-2 (Azumi et al. 1990)
p53 tumor suppressor gene product (Kawasaki et al. 1992)

Table 1 *(continued)*

Fast, primary microwave-chemical fixation vs. freeze fixation

Leu-1 (Login and Dvorak 1994b)
Ia (Login and Dvorak 1994b)

Microwave irradiation followed by chemical fixation vs. immersion fixation in aldehyde

Carcinoembryonic Antigen (Ohtani et al. 1990)
Cytokeratins: 40,46,50,52, 56.5,58,65–67 kd (Ohtani et al. 1990)
LN-1,3 (Ohtani et al. 1990)
Rust fungus surface-associated proteins (Kang et al. 1991)
Vimentin (Ohtani et al. 1990)
von Willebrand Factor (Ohtani et al. 1990)

Freeze fixation followed by microwave irradiation in polyethylene glycol vs. freeze fixation

Desmin (Suurmeijer et al. 1990)
Keratins 1,4,5,18 (Suurmeijer et al. 1990)
Vimentin (Suurmeijer et al. 1990)

Fast, primary microwave-chemical fixation in aldehydes or carbodiimide followed by freeze fixation vs. immersion fixation in aldehyde

Amylase (Yamashina et al. 1990)
Chymotrypsinogen (Yamashina et al. 1990)
Collagen IV (Yamashina et al. 1990)
Glutathione peroxidase (Utsunomiya et al. 1991)
Ia (Login and Dvorak 1994b)
Immunoglobulin Kappa light chain (Login and Dvorak 1994b)
Laminin (Yamashina et al. 1990)
Leu-3A (Login and Dvorak 1994b)
T-8 (Login and Dvorak 1994b)
T-9 (Login and Dvorak 1994b)

10.2.4 *Primary Chemical Fixation Followed by Microwave Irradiation*

This fixation approach defines a process by which microwave energy is used to augment primary fixation by chemicals. In other words, specimens are immersed at room temperature in chemicals for minutes (Leong et al., 1985; Leong, 1988; Smid et al., 1990) to hours (Marani et al., 1987; Boon et al., 1988) *before* microwave irradiation. Microwave acceleration of chemical fixation has been used in immunohistochemical studies to evaluate antigens with success equal to routine aldehyde immersion fixation. However, no studies have shown superior immunohistochemical labeling using this microwave fixation approach compared to alternate fixation methods.

10.2.5 *Combinations of Freeze and Microwave Fixation*

Several approaches have been described: (1) irradiation of frozen sections in a variety of chemical fixatives (e.g., aldehyde, alcohol, or polyethylene glycols) (Kok et al., 1987; Nakae et al., 1990; Therkildsen and Pilgaard, 1990); (2) cryosectioning of specimens previously fixed by microwave stabilization (Marani et al., 1987) or by fast, primary microwave-chemical fixation (Hopwood et al., 1984; Ohtani et al., 1989; Nakae et al., 1990; Yamashina et al., 1990; Login and Dvorak, 1994b), or by microwave irradiation followed by chemical fixation (Ohtani et al., 1989; Naganuma et al., 1990; Utsunomiya et al., 1991); and (3) most recently by freeze fixation simultaneous with microwave irradiation (Hanyu et al., 1992). In the first approach, microwave irradiation is said to enhance the diffusion and reaction rate of chemical fixatives without evidence of damage due to thawing (Kok et al., 1987). In the second set of approaches, microwave fixation provides sufficient structural preservation to limit freezing artifacts, however preservation is reportedly not uniform (Nakae et al., 1990). Alternatively, in the third approach, microwave irradiation is used to suppress ice crystal formation during rapid freezing (Hanyu et al., 1992). Antigens that are better preserved for immunohistochemistry by combinations of microwave and freeze fixation methods than by aldehyde immersion or by freeze alone are reported in Table 1. In addition, superior immunocytochemical detection of several antigens were reported in rat tissues irradiated in polyethylene glycol (Suusmeijer et al., 1990) a mixture of aldehyde and carbodiimide (Yamashina et al., 1990) or in 0.1% glutaraldehyde and 4% formaldehyde (Utsunomiya et al., 1991) followed by freeze fixation compared to routine chemical fixation (Table 1).

10.3 General Methods for Microwave Fixation

Microwaves are electromagnetic waves (i.e., oscillating electric and magnetic fields at right angles to one another) in a frequency range between 1 gigahertz (10^3 MHz) and 1 terahertz (10^6 MHz) corresponding to wavelengths in air between 30 cm and 0.3 mm, respectively (Neas and Collins, 1988) (Figure 1). They are non-ionizing radiation that cause molecular motion by migration of ions and rotation of dipolar molecules (e.g., water) without disruption of chemical bonds (Neas and Collins, 1988). The physical change easiest to measure in an aqueous sample irradiated by microwave energy is an increase in temperature. Microwave heating results from the release of energy following the disruption of hydrogen bonds primarily between water molecules (Mudgett, 1990).

Nonuniform heating of foods in household microwave ovens is common and is often the result of a variety of factors such as inadequate production and distribution of microwave power by inexpensive electronic components, microwave cavities designed for specific food volumes, differences in the conductivity of the foods, and the location of the food within the cavity (Kok and Boon, 1992; Login and Dvorak, 1994b). While the presence of microwave-induced hot and cold spots may be tolerated in foods they pose difficulties in biopsy specimens.

An important property of the microwaves produced in household microwave ovens is that they do not penetrate much more than 1.5 cm into water or into biological samples. Because microwave energy is absorbed as it enters a specimen with a high water content, less energy is deposited in the interior than at the exterior. Therefore, it is important to use small sample containers (Login and Dvorak, 1994b). We recommend using containers with at least one of its dimension being less than 1.5 cm.

Many published microwave procedures simply do not contain the details needed to reduce the variables introduced by microwave ovens. Even if the details were provided, they probably would not be the best conditions for all microwave ovens. Understandably, several questions must be answered to adapt a published microwave procedure to a particular microwave oven. For example, where should samples be placed in a microwave oven cavity? What size and shape container should be used for the sample? Is a water bath necessary and if so what volume should it be for a particular microwave oven? Fortunately, simple and inexpensive techniques are available to show the best locations for microwave irradiation for a specific set of microwave fixation conditions. We next describe the materials needed and a calibration procedure that can be followed to improve microwave fixation results in household or laboratory microwave ovens.

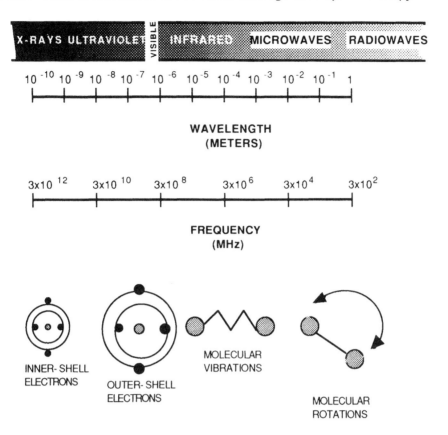

Figure 1   Electromagnetic spectrum. At frequencies below the microwave region, conductive effects associated with electrophoresis of dissolved salts in ionic solutions become increasingly more pronounced. At frequencies in the microwave region, molecular and ionic displacement occur and are expressed as orientation and ionic polarization effects. At frequencies above the microwave region but below the visible regions, atomic polarization mechanisms are seen which involve the displacement of atomic nuclei with respect to each other. In the visible region, polarization occurs by the displacement of electrons. At frequencies above the visible region, ionization effects occur, which result in the formation of free radicals. (Adapted with permission from Neas and Collins, 1988; Mudgett, 1986).

10.3.1 *Materials for Calibration of Microwave Ovens and for Microwave Fixation Procedures*

1. Neon bulb array: The device is made by placing neon bulbs (available from electronic stores) in a styrofoam base equally spaced 2.5 cm apart (Login and Dvorak, 1994b). The styrofoam base is made large enough to fit on the floor of the microwave oven. A square array will typically contain 25, 36, or 49 bulbs (Login and Dvorak, 1994b). (NOTE: This is a simple but valuable device that shows the distribution of high and low microwave power in a microwave oven. The neon gas glows orange when a high enough density of microwaves energizes the gas inside the glass bulbs.)
2. Alphanumeric grid: This is a plastic sheet (i.e., 25 cm × 8 cm) with ruled lines spaced 1 cm apart. The alphanumeric grid is centered on the floor of the microwave oven and secured in place with cellophane tape. (NOTE: The ruled sheet simplifies reproducible positioning of the specimen container [Login and Dvorak, 1990].)

3. Agar-Saline-Giemsa (ASG) blocks: ASG blocks are small blocks of agar (~0.5 cm^3) that contain a mixture of dyes that react to microwaves by showing different colors at different temperatures. ASG blocks are made by melting 2% agar (Difco Laboratories, Detroit, MI) in 0.9% saline (Travenol Laboratories, Deerfield, IL) followed by adding 0.1% Giemsa stain (Aldrich Chemical Co., Milwaukee, WI). The molten mixture is gelled in flat silicon embedding molds (i.e., 20 wells, each 5 mm^2 × 10 mm, Ted Pella, Redding, CA) and allowed to cool to room temperature. The Giemsa stain will give the gelled ASG blocks a translucent purple appearance at room temperature. Color changes of the Giemsa in the ASG blocks correspond to the following temperatures: light purple ~20°C, dark purple to dark blue = 40 to 55°C, light blue to clear = 55 to 60°C, melt >65°C. (NOTE: These blocks are easy to make. They are used to show the uniformity of microwave heating within a small sample for the microwave oven power and time conditions selected. ASG blocks in the calibration procedure are used to select the best microwave conditions before using a biological specimen.)
4. A 300 ml beaker and distilled water (20°C).
5. Thermocouple thermometer (i.e., Digisense model 8528-20; sensor type MT-4 copper/constantan: 0.025 sec time constant, Cole Parmer Instrument Co., Chicago, IL) for measuring solution temperature *following* microwave irradiation.
6. Standardized specimen containers such as a polystyrene tissue culture dish (3.5 cm diameter × 1 cm high; Becton Dickinson, Lincoln Park, NJ) or a glass vial (1.7 cm diameter × 4.5 cm high, Electron Microscopy Sciences, Fort Washington, PA).
7. A chemical fixative for transmission electron microscopy studies. We recommend a mixed aldehyde fixative, containing 2% formaldehyde, 2.5% glutaraldehyde, and 0.025% calcium chloride in 0.1 M sodium cacodylate buffer (pH 7.4).

10.3.2 A Calibration Procedure Recommended for Setting up a Microwave Oven for Microwave Fixation and for Periodic Checks of Oven Performance

The key components of the calibration procedure include: (1) determination of the warmup time needed for a microwave oven power tube to produce microwave energy, (2) determination of the appropriately sized water load for the microwave oven, (3) mapping the major hot spots in a loaded microwave oven using the neon bulb array, and (4) refining the hot spot map using the ASG blocks. In our experience, calibration does not need to be repeated if the following variables remain constant: water load volume, sample load volume, water load and sample placement in the oven, and initial solution temperature.

1. The microwave oven power is set on maximum.
2. The neon bulb array is used to measure the time delay between turning on the microwave oven (i.e., bulbs not illuminated) and production of microwaves by the magnetron (i.e., bulbs illuminated). The array is placed on the floor of an empty microwave oven. The timer is set for 10 sec. After several trials an average delay time can be calculated. This value is subtracted from the preset irradiation time to yield an actual irradiation time.
3. Determination of the water load volume (Login and Dvorak, 1985, 1988). Water loads (initial temperature, 20°C) ranging in volume from 100 to 300 ml are prepared in 300 ml beakers and are tested one at a time. The water load is placed in a rear corner of the microwave oven. The neon bulb array is placed on the floor of the microwave oven. The water load volume should not be so large that none of the bulbs in the array fail to illuminate. A water load volume that allows a cluster of bulbs to remain steadily illuminated during irradiation for ~10 sec is ideal.

4. The areas of maximum microwave power distribution within the oven cavity that contains the water load are determined by using the neon bulb array. The locations are recorded on the alphanumeric grid. The bulb array is removed.
5. The gelled ASG blocks are immersed individually in the 35 mm diameter culture dish or in the 17 mm diameter glass vial containing 3 ml of the same type of solution that will be used for fixation and placed on the alphanumeric sheet at one of the grid locations recorded in Step 4.
6. The best location for specimen fixation is identified by determining the region of the microwave cavity that results in the most uniform color change of the ASG block during the shortest irradiation interval (i.e., ~10 sec) needed to raise the solution temperature around the agar block from 20°C to 50°C.

10.3.3 A Standardized Method for Microwave Fixation of Samples in Microwave Ovens

We use the following standardized protocol before each microwave fixation series to improve reproducibility of sample preservation for light and electron microscopy. The key components of the standardization protocol are (1) magnetron warmup, (2) use of ASG blocks to check irradiation conditions in the loaded microwave cavity (Login and Dvorak, 1990), (3) use of specimen containers with one dimension of the container less than 1.5 cm, and (4) fast specimen handling to prevent conductive heating artifacts after irradiation. In practice, large cavity microwave ovens are set to maximum power (i.e., no cycling of the magnetron).

1. The water load volume determined in the calibration procedure is irradiated for 2 min.
2. The water load is replaced with fresh water at 20°C and at the same volume determined in Step 3 of the calibration protocol. The reproducibility of the micro-wave fixation method depends on proper use of the water load.
3. An ASG block is used as described in Step 5 of the calibration protocol to check the uniformity of irradiation. If the block does not show an irradiation pattern similar to that obtained during the calibration procedure then the following variables should be checked: water load volume, placement, and initial temperature, ASG block size, immersion fluid type and volume, type of sample container, location of the sample container on the alphanumeric grid, and actual irradiation time.
4. Specimen fixation should be completed as soon after the warmup procedure as possible (e.g., within 2 min) (Login and Dvorak, 1990).
5. Tissues are trimmed to dimensions <1 cm³. Next, they are immersed in a 35 mm diameter culture dish or in a 17 mm diameter glass vial containing 3 ml of the irradiation solution. Finally, the tissues are irradiated in the microwave oven at the coordinates used in Step 6 of the calibration procedure.

10.3.4 Recommendations for Processing and Embedding Microwave Fixed Specimens for Light and Electron Microscopy

The appearance of microwave fixed specimens is improved by using room temperature processing (Login and Dvorak, 1994b) and plastic embedding media (Notoya et al., 1990; van de Kant et al., 1990a; Login and Dvorak, 1994b). Processing microwave stabilized specimens at room temperature eliminates artifacts such as red blood cell lysis (Login and Dvorak, 1994b). Sensitivity to processing temperature may be the result of limited or absent protection of chemical cross-linking agents in microwave fixed specimens (Login and Dvorak, 1994b).

Embedding media affect the microscopic morphology of microwave fixed specimens. van de Kant et al. (1990a) showed that erythrocytes were well preserved in microwave stabilized specimens embedded in glycolmethacrylate but not in paraffin. In addition, they showed that embedding microwave stabilized Sertoli cells in paraffin resulted in 37% smaller nuclear areas (p <0.001) than embedding in glycolmethacrylate. Notoya et al. (1990) showed equally good morphology of microwave fixed tissues embedded in Epon or hydroxyethylmethacrylate. van de Kant et al. (1990a) suggest that plastic embedding media cross-links free amino groups of peptide chains, thus improving tissue stability.

10.4 Immunogold-Silver Labeling of Histamine and Serotonin in Rat Mast Cells Fixed by a Fast Primary Microwave-Chemical Method Followed by Freeze Fixation for Study by Light Microscopy

We have successfully localized histamine and serotonin in the cytoplasm of rat peritoneal mast cells using the combination of rapid microwave-aldehyde fixation and freeze fixation and an immunogold-silver staining protocol (Login and Dvorak, unpublished data). The experimental protocol for mast cell fixation and immunogold-silver staining illustrates how we use the calibration procedure for microwave fixation.

10.4.1 Microwave Fixation and Handling Rat Peritoneal Mast Cells

1. Rat peritoneal mast cells were purified to 79% through a Percoll solution containing 9 ml Percoll (Pharmacia Fine Chemicals; Uppsala, Sweden), 1 ml of 10× Hank's balanced salt solution, 0.035% $NaCHO_3$ and 0.1% bovine serum albumin (BSA, Pentex grade: Miles, Kankakee, IL) (Login et al., 1992a).
2. Cells were microwave fixed in suspension in a mixed aldehyde solution containing 2.0% formaldehyde, 0.1% glutaraldehyde, 0.025% calcium chloride, in 0.1 M sodium cacodylate buffer, pH 7.4, 20°C with a 700 watt microwave oven (Amana RR8B, Amana, IA). We used the standardized microwave fixation method described in Section 10.3.3.
3. The microwave fixed cells were pelleted through molten agar in plastic microfuge tubes in a Beckman microfuge (Beckman Instruments, Palo Alto, CA) for 1 min at 1000 rpm, 20°C, gelled in a 4°C ice slurry for 30 min, trimmed to 12 mm × 0.5 cm² blocks and frozen in O.C.T. compound (Miles Scientific, Naperville, IL).
4. 4-μm sections were cut on a Tissue Tek II cryomicrotome (Miles Scientific) mounted on glass slides coated with poly-L-lysine (Sigma Chemical Co., St. Louis, MO).

10.4.2 Immunogold Labeling Procedure for Serotonin and Histamine

(NOTE: A 100-μl aliquot of each reagent listed below was added in sequence at 20°C to the 4 μm sections):

1. Wash with 20 mM Tris buffer containing 0.9% saline, 0.2% nonimmune goat IgG (TBS-Ig) for 20 minutes (blocking).
2. Primary antibodies: monoclonal rat IgG anti-serotonin (Seralab clone YC5/45, Accurate Chemical & Scientific Corp., Westbury, NY) or polyclonal rabbit IgG antihistamine (Milab B80-100, Accurate Chemical & Scientific Corp., Westbury, NY) (used neat or diluted 1:25, 1:50, or 1:250) in TBS-Ig, 0.4% cold water fish gelatin (Sigma), and 1% normal goat serum for 1 h at 20°C in a humid chamber.

3. Wash 3 times, 10 min each in TBS-Ig.
4. Secondary antibodies: 5-nm gold-conjugated goat IgG directed against rabbit or rat IgG (Janssen Life Science Products, Beerse, Belgium) (diluted 1:50 in TBS-Ig, 0.4% cold water fish gelatin, and 1% normal goat serum) for 1 h at 20°C.
5. Wash twice, 10 min each in TBS-Ig.

10.4.3 Silver-Enhancement Procedure

(NOTE: A 100-µl aliquot of each reagent listed below was added in sequence at 20°C to the 4 µm sections.)

1. Wash twice, 3 min each in 10 mM phosphate buffered saline containing PBS, 100 ml distilled water, 0.8 g NaCl, 0.02 g KCl, 0.14 g $Na_2HPO_42H_2O$ and 0.02 g KH_2PO_4.
2. Washed twice, 3 min each in distilled water.
3. Blot excess distilled water around the section.
4. Prepare silver enhancer solution fresh in a small plastic tube: 2 drops of reagent A (enhancer) from IntenSE M™ kit (Janssen) added to 2 drops reagent B (initiator) of IntenSE M™ kit for each section.
5. Cover the section with the silver reaction mixture for 3 to 6 min at 20°C.
6. Wash twice, 3 min each in distilled water.
7. Counter-stain with Harris' hematoxylin, 5 min at 20°C.
8. Permanently mount and cover slipped for light microscopic examination.

The immunogold-silver staining revealed histamine (111 Da) and serotonin (173 Da) in the cytoplasm of rat peritoneal mast cells. Our light microscopy data for histamine localization are consistent with our more recent postembedding immunogold electron microscopic results where we localized histamine to rat mast cell granules (Figure 2) (Login et al., 1992a).

10.5   A Summary of Microwave Procedures that Improve Immunogold-Silver Staining by Enhancing Section Adherence to the Slide, Antigen Retrieval, and Stimulation of Primary and Secondary Antibody Reactions

In this section we discuss how microwave energy can be used to improve the handling of tissue sections for immunogold-silver staining. We also reference some of the microwave literature for immunoperoxidase labeling because of the similarity between microwave methods and immunogold-silver staining. We do not list detailed microwave methods used by original investigators because of the many differences in performance among brands of microwave ovens. We have learned that the methods optimized for one microwave oven may not be applicable to another microwave oven. Therefore, we present a survey of microwave applications in Section 10.5 followed by a microwave staining calibration protocol in Section 10.6. The microwave staining calibration protocol can be used to optimize the function of a microwave oven for microwave-enhanced section adhesion, antigen retrieval, and accelerated primary and secondary immunogold-silver staining reactions.

10.5.1 Microwave Enhanced Adherence of Tissue Sections to Glass Slides

Several investigators have successfully used microwave energy to improve the adherence of frozen or paraffin embedded tissues sections to glass slides. Sections are placed on slides

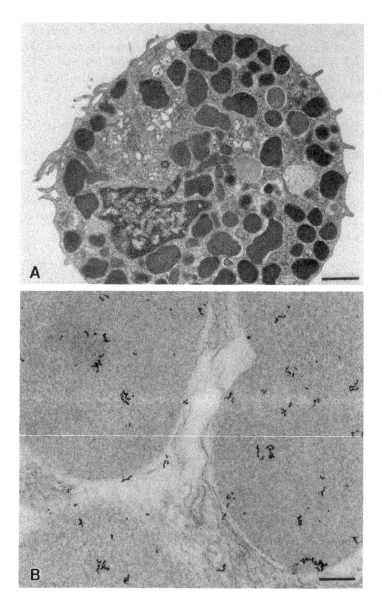

Figure 2   Electron micrographs of rat mast cells fixed in a microwave oven (Amana RR8B, 700 W) by fast, primary microwave-chemical fixation in a mixed-aldehyde fixative. In (A) and (B) the cells were fixed in an aldehyde mixture containing 2% paraformaldehyde, 2.5% glutaraldehyde, 0.025% calcium chloride, 0.1 M sodium cacodylate buffer, pH 7.4 and irradiated for 7 sec from ~20 to 48°C, processed routinely for electron microscopy, and embedded in Epon. (A) The mast cell shows excellent preservation of nuclear, cytoplasmic, and plasma membranes as well as secretory granules and other subcellular organelles. (B) The mast cell granules show 5-nm immunogold staining with specific guinea pig antihistamine-BSA antiserum diluted (Peninsula Labs, Belmont, CA) 1:30. Little or no label is present over the cytoplasm. (Note: the secondary immunogold reagent was custom prepared and was known to contain a high percentage of gold triplets.) Bars: A = 1 μm and B = 0.15 μm. (Reproduced with permission from Login et al., *J. Histochem. Cytochem.*, 40, 1247, 1992a.)

coated with 1% fish gelatin and irradiated for seconds (Leong and Milios, 1986; Cathieni and Taban, 1992). Cathieni and Taban (1992) further showed that fish gelatin resulted in less nonspecific precipitation of metallic silver from the IntenSE M™ mixture than other adhesives such as gelatin-chromalin, poly-L-lysine, or albumin.

In principle, microwave irradiation improves the cross-linking between the adhesive and the tissue section. In addition, microwave energy provides better control over heating samples than can be achieved in a convection oven or on a hot plate. This phenomenon can be conceptualized by recalling from Section 10.3 that microwaves are absorbed primarily by water. Therefore, when water in or on the tissue section has evaporated the section no longer absorbs microwave energy and the heating stops.

10.5.2 Microwave Stimulated Antigen Retrieval in Aldehyde Fixed Specimens

Microwave methods are used to improve the immunoreactivity of antigens in formalde-hyde-fixed, paraffin-embedded tissue sections (Sharma et al., 1990; Shi et al., 1991). Slides are placed in a humid chamber (e.g., a wooden slide box containing a few drops of water (Sharma et al., 1990)). Alternatively, they are placed in a Coplin jar filled with distilled water (Shi et al., 1991), 1% zinc sulfate solution (Shi et al., 1991), a saturated lead thiocy-anate solution (Shi et al., 1991), or with a dilute citrate buffer solution (McCormick et al., 1993) and irradiated for 1 to 10 min until the solution boils. A wide range of commonly used commercial antibodies show noticeably better immunoreactivity on slide sections treated by microwave heating than by conductive heating methods (Sharma et al., 1990; Shi et al., 1991). In theory, rapid microwave heating reduces the number of binding sites between formaldehyde and tissue proteins as compared to conductive heating methods (Sharma et al., 1990; Shi et al., 1991).

van de Kant et al. (1993) emphasize that tissues sections that will subsequently be used for silver-enhancement should not be exposed to metal solutions during antigen retrieval. The persistence of some heavy metals (i.e., lead) in the tissue sections may cause a nonspecific reduction and deposition of silver during silver-enhancement.

10.5.3 Microwave Stimulated Immunogold and Affinity Gold Silver Staining

A variety of microwave methods are used to enhance the reaction speed of each of the steps in various staining protocols. Microwaves have been used to accelerate protein blocking (Leong and Milios, 1990; Takes et al., 1991), primary and secondary antibody incubations (Jackson et al., 1988; Boon et al., 1989; Leong and Milios, 1990; van de Kant et al., 1990b; Takes et al., 1991), secondary protein A incubation (van de Kant et al., 1988, 1990b), gold-protein-ligand complex incubations (Cathieni and Taban, 1992), and silver-enhancement (van de Kant et al., 1990b). A major benefit of using microwave irradiation in these applications is the time factor. For example, incubation time with specific primary antibod-ies and gold-labeled secondary antibodies can be reduced from 1 h at 20°C to ~5 min each in a microwave oven. Microwave-assisted staining also offers the potential for increasing the positive reaction product and decreasing the nonspecific background label (Kok and Boon, 1992).

A brief review of the literature shows specific examples of how microwaves have been used in immunogold- and affinity-gold-silver staining protocols. Jackson et al. (1988) were the first to apply microwave-stimulated staining procedures to an immunogold-silver staining protocol. Using microwave irradiation, they completed within minutes, incuba-tions of the specific primary antibodies and gold-labeled secondary antibodies to detect IgA,G,D,M and Kappa and Lambda light chains in paraffin sections of human tonsil. Boon et al. (1989) showed superior immunogold-silver staining of beta-human chorionic gona-dotropin in paraffin sections of the syncytiotrophoblast of the first trimester placenta with their microwave method. van de Kant et al. (1990b) used microwave-accelerated immunogold-silver staining to improve the speed and reliability of identifying bromodeoxyuridine labeling on plastic sections of nuclei of mouse spermatogonia and preleptonene spermatocytes. Recently, Cathieni and Taban (1992) showed improved gold-protein-ligand, silver staining of specific cell binding sites using the ligands substance P, and delta-sleep-inducing peptide, oxytocin, and dopamine on frozen sections of rat brain using a microwave-accelerated protocol.

Two microwave-accelerated immunological staining protocols have emerged from the literature: (1) irradiating slides immersed in large volumes of reagents (van de Kant et al., 1990b) and (2) irradiating reagent droplets covering only the slide sections (Leong and Milios, 1986; Jackson et al., 1988; van de Kant et al., 1988; Boon et al., 1989; Kok and Boon, 1992). There are advantages to each of these microwave-accelerated staining approaches. Large reagent volumes simplify temperature measurement and regulation and prevent section drying during irradiation (Jackson et al., 1988; van de Kant et al., 1990b). The primary reason to use the droplet approach is to conserve limited quantities of the antibody solutions. Importantly, van de Kant et al. (1990b) showed that a 200 ml volume of monoclonal antibody solution could be reused up to five times provided the antibody solution was cooled to 12°C at the start of each incubation.

Most microwave staining protocols recommend setting the microwave oven power between 25 and 50%. In actuality, microwave oven power tubes in household microwave ovens produce only full power. Microwave oven power controls do not produce less power the way a dimmer switch decreases a light bulb's intensity. Instead, microwave oven power is controlled by turning the power tube on and off in a cyclical fashion to produce an average power over time. For example, setting a household microwave oven at 50% power for 1 min could result in a 30 sec burst of full power followed by a 30 sec period of no power (Login and Dvorak, 1994b).

A microwave oven's power cycle may range between 1 and 60 sec. The cycle time in household microwave ovens is predetermined by the manufacturer. The advantage of using microwave ovens with shorter cycle times (the number of times a magnetron cycles on and off per minute) is that the temperature rise in the sample is more gradual (Kok and Boon, 1992). Login and Dvorak (1994b) published a technique using a neon bulb array to determine the preset cycle time in household microwave ovens. Cycle time can be set on several models of microwave ovens designed for the laboratory.

Understanding how to control microwave oven power is important with respect to understanding how to limit heating of specimens during microwave enhanced immunogold-silver staining. Controlling the temperature rise in a sample will prevent immunological reagents from being denatured (Jackson et al., 1988). Generally the incubation temperature of the antibody solution is maintained below 42°C (Boon et al., 1989). Next, we discuss how to systematically and reproducibly determine the best power settings and irradiation times for immunogold-silver staining protocols.

10.6 Calibration Method for Microwave Staining Procedures

Published protocols on microwave-accelerated immunogold-silver staining will often report the temperature, solution, volume, microwave power, irradiation time, and cycle time. Unfortunately, it is difficult to reproduce microwave conditions in dissimilar microwave ovens. The single most important parameter to monitor is the final irradiation temperature of the incubation solutions. The slide calibration procedure is used to systematically select the appropriate functions of a microwave oven to achieve the best immunogold-silver staining results.

The calibration procedure for microwave staining is very different from the microwave calibration procedure for fixation. The key calibration tools needed for microwave staining are glass slides with liquid crystal temperature strips. The liquid crystal temperature strips simulate the position of tissue sections on the glass slides and directly measure the temperature of the incubation medium.

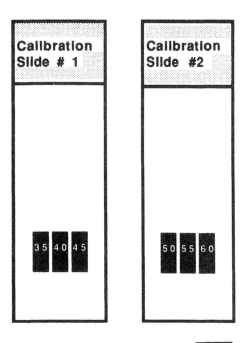

Figure 3 Diagram of calibration slides for microwave-accelerated staining. The black squares with white numerals printed on them correspond to the temperature to which the liquid crystal squares respond. The three squares are stuck to the bottom third of the slide in a group to simulate the placement of the tissue section. Slide 1 measures the temperature range between 35 and 45°C. Slide 2 measures the higher temperature range between 50 and 60°C. The black liquid crystal window will appear green or blue when the solution temperature is the same as the temperature printed on the liquid crystal. These calibration slides are safe to use in a microwave oven and they are reusable. Bar = 1 cm.

10.6.1 Materials for the Calibration Procedure for Microwave Staining

1. Glass slides with liquid crystal temperature squares (Login and Dvorak, 1994b): 24 precleaned glass slides (75 × 25 mm and ~1 mm thick) (Corning Glass Works, Scientific Glassware Dept., Corning, NY) and 6 liquid crystal temperature strips (LCTS) (Cat # TEMP-5, Owl Scientific Plastics, Woburn, MA) are needed. The strips contain black squares with white numerals printed on them corresponding to the temperature to which the liquid crystal squares respond. The LCTS show temperatures between 35 and 60°C in 5°C increments. Remove the protective paper from the back of the strips to expose the adhesive. Cut the LCTS into squares. Stick 3 squares to the bottom third of the slide in a cluster as shown in Figure 3. Group the individual liquid crystal squares corresponding to the following temperatures on each of the calibration slides:

    Slide 1: 35, 40, 45°C
    Slide 2: 50, 55, 60°C

    (NOTE: We recommend making 6 replicates of each calibration slide. These slides are reusable. Place the slides in a slide box labeled calibration slide set for safe keeping. The adhesive backing on the LCTS will melt if the temperature of the solution exceeds 70°C.)
2. Three 50 ml Coplin jars
3. 150 ml distilled water at room temperature
4. Neon bulb array (see p. 169)
5. Alphanumeric grid (see p. 169)

10.6.2 Calibration Procedure for Microwave-Accelerated Staining

1. Tape the alphanumeric grid (see p. 169) to the floor of the microwave oven.
2. Center the neon bulb array on the alphanumeric grid. Remove some of the bulbs in the array so that the bulb spacing is now 5 cm. (NOTE: In Section 10.3.1 a bulb spacing of 2.5 cm was used in the array).

3. Place 1 to 3 Coplin jars each filled with 50 ml distilled water on the styrofoam base of the neon bulb array in the microwave oven.
4. Turn the microwave oven on at full power for 20 sec and observe the areas where the neon bulbs are steadily illuminated. Notice that only a few bulbs are illuminated when the oven is loaded with the Coplin jars.
5. Open the microwave oven door and reposition 1 or more of the Coplin jars near clusters of bulbs that were illuminated in Step 4. Repeat this step until the bulbs all around the Coplin jar(s) are illuminated when the microwave power is on. (NOTE: Placing more than 1 Coplin jar filled with solution into the microwave oven at the same time may overload the oven (i.e., the neon bulbs do not illuminate around the jars). Overloading the microwave oven will result in uneven power absorption by the solution in the Coplin jars.)
6. Based on the results from Step 5, mark the ideal positions for the Coplin jar placement on the alphanumeric grid.
7. Remove the neon bulb array.
8. Replenish the water in the Coplin jars with 50 ml room temperature, distilled water.
9. Place one set of calibration slides labeled #1 or #2 (depending on your target incubation temperature) into each of the Coplin jars. Use the maximum number of Coplin jars that was determined in Step 5.
10. Position the jars on the alphanumeric grid corresponding to the position(s) marked in Step 4.
11. Turn on the microwave oven at the power setting and for the time recommended in a published microwave staining protocol. (Caution: Heating the solution above 70°C may melt the adhesive holding the LCTS on the slides.)
12. Remove the jars from the microwave oven. Record the temperature change of the liquid crystal squares for each of the calibration slides in each of the Coplin jars.
13. Repeat Steps 8 through 12 until the microwave power and irradiation time settings result in the final irradiation temperature reported in the published staining protocol.
14. Next, determine if multiple slides in a Coplin jar will reach the same temperature for the conditions selected in Step 13. First, place a set of 6 calibration slides with liquid crystal temperature squares that are all the same temperature range (i.e., 6 slides labeled calibration slide #2) into a Coplin jar. Repeat Steps 8 through 13. Record the slide positions within the Coplin jars containing calibration slides that reach the same temperature needed for the staining protocol. (NOTE: This is important because nonuniform microwave irradiation within a Coplin jar may lead to hot and cold spots. Slides that are exposed to too high a temperature may lose their sections. Slides that are in a cold spot may not have been exposed to sufficient microwave power for adequate staining.)
15. Be sure to use only those positions in the Coplin jar that were determined in Step 14 to yield the appropriate temperature for doing microwave-accelerated incubations.

10.6.3 *Additional Applications of the Liquid Crystal Calibration Slides in Microwave-Accelerated Staining*

The calibration slides made in Step 1 of the calibration procedure are ideal for monitoring the temperature of reagent droplets. The calibration procedure can be adapted to do microwave-stimulated immunogold-silver staining with the droplet technique (see p. 176). For example, microwave-accelerated staining of electron microscope grids can be done on calibration slides by floating the grids on or in the droplets located on the liquid crystal temperature square.

Standardization of droplet size and shape for the droplet technique should be done on slides with standardized slide wells. van de Kant et al. (1993) report that the shape of the droplet affects microwave energy absorption and the rate of microwave heating. We highly recommend using the printed slides available from Roboz Surgical Instrument Company, Inc. (Washington, D.C.) for standardizing droplet size and shape (Login et al., 1992a).

Do not store metal instruments used for immunogold labeling of electron microscopic grid sections on or near microwave ovens. We have learned that placing metal forceps near microwave ovens may magnetize the forceps in minutes. Magnetized forceps make handling of nickel grids very difficult.

10.7 Concluding Remarks

Microwave-accelerated procedures are useful in almost every step in the spectrum of sample handling for detection of antigens. At the very least, microwave procedures speed up reaction processes and save time. Even more importantly, microwave procedures improve the retention of soluble antigens and preserve antigen immunoreactivity often better than conventional fixation methods (Login and Dvorak, 1994b). Microwave-accelerated staining provides a new degree of control to improve the efficiency of antigen-antibody staining reactions. Judging the level of interest in laboratory microwave methods by the increasing number of articles written every year, we see a very promising future for this technology (Login and Dvorak, 1994b). The standardized methods in this chapter allow published procedures to be readily customized to a particular microwave oven. These standardized methods will greatly facilitate the use of microwave procedures in laboratories.

References

Abbate, M., Bachinsky, D., Zheng, G., Stamenkovic, I., McLaughlin, M., Niles, J. L., McCluskey, R. T. and Brown, D. 1993. Location of GP330/α2m-receptor associated protein (α2-MRAP) and its binding sites in kidney: distribution of endogenous α2-MRAP is modified by tissue processing. *Eur. J. Cell Biol.* 61: 139.

Azumi, N., Joyce, J., and Battifora, H. 1990. Does microwave fixation improve immunohistochemistry. *Mod. Pathol.* 3: 368.

Barsony, J. and Marx, S. J. 1990. Immunocytology on microwave-fixed cells reveals rapid and agonist-specific changes in subcellular accumulation patterns for cAMP or cGMP. *Proc. Natl. Acad. Sci. U.S.A.* 87: 1188.

Barsony, J. and Marx, S. J. 1991. Rapid accumulation of cGMP near activated vitamin D receptors. *Proc. Natl. Acad. Sci. U.S.A.* 88: 1436.

Barsony, J. and McKoy, W. 1992. Molybdate increases intracellular 3′,5′-guanosine cyclic monophosphate and stabilizes vitamin D receptor association with tubulin filaments. *J. Biol. Chem.* 267: 24457.

Barsony, J., Pike, J. W., DeLuca, H. F., and Marx, S. J. 1990. Immunocytology with microwave fixed fibroblasts shows 1-alpha,25-dihydroxyvitamin D3 dependent rapid and estrogen dependent slow reorganization of vitamin D receptors. *J. Cell Biol.* 111: 2385.

Benhamou, N., Noel, S., Grenier, J., and Asselin, A. 1991. Microwave energy fixation of plant tissue: an alternative approach that provides excellent preservation of ultrastructure and antigenicity. *J. Electron Microsc. Tech.* 17: 81.

Bernard, G. R. 1974. Microwave irradiation as a generator of heat for histological fixation. *Stain Technol.* 49: 215.

Boon, M. E., Gerrits, P. O., Moorlag, H. E., Nieuwenhuis, P., and Kok, L. P. 1988. Formaldehyde fixation and microwave irradiation. *Histochem. J.* 20: 313.

Boon, M. E., Kok, L. P., Moorlag, H. E., and Suurmeijer, A. J. 1989. Accelerated immunogold silver immunoperoxidase staining of paraffin sections with the use of microwave irradiation. *Am. J. Clin. Pathol.* 91: 137.

Bullock, G. R. 1984. The current status of fixation for electron microscopy: a review. *J. Microsc.* 133: 1.

Cathieni, M. M. and Taban, C. H. 1992. Microwave-aided binding of gold-protein-ligand (GPL) complexes. Light microscopic observations in the rat brain. *J. Histochem. Cytochem.* 40: 387.

Chew, E. C., Riches, D. J., Lam, T. K., and Hou Chan, H. J. 1983. A fine structural study of microwave fixation of tissues. *Cell. Biol. Int. Rep.* 7: 135.

Eggli, P. S., Lucocq, J., Ott, P., Graber, W., and van der Zypen, E. 1992. Ultrastructural localization of hyaluronan in myelin sheaths of the rat central and rat and human peripheral nervous systems using hyaluronan-binding protein-gold and link protein-gold. *Neuroscience* 48: 737.

Hanyu, Y., Ichikawa, M., and Matsumoto, G. 1992. An improved cryofixation method: cryoquenching of small tissue blocks during microwave irradiation. *J. Microsc.* 165 Pt 2: 255.

Hopwood, D., Coghill, G., Ramsay, J., Milne, G., and Kerr, M. 1984. Microwave fixation: its potential for routine techniques, histochemistry, immunocytochemistry, and electron microscopy. *Histochem. J.* 16: 1171.

Hsu, H. C., Peng, S. Y., and Shun, C. T. 1991. High quality of DNA retrieved for southern blot hybridization from microwave-fixed, paraffin-embedded liver tissues. *J. Virol. Methods* 31: 251.

Ikarashi, Y., Okada, M., and Maruyama, Y. 1986. Tissue structure of rat brain in microwave irradiation using maximum magnetic field component. *Brain Res.* 373: 182.

Jackson, P. 1991. Microwave fixation in molecular biology. *European J. Morphol.* 29: 57.

Jackson, P., Lalani, E. N., and Boutsen, J. 1988. Microwave-stimulated immunogold silver staining. *Histochem. J.* 20: 353.

Jensen, F. E. and Harris, K. M. 1989. Preservation of neuronal ultrastructure in hippocampal slices using rapid microwave-enhanced fixation. *J. Neurosci. Methods* 29: 217.

Kang, Z., Rohringer, R., Chong, J., and Haber, S. 1991. Microwave fixation of rust-infected wheat leaves. Preservation of fine structure and detection of cell surface antigens, lectin- and sugar-binding sites. *Protoplasma* 162: 27.

Kawasaki, Y., Monden, T., Morimoto, H., Murotani, M., Miyoshi, Y., Kobayashi, T., Shimano, T., and Mori, T. 1992. Immunohistochemical study of p53 expression in microwave-fixed, paraffin-embedded sections of colorectal carcinoma and adenoma. *Am. J. Clin. Pathol.* 97: 244.

Kok, L. P. and Boon, M. E. 1992. *Microwave Cookbook for Microscopists*, 3rd ed. Coulomb Press, Leyden.

Kok, L. P., Boon, M. E., and Suurmeijer, A. J. 1987. Major improvement in microscopic-image quality of cryostat sections. Combining freezing and microwave stimulated diffusion. *Am. J. Clin. Pathol.* 88: 620.

Leong, A. S.-Y. 1988. Microwave irradiation in histopathology. In: *Pathology Annual Part 2, Vol 23.* Rosen, P. P. and Fechner, R. E. (Eds.), Appleton & Lange, Norwalk.

Leong, A. S.-Y., Daymon, M. E., and Milios, J. 1985. Microwave irradiation as a form of fixation for light and electron microscopy. *J. Pathol.* 146: 313.

Leong, A. S.-Y. and Milios, J. 1986. Rapid immunoperoxidase staining of lymphocyte antigens using microwave irradiation. *J. Pathol.* 148: 183.

Leong, A. S.-Y. and Milios, J. 1990. Accelerated immunohistochemical staining by microwaves. *J. Pathol.* 161: 327.

Leong, A. S.-Y., Milios, J., and Duncis, C. G. 1988. Antigen preservation in microwave irradiated tissues: a comparison with formaldehyde fixation. *J. Pathol.* 156: 275.

Login, G. R. 1978. Microwave fixation versus formalin fixation of surgical and autopsy tissue. *Am. J. Med. Technol.* 44: 435.

Login, G. R. and Dvorak, A. M. 1985. Microwave energy fixation for electron microscopy. *Am. J. Pathol.* 120: 230.

Login, G. R. and Dvorak, A. M. 1988. Microwave fixation provides excellent preservation of tissue, cells and antigens for light and electron microscopy. *Histochem. J.* 20: 373.

Login, G. R. and Dvorak, A. M. 1990. Rapid microwave fixation. I. A model system and standardization protocol for large cavity microwave ovens. *Proc. Microsc. Soc. Can.* 17 (abstract): 12.

Login, G. R. and Dvorak, A. M. 1994a. Application of microwave fixation techniques in pathology to neuroscience studies: A review. *J. Neurosci. Meth.* 55: 173.

Login, G. R. and Dvorak, A. M. 1994b. Methods of microwave fixation for microscopy. A review of research and clinical applications: 1970–1992. *Prog. Histochem. Cytochem.* 27/4: 1.

Login, G. R. and Dvorak, A. M. 1994b. *The Microwave Toolbook. A Practical Guide with Laboratory Exercises for Microscopists.* Beth Israel Hospital Press, Boston.

Login, G. R., Galli, S. J., and Dvorak, A. M. 1992a. Immunocytochemical localization of histamine in secretory granules of rat peritoneal mast cells with conventional or rapid microwave fixation and an ultrastructural post-embedding immunogold technique. *J. Histochem. Cytochem.* 40: 1247.

Login, G. R., Galli, S. J., Morgan, E., Arizono, N., Schwartz, L. B., and Dvorak, A. M. 1987a. Rapid microwave fixation of rat mast cells. I. Localization of granule chymase with an ultrastructural postembedding immunogold technique. *Lab. Invest.* 57: 592.

Login, G. R., Kissell, S., Dwyer, B. K., and Dvorak, A. M. 1991. A novel microwave device designed to preserve cell structure in milliseconds. In: *Microwave Processing of Materials II.* Snyder, W. B., Jr., Sutton, W. H., Iskander, M. F., and Johnson, D. L. (Eds.), Materials Research Society, Pittsburgh.

Login, G. R., Quissell, D. O., and Dvorak, A. M. 1992b. Immunocytochemical detection of amylase in microwave fixed parotid acinar cells. *J. Dent. Res.* 71: 304.

Login, G. R., Schnitt, S. J., and Dvorak, A. M. 1987b. Rapid microwave fixation of human tissues for light microscopic immunoperoxidase identification of diagnostically useful antigens. *Lab. Invest.* 57: 585.

Login, G. R., Stavinoha, W. B., and Dvorak, A. M. 1986. Ultrafast microwave energy fixation for electron microscopy. *J. Histochem. Cytochem.* 34: 381.

Login, G. R., Stavinoha, W. B., and Dvorak, A. M. 1989. Fast and ultrafast microwave energy fixation techniques demonstrated by immunofluorescence, light and electron microscopy. In: *Microwave Irradiation For Histological And Neurochemical Investigations.* Blank, C. L., Howard, S., and Maruyama, Y. (Eds.), Soft Science Publications, Tokyo.

Marani, E., Boon, M. E., Adriolo, P. J., Rietveld, W. J., and Kok, L. P. 1987. Microwave-cryostat technique for neuroanatomical studies. *J. Neurosci. Meth.* 22: 97.

Mayers, C. P. 1970. Histological fixation by microwave heating. *J. Clin. Pathol.* 23: 273.

McCormick, D., Chong, H., Hobbs, C., Datta, C., and Hall, P. A. 1993. Detection of the Ki-67 antigen in fixed and wax embedded sections with the monoclonal antibody MIB1. *Histopathology* 22: 355.

Mizuhira, V., Notoya, M., and Hasegawa, H. 1990. New tissue fixation method for cytochemistry using microwave irradiation I. General remarks. *Acta. Histochem. Cytochem.* 18: 501.

Moran, R. A., Nelson, F., Jagirdar, J., and Paronetto, F. 1988. Application of microwave irradiation to immunohistochemistry: preservation of antigens of the extracellular matrix. *Stain Technol.* 63: 263.

Mudgett, R. E. 1986. Electrical properties of foods. In: *Engineering Properties of Foods.* Rao, M. and Rizvi, S. S. H. (Eds.), Marcel Dekker, Inc., New York.

Mudgett, R. E. 1990. Developments in microwave food processing. In: *Biotechnology and Food Process Engineering.* Schwartzberg, H. G. and Rao, M. (Eds.), Marcel Dekker, Inc., New York.

Naganuma, H., Ohtani, H., Harada, N., and Nagura, H. 1990. Immunoelectron microscopic localization of aromatase in human placenta and ovary using microwave fixation. *J. Histochem. Cytochem.* 38: 1427.

Nakae, T., Hosokawa, Y., Sugihara, H., Fushiki, S., and Tsuchuhashi, Y. 1990. Application of microwave irradiation in surgical pathology; improvement of microscopic image of cryostat sections and exploration in rapid metallic stainings. *Acta Histochem. Cytochem.* 23: 573.

Neas, E. D., and Collins, M. J. 1988. Microwave heating: Theoretical concepts and equipment. In: *Introduction to Microwave Sample Preparation: Theory and Practice.* Kingston, H. M. and Jassie, L. B. (Eds.), American Chemical Society, Washington, D.C.

Notoya, M., Hasegawa, H., and Mizuhira, V. 1990. New tissue fixation method for cytochemistry by the aid of microwave irradiation. II. Details. *Acta Histochem. Cytochem.* 23: 525.

Ohtani, H., Maruyama, I., and Yonezawa, S. 1989. Ultrastructural immunolocalization of thrombomodulin in human placenta with microwave fixation. *Acta Histochem. Cytochem.* 22: 393.

Ohtani, H., Naganuma, H., and Nagura, H. 1990. Microwave-stimulated fixation for histochemistry: Application to surgical pathology and preembeding immunoelectron microscopy. *Acta Histochem. Cytochem.* 23: 585.

Patterson, M. K., Jr., and Bulard, R. 1980. Microwave fixation of cells in tissue culture. *Stain Technol.* 55: 71.

Sharma, H. M., Kauffman, E. M., and McGaughy, V. R. 1990. Improved immunoperoxidase staining using microwave slide drying. *Lab. Med.* 21: 658.

Shi, S.-R., Key, M. E., and Kalra, K. L. 1991. Antigen retrieval in formalin fixed, paraffin-embedded tissues: an enhancement method for immunohistochemical staining based on microwave oven heating of tissue sections. *J. Histochem. Cytochem.* 39: 741.

Smid, H. M., Schooneveld, H., and Meerloo, T. 1990. Microwave fixation of water-cooled insect tissues for immunohistochemistry. *Histochem. J.* 22: 313.

Stavinoha, W. B., Pepelko, B., and Smith, P. W. 1970. Microwave radiation to inactivate cholinesterase in the rat brain prior to analysis for acetylcholine. *The Pharmacologist* 12: 257.

Suurmeijer, A. J. H., Boon, M. E., and Kok, L. P. 1990. Notes on the application of microwaves in histopathology. *Histochem. J.* 22: 341.

Takes, P. A., Kohrs, J., Krug, R., and Kewley, S. 1991. Microwave technology in immunohistochemistry: application to avidin-biotin staining of diverse antigens. *J. Histotechnol.* 12: 95.

Therkildsen, M. H. and Pilgaard, J. 1990. Microwave-assisted frozen section diagnosis. *APMIS.* 98: 200.

Utsunomiya, H., Komatsu, N., Yoshimura, S., Tsutsumi, Y., and Watanabe, K. 1991. Exact ultrastructural localization of glutathione peroxidase in normal rat hepatocytes: advantages of microwave fixation. *J. Histochem. Cytochem.* 39: 1167.

van de Kant, H. J. G., Boon, M. E., and de Rooij, D. G. 1988. Microwave-aided technique to detect bromodeoxyuridine in S-phase cells using immunogold-silver staining and plastic embedded sections. *Histochem. J.* 20: 335.

van de Kant, H. J. G., Boon, M. E., and de Rooij, D. G. 1993. Microwave applications before and during immunogold-silver staining. *J. Histotechnol.* 16: 209.

van de Kant, H. J. G., De Rooij, D. G., and Boon, M. E. 1990a. Microwave stabilization versus chemical fixation. A morphometric study in glycolmethacrylate- and paraffin-embedded tissue. *Histochem. J.* 22: 335.

van de Kant, H. J. G., Van Pelt, A. M. M., Vergouwen, R. P. F. A., and Rooij, D. G. 1990b. A rapid immunogold-silver staining for detection of bromodeoxyuridine in a large number of plastic sections, using microwave irradiation. *Histochem. J.* 22: 321.

Yamashina, S., Katsumata, O., and Sekine, R. 1990. Evaluation of microwave irradiation in immunohistochemical reactions. *Acta Histochem. Cytochem.* 23: 553.

Yasuda, K., Yamashita, S., and Shiozawa, M. 1989. An experimental study on the application of microwave fixation to immunohistochemical studies. *Acta Histochem. Cytochem.* 22: 91.

Zimmerman, G. R. and Raney, J. A. 1972. Fast fixation for surgical pathology specimens. *Lab. Med.* 3 (12): 29.

Chapter 11

Microwave-Aided Binding of Colloidal Gold-Protein-Substance P

Charles H. Taban and Maria M. Cathieni

Contents

11.1 Introduction

The Gold-Protein-Ligand complex (GPL complex) method was developed for specific labeling of binding sites without requiring the use of antibodies or of radioactive molecules. Colloidal gold was chosen as a label because of its ability to form stable complexes with large protein molecules, usually without modifying their biological or physiological activities. However, when colloidal gold particles are conjugated with a low molecular weight protein, the rate of label aggregation is increased. To overcome this problem, the size of the protein can easily be increased by cross-linking it with bovine serum albumin in the presence of glutaraldehyde (Horisberger and Rosset, 1977; Larsson 1979, 1981; Ackermann and Wolken, 1981). For binding studies, the biological activity of the conjugated ligand must be retained and, ideally, must be the same as that of the native ligand molecule itself.

In the first section we will describe the preparation of one of the several possible GPL complexes, namely the Gold-Protein-Substance P complex (GPSP complex), and the procedure for binding studies of unfixed, cryostat sections of the rat brain. For the labeling procedure, a microwave oven (MWO) is employed for two main purposes: (1) to achieve satisfactory stabilization of cryostat sections, and (2) to obtain more rapid preincubation, incubation, and fixation. The theory and application of microwaves for histologists is well documented (Boon and Kok, 1989; Kok and Boon, 1990, and Login and Dvorak in this volume).

In the second section we will show that the concentration of GPSP complex can be evaluated in molar-equivalent and that the biological activity of GPSP complex can be measured in a well-recognized pharmacological test (Cathieni et al., 1995). Preliminary results from samples subjected to the same conditions as those used in the pharmacological test show the potential for GPSP complex techniques for electron microscopy.

11.2 Preparation of GPSP Complex

For specific binding site studies with GPL complexes, the biological activity of the conjugated ligand has to be retained in order to localize and eventually to activate the correct sites. Different coupling procedures have been thoroughly investigated by immunologists to solve the problem of enhancing the immunogenicity of small molecules in order to produce antibodies. As reported by Van Regenmortel et al. (1988), most procedures for preparing peptide-protein conjugates are based on the use of symmetrical or asymmetrical bifunctional reagents which either become incorporated into the final conjugate or activate certain reactive sites of the carrier protein molecule for subsequent linkage with the peptide. Glutaraldehyde is the most extensively used coupling reagent and yet the mechanism of reaction of glutaraldehyde with proteins is not completely understood (Hayat, 1981). The reaction of glutaraldehyde with proteins involves mainly lysine residues (Korn et al., 1972) as well as the alpha amino group and the sulfhydryl group of cysteine (Habeeb and Hiramoto, 1968). Moreover, Avrameas (1969) and Jeanson et al. (1988) have shown that glutaraldehyde was by far the most effective reagent for producing enzyme-protein complexes which retained a part of their enzymatic and immunological specificity.

The serum albumins are widely used as carrier proteins because they are inexpensive and liable to yield soluble conjugates. Therefore, GPSP complexes are prepared in the following way with substance P (SP) as ligand, bovine serum albumin (BSA) as carrier molecule, and glutaraldehyde as coupling agent.

The protein-peptide is mixed at a molar ratio of 1:5. We also tested a molar ratio of 1:20, but since the final histological result appeared unchanged, the 1:5 molar ratio is used. A BSA solution of 20 mg/ml is prepared and filtered through a 0.22 μm millipore filter. Substance P is dissolved in 1 ml bidistilled H_2O at a concentration of 1.5 mM. At room temperature, 1 ml of BSA solution is added to 1 ml of SP solution. Under constant stirring, 240 μl of 0.5% glutaraldehyde solution and between 100 and 200 μl of 500 mM Tris buffer (pH 8.0) are then added to adjust the pH of the solution to approximately 7.5 to 7.8. During the coupling reaction, a yellow chromophore develops and the pH drops; a stable pH indicates the completion of the reaction which takes between 30 min and 1 h. To saturate the double bonds, a freshly prepared sodium borohydride solution at a concentration of 20 mg/ml is added until the solution becomes colorless; 400 μl of borohydride solution is generally adequate. The final volume of the solution is adjusted to 4 ml with distilled water to give a final BSA-SP conjugate concentration of 5 mg/ml. It is then placed at 4°C until disappearance of the foam which results from the addition of the NaH$_4$Bo solution. The conjugate solution is then dialyzed at 4°C against 50 mM Tris buffer (pH 7.0) with three changes in 24 h, filtered through a 0.22 μm millipore filter and stored in aliquots at –20°C until use.

Various methods for synthesis of colloidal gold particles of different sizes have been reviewed by Handley (1989). For the preparation of 5-nm diameter gold particles, we used the method of Slot and Geuze (1985) based on a temperature controlled reduction of chloroauric acid in a mixture of sodium citrate and tannic acid. A stable temperature of 60°C during the reaction is an important parameter for obtaining uniformly sized colloidal gold particles. The size of the prepared gold particles can be measured on electron micrographs by the procedure described by De Mey (1984). The colloidal gold suspension is cooled to room temperature (20°C) and used immediately for the preparation of the gold-protein complex without adjustment of the pH (pH 5.7).

The minimum concentration of BSA protein necessary to stabilize the gold sol is 20 µg/ml, as determined by the NaCl-induced flocculation test (Geoghegan and Ackerman, 1977; Wang and Larsson, 1985). The gold-protein complex itself is prepared by adding ten times the minimum amount of protein needed to stabilize the colloid. For adsorption on colloidal gold, the BSA-SP conjugate is added dropwise to the colloidal gold suspension under constant stirring with a magnetic stirrer. We find it unnecessary to add any extra stabilizer. The GPL complex is then centrifuged at 54,000 g for 1 h at 4°C in a centrifuge with a fixed angle rotor. The dark red sediment formed at the bottom of the tube corresponds to the GPSP complex. When a black spot is formed on the side near the bottom of the tube, it corresponds to metallic gold not stabilized by the protein and has to be discarded. The clear supernatant that contains the free BSA-SP conjugate is very carefully discarded, except the aliquot collected for spectrophotometric determination of absorbance ($A_{525\,nm}$) of the crude suspension. The measure of absorbance is a prerequisite especially for pharmacological measures where the calculation of the concentration of the colloid in molar equivalents (M-e) is needed (Cathieni et al., 1995). To eliminate any possible residual, unadsorbed free conjugate, the pellets can be resuspended in 50 mM Tris buffer (pH 7.0) and centrifuged again, as before. The pellets are pooled and stored at 4°C with 0.02% Na N_3. In this purified and concentrated form, the GPL complex is stable for several months.

11.3 Binding Site Visualization

11.3.1 Slide Preparation

Some conventional media used for coating slides appear inappropriate for the application of colloidal gold technique. Coating slides with albumin does not firmly retain the preparations. Coating slides with gelatin-chromalum or poly-L-lysine retains the preparations; however, excessive background is developed around these preparations during the process of gold particle enlargement by silver deposition during the silver development step. Metal ions of chromalum produce silver deposits and poly-L-lysine itself retains a number of gold particles. In both cases, the result is an excess of background staining around the preparations as a consequence of silver intensification. For these reasons the slide-coating medium should be free of metal ions and able to retain the preparations firmly, without trapping too many gold particles.

We propose the use of IGSS quality gelatin (20% solution, Amersham) further diluted in bidistilled water to a 1% solution. Thoroughly cleaned glass slides are dipped three times in this 1% gelatin solution and dried at room temperature in a dust-free place. The slides are then immersed in a 10% formaldehyde solution and irradiated for 1 min in a microwave oven at 600 W. The aim of this procedure is to render gelatin insoluble by polymerization. In this way the gelatin is firmly fixed to the glass support and the histological section holds well with minimal background around the preparations. The quality of the glass slide may be a problem and some glass may be the cause of undesirable silver autonucleation.

11.3.2 Tissue Processing

Unfixed tissues give the best results for GPSP complex marking. Each tissue may present some peculiarities that have to be taken into account in choosing the type of preparation; minor changes from the method described here for rat brain tissue may be required.

Anesthesia of animals by intraperitoneal injection of sodium pentobarbital (0.4 ml/100 g body weight) followed by perfusion with an intra-aortic catheter of a cold 0.9% NaCl (or cold PBS solution) containing 5.5% sucrose (w/v) (pH 7.4) allows easy treatment of tissues and good retention of reactivity. The addition of only 0.1% formaldehyde solution to the perfusion medium reduces the specific marking, especially in tissues in the immediate vicinity of vessels or on endothelial cells.

The brain can be quickly removed by using small dental forceps, allowing dissection of the calvaria without causing brain damage. Two or three drops of IGSS gelatin can be poured over the brain surface, forming a very thin film protecting it from dessication during the subsequent quick freezing with carbon dioxide and cryostat sectioning.

Cryostat sections that are 6 to 20 µm thick are cut. The preparations are placed on the polymerized IGSS gelatin coated slides, immediately dried for 2 min under a stream of cold air then irradiated for 15 sec in the microwave oven (MWO) at 240 W, the slides being placed 6 cm from the center of the carrousel. Two to four preparations can be irradiated at a time. Exposure of the sections in the MWO gives better tissue preservation and more standardized, reproducible results than when air drying is used alone, especially when the tissues are not further processed the same day.

Experimentally we found that humidity is detrimental for the preservation of binding sites. Therefore, the beneficial effect of irradiation may be due to microwave enzyme inactivation and to stabilization of the brain tissues; the term "stabilization" meaning that no chemical fixative, but only the physical effect of microwaves, is involved for tissue preservation (Marani et al., 1987). The irradiation time is also critical (Leong and Milios, 1990), an overexposure to MW results in excessive nonspecific labeling.

After drying and irradiation, the slides are collected in boxes containing silica gel and stored at room temperature until use, in a tightly closed dessicator also containing silica gel. Replacement of silica gel by P_2O_5 must be avoided, P_2O_5 being the cause of an important unspecific labeling produced during the silver intensification step. When immediately placed in a dry box after drying and irradiation, the preparations can be kept undamaged for at least 7 d. Thereafter, the amount of binding may progressively diminish.

11.3.3 Remarks on Microwave Irradiation

The theoretical basis for using microwave ovens and their application for laboratory work are well documented in Boon and Kok (1989), Kok and Boon (1990) and Boon et al. (1991).

We used a commercially available microwave oven without built-in temperature control (Mio-Star Hi-speed 2'100; Brother Industries, Nagoya, Japan), equipped with a platform rotating at 4.75 rpm. It operates at 2.45 gHz and has a maximum output of 600 W at "high" and 240 W at "low" (defrost) power. The radiation time at "low" setting is 8 sec (emitted at 600 W) for each block of 20 sec.

The use of MWO for slide preparation has been mentioned above. For preincubation and incubation we find that MW irradiation provides excellent tissue preservation, short reaction rates, and more regular marking than in the absence of tissue irradiation. The intensity of the marking depends on the volume and temperature of the reaction media and also on the power and length of irradiation in the MWO; any change in one of these parameters modifies the final quality of the marking and reproducibility of results.

To counteract the temperature increase of the reaction media due to the action of microwaves, the jars accommodating the slides are plunged in a vial containing a 500 ml water load, as described by Cathieni and Taban (1992). The containers used are ordinary glassware previously checked for temperature rise when irradiated. As stated by Boon and Kok (1989), glassware which heats up appreciably when irradiated for 1 to 2 min in the MWO should be avoided in order to obtain standardized results. However, plastic tanks and vessels can probably replace glass ones, avoiding possible undesired temperature shifts due to variations in the glass composition. Using MWO, it is possible to process several preparations from tissue sampling to microscopic observation within a day.

11.3.4 Visualization Procedure

The visualization procedure for GPSP complex is realized in five steps: preincubation, incubation, fixation, silver-enhancement, and counterstaining.

11.3.4.1 Media Composition

Preincubation medium: 15% BSA (w/v) is dissolved in 50 mM Tris buffer (pH 7.4) and filtered. Incubation medium: 1% gelatin (w/v) (Merck, 60–100 Bloom) is dissolved by gentle heating (40°C) in 50 mM Tris buffer (pH 7.4) and filtered. The two media are supplemented by 40 μg/ml bacitracin, 10 μg/ml leupeptin, and 10 μg/ml chymostatin.

A stock solution of 10 mM leupeptin in bidistilled water is prepared, aliquoted, and stored at –20°C. This solution is stable for one month at –20°C, one week at 4°C, and only for several hours at room temperature. The effective concentration is 10 to 100 μM.

A stock solution of 10 mM chymostatin is prepared in DMSO and aliquoted. This stock solution remains stable for one month at –20°C and several hours at room temperature. The effective concentration is 5 to 100 μM and the half-life 20 min at pH 7.5. Experiment shows that 10 μg/ml of leupeptin and 10 μg/ml of chymostatin are adequate for good labeling of brain tissue. In the absence of chymostatin and leupeptin the labeling is diminished. Fixation medium is 5% formaldehyde (pH 7.4).

11.3.4.2 Preincubation and Incubation

To prevent silver self-precipitation during the silver-enhancement step, heavy metal contamination of glassware and contact of metallic objects with washing and incubation media should be avoided. To prevent excessive heating of the preparations during microwave irradiation, the staining jar used is placed in a 500 ml waterbath at 15 ± 0.5°C. After 50 sec irradiation at 240 W, the temperature of the water bath rises from 15 to 21 to 22°C while the temperature of the medium in the staining jar rises only from 22 to 23 to 24°C. This procedure is used for the preincubation and incubation steps and the water load renewed each time.

11.3.4.3 Silver-Enhancement

Four to six slides are placed flat in plastic boxes and tissue sections are covered with IntenSE M (Cathieni and Taban, 1992). IntenSE M (Amersham) gives satisfactory, reproducible results; however, with this mixture 5-nm gold particles have to be enhanced twice to become visible with the light microscope. Although IntenSE M is rather light insensitive and does not usually need a silver fixation step, we find that enhancing the gold signal in the dark under a safe light and fixing the preparations in a 5% sodium thiosulfate solution are additional precautions for preventing any autocatalytic silver precipitation.

11.3.4.4 Counterstaining

Due to the nature of the technique, classical methods used for histological counterstaining cannot readily be applied; silver-enhanced gold granules are damaged by oxidative staining steps. However, GPL complex-labeled sections of rat brain can be satisfactorily counterstained with 0.025% toluidine blue solution in 0.01 N acetate buffer (pH 4.6).

11.3.4.5 Protocol for Binding Site Visualization

This protocol applies for 12 μm thick cryostat sections of rat brain.

1. Immediately dry each section for 2 min under a stream of cold air.
2. Irradiate for 15 sec in the MWO at 240 W, the sections being placed 6 cm from the center of the platform.
3. Collect slides in boxes containing silica gel. Store until use at room temperature in a tightly closed dessicator also containing silica gel.
4. Transfer the slides into the preincubation medium (22°C) and preincubate for 50 sec in the MWO at 240 W.
5. Transfer the slides to the incubation medium (22°C) and incubate 50 sec in the MWO at 240 W.
6. Wash in four changes (45 sec each) of ice cold 50 mM Tris buffer (pH 7.4).
7. Fix in 5% formaldehyde solution (pH 7.4) for 30 sec in the MWO at 600 W.
8. Wash with three changes of distilled water, 5 min each.
9. Change water and proceed to the silver-enhancement step. Under photographic safe light and just before use, mix equal volumes of silver enhancer and initiator in a polystyrene vial. Dry the back of the slide with a towel, eliminate excess water around the preparation and cover it with the IntenSE mixture.
10. Enhance the gold signal for 17 to 20 min, depending on room temperature.
11. Wash thoroughly with distilled water and repeat Step 10.
12. Wash thoroughly with distilled water and fix for 5 min in 5% sodium thiosulfate.
13. Wash in distilled water.
14. Counterstain for 30 to 50 sec in 0.025% toluidine blue solution.
15. Rinse twice in acetate buffer.
16. Rinse twice in distilled water.
17. Dry under a stream of cold air.
18. Dip the slides in xylene for 5 min and mount with Merckoglas.

Application of the method yields permanent histological preparations exhibiting binding sites labeled with black granules well distinguished on the pale blue color of the counterstain (Figures 1 and 2).

11.3.4.6 Specificity Tests

The labeling specificity of the histological preparations can be verified by:

1. addition of the specific conjugate BSA-SP or the nonspecific conjugate BSA-Gly-Gly or BSA-BSA to the preincubation medium. Following incubation with GPSP, this should result in an absence of labeling after application of the first conjugate, and in a correct labeling after addition of BSA-Gly-Gly or BSA-BSA;
2. addition of BSA-SP or of the nonspecific conjugate BSA-Gly-Gly or BSA-BSA to the incubation medium should result in a diminished labeling after application of BSA-SP and in a correct labeling after addition of BSA-Gly-Gly or BSA-BSA.
3. addition of the native ligand (in this case SP) or of the nonspecific ligand (Gly-Gly) to the preincubation medium should result in an absence of further labeling after application of SP, and a normal labeling after addition of Gly-Gly;
4. addition of the native ligand (SP) or of an unspecific ligand (Gly-Gly) to the incubation medium. In this case, a ligand concentration-related reduction of the labeling should be observed only after addition of the native ligand. It is evident that the silver intensification step applied to untreated preparations should not by itself develop any labeling.

When carried out as described in the protocol, the specificity tests require large and expensive amounts of SP. To circumvent this problem, it is possible to carry out specificity

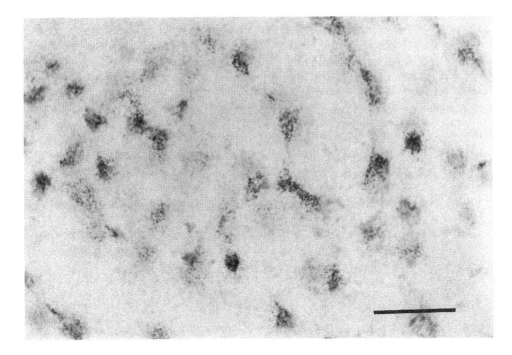

Figure 1 Rat brain horizontal section at the level of the mitral cell layer of the olfactory lobe labeled with 2 nM-e GPSP complex. Most neurons exhibit a distribution of black granules. Bar = 20 μm.

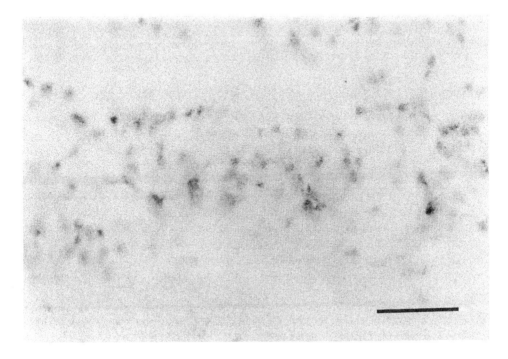

Figure 2 Rat brain horizontal section at the level of the dentate gyrus labeled with 2 nM-e GPSP complex. Several neurons exhibit the presence of black granules. Bar = 40 μm.

tests by the droplet incubation method as described by Boon et al. (1991) with a water load placed in the MWO. The slides can also be placed in an incubation box containing a water load in the floor of the tray (Leong and Milios, 1990). The volume and location of water load, the volume of the droplets covering the tissue sections, and the location of the sections in the MWO must be standardized.

For specificity tests we routinely use an incubation box containing 800 ml tap water (t = 15°C) placed in the center of the rotating platform. Brain sections covered with 500 μl of preincubation medium (t = 22°C) are irradiated for 50 sec at 240 W, then the medium is drained, replaced by 500 μl of incubation medium and irradiated for 50 sec at 240 W. The heated water load is replaced after each run. As some irregularities in the marking may occur despite standardized parameters, we process slides simultaneously by pairs, one control and one subjected to blocking or competitive agents. Slides are then further processed as described.

The most important test of GPL complex specificity is the verification that, in their conjugated form, the ligands do not lose their biological activity. This verification for GPSP complex is carried out in the second part of this contribution.

11.4 Biological Activity of GPSP Complex

The activity of the ligand should be retained in the carrier conjugated form, and this can be verified by pharmacological testing. However, for quantitative comparison between the biological effect elicited by molar concentration of native SP and that produced by GPSP, the molar equivalent concentration of GPSP complex must be calculated. The molar-equivalent calculation of GPSP complex would correspond to reality only if each colloidal gold particle carries at least one SP molecule. That at least one molecule is, indeed, bound can be verified by application to the GPSP complex of an antibody recognizing SP, which itself is detected by a second fluorescent antibody. Analysis is done first by choosing an area under the fluorescent microscope, counting the fluorescent particles seen with the microscope equipped with an HBO 100 lamp, BP 38, KP 500 excitation and K 510 stop filters, and then by counting the same particles again under dark light illumination.

11.4.1 Immunological Detection of SP in GPSP Complex

For this test, poly-L-lysine (PLL) coated slides are used, the gold particles adhering more strongly to PLL than to gelatin coated slides. The test is carried out in 10 mM phosphate buffer (pH 7.4) containing 0.2% protein-BSA (PBSA) and 0.01% NaN_3. Phosphate buffer is used instead of Tris buffer, since phosphate ions sustain the antigen-antibody reaction. However, the phosphate concentration should remain low since phosphate ions may impair the stability of the colloid by lowering the energy barrier (Horisberger, 1981). We observed that when GPSP complex associated with SP-antibody is submitted to centrifugation forces, a dark spot of metallic gold appears on the side near the bottom of the centrifugation tube, indicating that GPSP-SP antibody-complex may dissociate during centrifugation and that addition of 0.2% of PBSA to the suspension stabilizes the complex. The conjugate PBSA is prepared as described for PSP, but omitting SP from the conjugation reaction.

11.4.1.1 Procedure

1. Mix 50 μl of concentrated GPSP complex (control: 50 μl of GPBSA) with 50 μl of rabbit anti-SP serum diluted 1/100 with 20 mM phosphate buffer-PBSA. Incubate at least 1 h at room temperature.
2. Add 100 μl of goat-antirabbit fluorescent antiserum diluted 1/100 with 10 mM phosphate buffer-PBSA. Incubate 5 min in the dark at room temperature.

3. Add 1 ml of 10 mM phosphate buffer-PBSA, centrifuge at 20,000 g, 45 min at 12°C. Carefully discard the supernatant.
4. Transfer the pellet into a clean centrifuge tube (to eliminate the free antibodies possibly retained on the wall of the tube). Repeat Step 3.
5. Mount the pellets with phosphate buffer on PLL coated slides.

11.4.2 Evaluation of GPL-Complex Molar Equivalent (M-e) Concentration

In contrast to qualitative histological observations, the quantitative pharmacological studies require known concentrations of GPL complex. The mean particle diameter and the size distribution of the gold particles must be measured in order to determine GPSP complex concentration. Ideally, the gold particles should have the same volume. Determination of particle size from electron micrographs is satisfactory but time consuming; therefore, we are exploring the feasibility of particle sizes analysis carried out by photon correlation spectroscopy.

After the gold particle size has been established, the GPL complex molar equivalent (M-e), that is, the number of particles related to the Avogadro's number, is estimated by taking into account: (1) the number of colloidal gold particles of the freshly prepared suspension (Park et al., 1987), (2) the relationship between absorbance and gold particles concentration (Park et al., 1987; Horisberger et al., 1977), and (3) the Avogadro's constant.

1. The number of gold particles per ml is calculated as described by Park et al. (1987). For the proposed example, 100 ml of 5-nm colloidal gold suspension prepared according to Slot and Geuze (1985) contains 1 ml of 1% $HAuCl_4.3H_2O$ (mol. wt. = 393.83) solution, i.e., 5 mg of Au (mol. wt. = 196,97). Since the final volume of the GPSP complex suspension was 108.2 ml (100 ml gold suspension +4 ml BSA-SP conjugate + 4.2 ml H_2O), the concentration of Au was 5 mg/108, 2 ml = 4.62×10^{-5} g/ml. The volume of a 5-nm gold particle is (4/3) (3.14) $(2.5$ nm$)^3$ = 65.4 nm^3 = 6.54 $\times 10^{-20}$ ml. The density of Au being 19.32 g/ml, the weight of each 5-nm gold particle is $6.54 \times 10^{-20} \times 19.32 = 1.26 \times 10^{-18}$ g and the total number of gold particles in 1 ml is $(4.62 \times 10^{-5}$ g/ml$): (1.26 \times 10^{-18}$ g$) = 3.67 \times 10^{13}$.
2. The absorbance of freshly made GPSP complex at 525 nm was 0.935. Thus, absorbance of 0.935 corresponds to 3.67×10^{13} particles per ml. After centrifugation the absorbance of the GPSP-complex suspension reconstituted in 50 mM tris buffer was 0.584, corresponding to $(0.584/0.935) \times 3.67 \times 10^{13} = 2.29 \times 10^{13}$ particles per ml.

   When GPSP complex is too concentrated for direct spectrophotometric measures, absorbance can be measured by diluting an aliquot of GPSP complex suspension with Tris buffer without forgetting the dilution factor in the calculations. The correlation between the values given by calculated theoretical particle concentration and the values obtained by spectrophotometric measures has been discussed by Cathieni et al. (1995).
3. By analogy with a molar solution, the GPL complex M-e suspension contains 6.022 $\times 10^{23}$ (Avogadro's constant) particles per l. Thus, the M-e concentration of the GPSP complex suspension of A_{525nm} = 0.584 is $(2.29 \times 10^{13} \times 10^3): (6.022 \times 10^{23}) = 3.8 \times 10^{-8}$.

11.4.3 Pharmacological Test

The biological action of SP on isolated pig coronary artery reported by Bény et al. (1986) and by Gulati et al. (1987) has been used by Cathieni et al. (1995) to compare SP and GPSP complex activity. The pig coronary artery, which shows a high sensitivity and selectivity to SP, possesses NK_1 neurokinin receptors restricted to endothelial cells. The experimental

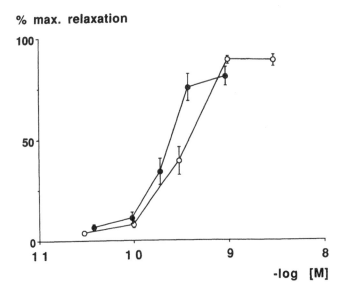

Figure 3 Dose-response curves of SP (O) and GPSP (●) induced relaxations measured in precontracted transverse strips of pig coronary arteries. Abscissa: molar (for SP) or molar-equivalent (for GPSP complex) concentration. Ordinate: percent of maximal relaxation. Results show the mean ± SEM of 8 experiments.

conditions provided by the pig coronary artery are particularly suitable in the sense that a direct contact can be managed between GPSP complex and endothelial NK_1 receptors, thus avoiding penetration problems of GPSP complex.

Exposure of strips of pig coronary artery to 10 μM of prostaglandin (PGF_2^α) produces a stable maximal contraction, sustained as long as the PGF_2^α is applied. Addition of SP to coronary artery strips precontracted with PGF_2^α produces dose-dependent relaxation comparable to that produced by different dilutions of the parent solution of GPSP complex calculated in M-e concentration. When expressed in terms of percent of the maximal relaxation, the data obtained with SP and with GPSP fit into dose-response curves which are similar (Figure 3). Moreover, 0.1 μM concentration of the neurokinin antagonist [D-Pro[4], D-Trp[7,9], Nle[11]] - SP (4–11) significantly reduces the vasodilation induced by SP and by GPSP.

The results show that, like native SP, GPSP (but not GPBSA) was able to stimulate dose-dependent relaxing effects on the pig coronary artery, which meant that endothelial NK_1 receptors could be activated by native SP as well as by SP bound to BSA and gold particles. Therefore, the process of BSA-SP conjugation and adsorption of the conjugate to 5-nm colloidal gold particles does not appear to modify the biological capacities of the SP C-terminal activating NK_1 endothelial cell receptors.

11.4.4 Visualization of GPSP Complex Binding

Small pieces of coronary artery are incubated under conditions identical to those used in the pharmacological tests, i.e., in a modified Krebs solution of the following composition (mM): 118.7 NaCl, 4.7 KCl, 2.5 $CaCl_2$, 1.2 KH_2PO_4, 24.8 $NaHCO_3$, 10.1 glucose, gassed with 95% O_2 –5% CO_2, pH 7.4, 35°C.

For light microscopy, a piece of coronary artery approximately 7 mm in length is incubated for 3 min in modified Krebs solution containing 1 nM-e GPSP complex, washed 1 min in modified Krebs solution, and then fixed in 5% formaldehyde (also in modified Krebs solution). After washing, the tissue is frozen with solid carbon dioxide, cut at 20 μm with a cryostat, placed on polymerized gelatin-coated slides and dried. The preparations

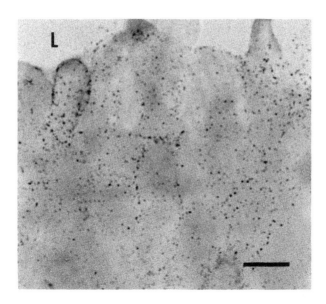

Figure 4   Pig coronary artery, 20 µm coronal section showing tangentially cut endothelial cells at the level of the beginning of a small effluent arterial branch. Incubation for 3 min with 1 nM-e GPSP complex. Accumulations of GPSP complex black dots are seen at the endothelial cell borders. L = vessel lumen. Bar = 10 µm.

are then rehydrated, the gold granules enlarged with IntenSE M as previously described, counterstained with hematoxylin-eosin and mounted with Merckoglas (Figure 4).

For GPSP complex ultrastructural localization, strips 4 mm in length and 1.5 mm in width are incubated as before, washed for 2 sec in cold modified Krebs solution and fixed in 2.5% glutaraldehyde in cacodylate buffer (pH 7.4) for 2 h at room temperature, then for 12 to 16 h at 4°C. The strips are then treated for electron microscopy. Following 3 min incubation, light microscopic examination reveals large accumulations of black dots on endothelial cells, but no marking is detected in connective or muscular tissues subjacent to the endothelium (Figure 4).

The gold particles are revealed by electron microscopic examination (Figure 5). Marking occurs in the endothelium; however, following 5 min incubation, GPSP complexes appear also in cell membranes of subjacent connective tissues.

11.5 Concluding Remarks

We have shown how a conjugated small peptide when adsorbed on colloidal gold can be used not only for detecting binding sites on tissue with EM or LM, but, in addition, for pharmacological tests, verifying the possible activation of the detected specific binding sites. For these purposes incubation with GPSP complex can be carried out either *in vitro* before cryostat sectioning or on cryostat sections of unfixed tissue. However, we are conscious that the results obtained so far with the GPSP complex method are preliminary, especially if one considers that labeling may reveal, in addition to the receptors stimulated *in vitro*, some binding sites possibly located in the elements of the extracellular matrix (ECM) for which a biological role has yet to be detected. Such ECM binding sites have been detected for the fibroblast growth factor (Yayon et al., 1991).

The final goal of this study is to understand what really happens *in vivo* for tissues in the presence of an active ligand. Pharmacological studies measure the effects produced by the ligand and histological observations detect the localization of binding sites or of cell receptors potentially able to be activated by the ligand. Importantly, since small ligands

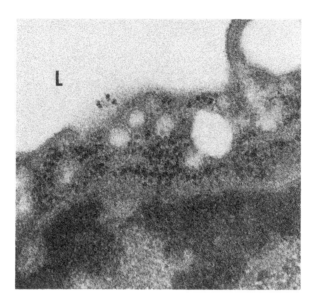

Figure 5   Pig coronary artery strip incubated for 5 min with 2 nM-e GPSP-complex and processed for electron microscopy. Black GPSP complex granules are present at the level of the endothelial cell. Magnification ×80,000.

other than SP can be conjugated with BSA, numerous different GPL complexes can be prepared and used for pharmacological and binding studies.

11.6 Summary

Conjugates of substance P with bovine serum albumin absorbed on colloidal gold (GPSP complex) are used for binding site visualization or activation. Cryostat sections of unfixed rat brain are preincubated for 50 sec in the microwave oven in a Tris buffered solution (pH 7.4) containing 1.5% BSA, then further incubated for 50 sec in the microwave oven in a Tris buffer solution containing 1% gelatin and the diluted colloidal gold complex. After washing, the preparations are postfixed for 30 sec in the microwave oven in 5% formaldehyde solution (pH 7.4). Finally, the cell-bound GPSP complexes are enlarged by a silver-enhancing process and counterstained.

The specificity of the biological activity of GPSP complex has been tested on strips of pig coronary artery which possess NK_1 neurokinin receptors restricted to endothelial cells. This assay shows a high sensitivity to SP. Like native SP, GPSP complex produces dose-dependent relaxation of pig coronary strips precontracted with 10 μM Prostaglandin F_2 alpha. Molar-equivalent concentrations of GPSP complex are determined to allow quantitative comparisons with native SP. The respective potencies and intrinsic activities of SP and GPSP complex do not show any significant differences. Histological samples subjected to the conditions of incubation used in the pharmacological testing show the presence of GPSP complex on endothelial cell membranes as confirmed by light and electron microscopy.

References

Ackerman, G. A. and Wolken, K. W. 1981. Histochemical evidence for the differential surface labeling, uptake, and intracellular transport of a colloidal gold-labeled insulin complex by normal human blood cells. *J. Histochem. Cytochem.* 29: 1137.

Avrameas, S. 1969. Coupling of enzymes to proteins with glutaraldehyde. Use of the conjugates for the detection of antigens and antibodies. *Immunochemistry* 6: 43.

Bény, J. L., Brunet, P., and Huggel, H. J. 1986. Effect of mechanical stimulation, substance P and vasoactive intestinal polypeptide on the electrical and mechanical activities of circular smooth muscle from pig coronary arteries contracted with acetylcholine: role of the endothelium. *Pharmacology* 33: 61.

Boon, M. E. and Kok, L. P. 1989. *Microwave Cookbook of Pathology: The Art of Microscopic Visualization.* 2nd revised edition. Coulomb Press, Leiden.

Boon, M. E., van de Kant, H. J. G., and Kok, L. P. 1991. Immunogold-silver staining using microwave irradiation. In: *Colloidal Gold: Principles, Methods, and Applications.* Vol. 3. pp. 347–367. Hayat, M. A. (Ed.), Academic Press, San Diego.

Cathieni, M. M. and Taban, C. H. 1992. Microwave-aided binding of gold-protein-ligand (GPL) complexes. Light microscopic observations in the rat brain. *J. Histochem. Cytochem.* 40: 387.

Cathieni, M. M., Mastrangelo, D., and Taban, C. H. 1995. Pharmacological evidence that gold-protein-substance P complex activates Neurokinin NK1 receptors. *Cell. Physiol. Biochem.* 5: 85.

De Mey, J. 1984. Colloidal gold as marker and tracer in light and electron microscopy. *EMSA Bull.* 14: 54.

Geoghegan, W. D. and Ackerman, G. A. 1977. Adsorption of horseradish peroxidase, ovomucoid and anti-immunoglobulin to colloidal gold for the indirect detection of concanavalin A, wheat germ agglutinin and goat antihuman immunoglobulin G on cell surfaces at the electron microscopic level: a new method, theory and application. *J. Histochem. Cytochem.* 25: 1187.

Gulati, N., Mathison, R., Huggel, H. J., Regoli, D., and Bény, J. L. 1987. Effects of neurokinins on the isolated pig coronary artery. *Eur. J. Pharmacol.* 137: 149.

Habeeb, A. F. S. A. and Hiramoto, R. 1968. Reaction of proteins with glutaraldehyde. *Arch. Biochem. Biophys.* 126: 16.

Handley, D. A. 1989. Methods for synthesis of colloidal gold. In: *Colloidal Gold: Principles, Methods, and Applications.* Vol. 1. pp. 13–32. Hayat, M. A. (Ed.), Academic Press, San Diego.

Hayat, M. A. 1981. *Fixation for Electron Microscopy.* Academic Press, San Diego.

Horisberger, M. 1981. Colloidal gold: a cytochemical marker for light and fluorescent microscopy and for transmission and scanning electron microscopy. *Scanning Electron Microsc.* II: 9.

Horisberger, M. and Rosset, J. 1977. Colloidal gold, a useful marker for transmission and scanning electron microscopy. *J. Histochem. Cytochem.* 25: 295.

Jeanson, A., Cloes, J.-M., Bouchet, M., and Rentier, B. 1988. Comparison of conjugation procedures for the preparation of monoclonal antibody-enzyme conjugates. *J. Immunol. Meth.* 111: 261.

Kok, L. P. and Boon, M. E. 1990. Microwaves for microscopy. *J. Microsc.* 158: 291.

Korn, A. H., Feairheller, S. H., and Filachione, E. M. 1972. Glutaraldehyde: nature of the reagent. *J. Mol. Biol.* 65: 525.

Larsson, L.-J. 1979. Simultaneous ultrastructural demonstration of multiple peptides in endocrine cells by a novel immunocytochemical method. *Nature* 282: 743.

Larsson, L.-J. 1981. Peptide immunocytochemistry. *Progr. Histochem. Cytochem.* 13: 73.

Leong, A. S. Y. and Milios, J. 1990. Accelerated immunohistochemical staining by microwaves. *J. Pathol.* 161: 327.

Marani, E., Boon, M. E., Adriolo, P. J. M., Rietveld, W. J., and Kok, L. P. 1987. Microwave cryostat technique for neuroanatomical studies. *J. Neurosci. Meth.* 22: 97.

Park, K., Scott, R. S., and Albrecht, R. M. 1987. Surface characterization of biomaterials by immunogold staining. Quantitative analysis. *Scanning Microsc.* 1: 339.

Slot, J. W. and Geuze, H. J. 1985. A new method for preparing gold probes for multiple-labeling cytochemistry. *Eur. J. Cell Biol.* 38: 87.

Van Regenmortel, M. H. V., Briand, J. P., Muller, S., and Plaué, S. 1988. Peptide-carrier conjugation. In: *Synthetic Polypeptides as Antigens. Laboratory Techniques in Biochemistry and Molecular Biology.* Vol 19. pp. 95–102. Burdon, R. H. and van Knippenberg, P. H. (Eds.), Elsevier, Amsterdam.

Wang, B. L. and Larsson, L. I. 1985. Simultaneous demonstration of multiple antigens by indirect immunofluorescence and immunogold staining. *Histochemistry* 83: 47.

Yayon, A., Klagsbrun, M., Esko, J. D., Lefer, P., and Ornitz, D. M. 1991. Cell surface, heparin-like molecules are required for binding of basic fibroblast growth factor to its high affinity receptor. *Cell* 64: 841.

Chapter 12

Increased Sensitivity for Detection of Immunogold-Silver Stain With Epipolarization Microscopy

Kuixiong Gao and Alvin Wei Gao

Contents

12.1 Introduction

The utilization of colloidal gold in the human history is far earlier than what most immunocytochemists thought. In ancient China, alchemists made the earliest synthetic drug for longevity — the red colloidal gold — called Chin-Yeh = Gold + Herbal juice, or an herbogolden complex, assuming that the red substance was as red as blood, the soul of life (Mahdihassan, 1984). However, it wasn't until the 1930s and 1940s that the study of colloidal gold with scientific base was started. (Weiser, 1949). The report entitled "An immunocolloid method for the electron microscope" (Faulk and Taylor, 1971) was considered as a landmark of the application of colloidal gold in immunocytochemistry. Later Ferns (1973) described a simple sodium citrate reduction method to produce colloidal gold solution of controllable and uniform size of gold particles. That method advocated the use of colloidal gold as a probe for biomedical research. During the first decade after Faulk and Taylor's publication, there were only a few articles published regarding the application of colloidal gold. However, after 1980 the number of publications relevant to the use of colloidal gold in histology and cytology increased tremendously (see a review by Handley, 1989). From 1981 to 1987, approximately 1000 articles related to colloidal gold for immunohistochemistry were published. In 1993, 892 articles were found under the search words colloidal gold and immunohistochemistry or immunocytochemistry (Medline [R] January-December 1993, Silver Platter 3.0).

Colloidal gold was initially used only as a marker for electron microscopy (EM), because of its electron dense nature and secondary electron emission feature (Horisberger, 1979). Direct visualization of colloidal gold in light microscopy (LM) was limited. The size of colloidal gold is too small to be detected at the light microscope level, although using highly concentrated immunogold cells may be stained red by this reagent (Geoghegan et al., 1978; Roth, 1982; Holgate et al., 1983).

Regardless of other immunostaining methods, there is always a need of a more sensitive way to detect much smaller amounts of antigen than presently possible. Since the main disadvantage for the colloidal gold marker at LM level was the small size of the colloidal gold itself, Danscher and Nörgaard (1983), and Holgate et al. (1983) reported the principle and practice of a very sensitive new immunostaining method such that gold particle was enhanced with silver in a physical developer. This method was called immunogold-silver staining (IGSS). After IGSS, the diameter of colloidal gold particle is increased by silver shields. Therefore, the final gold-cored silver granule (GSG) is discernible under LM. It is claimed that the IGSS method is up to 200-fold more sensitive than the standard immunoperoxidase method (Holgate et al., 1983).

Gao and Huang (1985) reported another method to amplify the invisible colloidal gold at LM level. They were able to produce microspherules encaptured with hundreds of colloidal gold particles (MS-Au). The size of MS-Au ranges from 0.5 to 10 µm, which is seen in a microscope. Besides, when specific protein-coated MS-Aus were fed to the cultured cells, those cells that recognize the specific protein could internalize gold labeled microbeads, which were rose-red and distinguishable in the bright field LM, even in the unfixed and unstained living cells. Endocytosis of these nontoxic microbeads by cultured cells was vividly photographed (Gao et al., 1991b).

Presently, the IGSS method has become one of the most important immunocytochemical techniques, and consequently constitutes the main focus of this book. Recently, the use of microscope with epi-illumination of polarized light, or, epipolarization microscopy (EPM; some of the authors used the abbreviation EPI) offers another opportunity to achieve even more sensitive detection of low abundance antigen signal after IGSS, especially when the high power of objective lens is used to examine semithin sections. This chapter aims to review these EPM observations. The potential for the IGSS-EPM method is great, but its use has not yet been widely spread. This situation seems to be similar to the early stage of utilization of colloidal gold. At the present time, the amount of literature on the IGSS-EPM method is rather limited. Some of the information presented here is based on personal experience.

12.2 Detection of Gold-Cored Silver Granules with Epipolarization Microscopy

12.2.1 Microscope Equipped with Epi-Illumination of Polarized Light

In an epipolarization microscope (EPM), the light from an epi-illumination light source, such as from an high intensity mercury lamp, is polarized before it reaches the specimen. The light reflected by GSG loses its polarization and passes a crossed analyzer, whereas most other reflected polarized light shall be cut off by the analyzer. A microscope with epi-illuminated fluorescence equipment has the potential to be converted into an EPM by replacing the fluorescence filter with a polarizing filter system, normally called IGS block. The quality of an EPM is mainly dependent on the quality of microscope itself, especially the quality of objective lens used.

The senior author has tested four different models of advanced EPMs manufactured by four leading microscope companies: the Zeiss, the Leica, the Nikon, and the Olympus. Basically, these EPMs were made under the same working principle, although the structure of each microscope has some modifications, and the price range varies widely. It is not our intention to evaluate which microscope is the best for EPM observation. Anyone who intends to purchase a set of EPM must balance the budget availability and the level of study required. Besides, a direct personal operation and comparison with different brand microscopes is important.

12.2.2 Type of Specimen

The IGSS method shows enhanced sensitivity as compared with standard immunoperoxidase and immunogold methods. Immunogold-silver staining has become the method of choice among many immunostaining methods for cultured cells, blood smears, cryosections, and semithin resin sections. In contrast with bright LM, which gives the image of GSG as brown-black granules, they look like very bright spots in EPM. The signal of GSG is so strong that it can be viewed with the image produced by transmission light to demonstrate the tissue profile simultaneously.

12.2.2.1 Blood Smear and Cultured Cells

De Waele et al. (1986, 1988, 1989) examined the potential of IGSS for the study of leukocyte subpopulations with monoclonal antibodies using EPM. The signal was strong enough to be seen against a background of transmitted light. Consequently, it was possible to visualize the May-Grunwald-Giemsa-stained cells and to make an accurate cell identification. The highest dilution of the monoclonal antibody for detecting all OKT3-positive cells in the cell suspension was 1:64 for bright LM, and 1:256 for dark field LM and EPM. The latter was eightfold greater than the highest dilution (1:32) for immunofluorescence microscopy (De Waele et al., 1988). The IGSS method was also used to quantify T- and B-lymphocytes and natural killer cells in buffy coated smears of normal adult blood. These lymphocytes subsets correlated well with those obtained in smears with the alkaline phosphatase-antialkaline phosphatase (APAAP) method (De Waele et al., 1989).

Hagèt et al. (1992) have used the IGSS-EPM method to study tissue type plasminogen activator (t-PA) (a protease that converts the plasma zymogen plasminogen into plasmin; the latter can degrade fibrin clots and extracellular matrix protein) in human mesangial cells. The kidney glomerulus contains three different cell types. Endothelial and mesangial cells are closely associated at the inner surface of the glomerular basement membrane, whereas visceral epithelial cells cover its outer shell. Mesangial cells are smooth muscle-like structures that arise from the glomeruli 8 to 14 days after attachment. They showed with IGSS-EPM that matrix-associated plasminogen activator inhibitor 1 (PAI-1; it plays a major role in fibrinolysis) is synthesized by spreading cultured human mesangial cells. The t-PA is present inside the cells, or at the cell surface, but is never associated with the extracellular matrix. Delarue et al. (1991) used the IGSS-EPM for phenotypic characterization of immortalized human glomerular visceral epithelial cells (VEC).

12.2.2.2 Tissue Sections of 5 to 10 μm

Tissue sections of 5 to 10 μm thickness are frequently used in basic research and clinical investigations, especially at the beginning of a project. Either paraffin sections, cryosections, or polyethylene glycol sections can be used for IGSS-EPM detection. However, only low or medium range of magnification, such as 10×, 20×, or 40× objective lens, is recommended, because there will be too many signals in the viewing field of the microscope if a higher power (above 40×) lens is used to observe thick sections. After specific IGSS, keratin, retinol-binding protein in embryonic membranes (Gao and Godkin, 1991; Gao et al., 1991a) and carbohydrate metabolizing enzymes in rat liver have been visualized clearly with bright field microscopy (Giffin et al., 1993), or by EPM (unpublished data).

Wei (personal communication) used cryosections (10 μm) of Bouin's solution-fixed pregnant mouse uterus and the IGSS method for detecting insulin-like growth factor binding protein (IGFBP4) distribution in the uterus and embryo at an early developmental stage. She found that antigenicity of IGFBP-4 was located on the surface of decidual cells that were adjacent to the embryo, but not in the embryo itself in Day 2 through 8 of pregnant mice. The signal of IGFBP-4 shown by GSG was stronger in EPM than in normal light trans-illumination. Godkin and associates observed transforming growth factors (TGFα and TGFβ2) distribution in the epithelial cells of uterus with IGSS and found that EPM was superior to normal LM with trans-illumination for signal detection (unpublished observations). Using the IGSS-EPM method (4 μm cryosections of rat liver), Hamilton et al. (1990) reported that apolipoprotein (apo) E in hepatic parenchyma cells was most intensely stained at the sinusoidal front with punctate deposits in cytoplasm of every hepatocyte.

12.2.2.3 High Resolution Detection with Semithin Sections
High resolution detection can be achieved with semithin sections using IGSS-EPM. The use of immunogold labels in combination with resins (which produce weak staining) for LM has been problematic. Shires et al. (1990) reported that they used LR gold and Lowicryl K4M two hydrophilic resins to study glomerular mesangial antigens. In their experimental conditions, the antigenicity was well preserved in tissue embedded in both resins. Blocks of normal rat kidney (1 mm³) were immersion-fixed for 4 h at 4°C in 2% paraformaldehyde or periodate-lysine-paraformaldehyde (PLP), dehydrated with ethanol and embedded in LR gold or Lowicryl K4M in cold. Sections of 0.5 μm thickness were immunostained with mouse Thy-1.1 monoclonal antibody that identifies a mesangial antigen, followed by 5-nm gold conjugated goat-antimouse IgG, and then silver enhanced. The sections were finally counterstained with eosin and mounted in Eukitt (BDH) containing 2.5% DABCO (Sigma, St. Louis, MO). The slides were viewed under a Leitz Dialux 20 microscope equipped with epipolarized illumination, with a 50/1.00 water immersion objective lens. Antigen was clearly shown in semithin sections embedded in both resins (Figures 1 and 2). K4M sections displayed a phenomenon of self-illumination when counterstained with eosin, which enabled morphological delineation without simultaneous trans-illumination (Figure 2). The pre-embedding IGSS method developed for detecting carbohydrate metabolic-related enzymes by Gao et al. (1993a,b) will be presented in the later part of this chapter.

12.2.2.4 Detection of Silver Granules After In Situ Hybridization
The routinely used approach of *in situ* hybridization for visualizing labeled gene probes involves the autoradiographic technique (Gupta et al., 1985). The silver grains as the final reaction products are usually counted under bright field LM. However, these silver grains can also be visualized by EPM (Ouchit et al., 1993). Jackson et al. (1989) introduced another method for in situ hybridization, in which 4% paraformaldehyde-fixed cell smears or paraffin sections (4 μm) were *in situ* hybridized with DNA probes (HPV 6b and PHY2.1) that had been biotinylated with biotin-11-deoxyuridine triphosphate by nick translation. The hybridized signal was detected by incubations of rabbit antibiotin and gold labeled goat-antirabbit IgG. Finally, the gold was silver-enhanced and detected by EPM. They claimed that the method is rapid, reliable, and economic. The principle may apply to other systems.

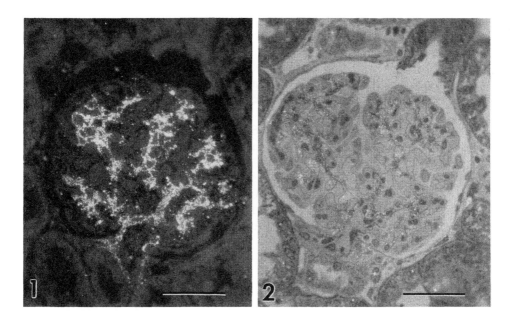

Figure 1 Light micrograph of a PLP-fixed, and LR Gold-embedded glomerular section (0.5 μm), showing the distribution of glomerular mesangial antigen (Thy-1.1) demonstrated by IGSS-EPM (bright dots). Bar = 25 μm. (Magnification ×680.) (From Shires et al., *J. Histochem. Cytochem.*, 38(2), 287, 1990. With permission.) **Figure 2** Light micrograph of a PLP-fixed, and Lowicryl K4M-embedded glomerular section (0.5 μm), showing the distribution of glomerular mesangial antigen (Thy-1.1) demonstrated by IGSS-EPM (bright dots). The eosin counterstain caused a self-illumination phenomenon of the tissue profile which was not seen with LR Gold embedding (Figure 1). Bar = 25 μm. (Magnification ×680.) (From Shires, et al., *J. Histochem. Cytochem.*, 38(2), 287, 1990. Permission granted by Histochemical Society of America.)

12.3 Optimized Condition for High Resolution Observation

Semithin sections are needed for high resolution detection of antigenic sites marked by GSG, since the out-of-focus signals in thick sections are problematic. The IGSS procedure can be performed either directly on semithin resin sections (Shires et al., 1990), or on thick sections which are then reembedded in resin to prepare semithin sections for observation (Gao et al., 1993a). A pre-embedding IGSS method is given below as an example.

12.3.1 Pre-Embedding Immunogold-Silver Staining

Immunostaining of properly fixed and cryosectioned tissues (e.g., rat liver) is carried out in 15 ml plastic slide boxes (Ted Pella, Inc., Redding, CA). For further details the reader is referred to Giffin et al. (1993) and Gao et al. (1993b).

1. Permeabilize the sections with 1% Surfact Amps X-100 (a purified 10% of Triton X-100 from Pierce, Rockford, IL) and 1% NH_4Cl (to quench the aldehyde residues in tissue) in 1/2 strength phosphate-buffered saline (PBS) (pH 7.4) at room temperature for 10 min.
2. Block the sections in a blocking solution, which consists of 5% normal calf serum, 1% bovine serum albumin (BSA), and 0.9% cold fish gelatin (Sigma Chemical Co., St. Louis, MO) in 0.1 M phosphate buffer (PB) (pH 7.4) containing 0.5% Triton X-100, for 1 h at 37°C.

3. Incubate the sections with a primary antibody (e.g., goat antirat PEPCK, or the normal serum from the same species as a control), diluted 1:1000 in blocking solution.
4. Wash the sections 6 times for 5 min each with PBS.
5. Incubate in a gold labeled secondary antibody (5 nm colloidal gold-labeled rabbit-antigoat IgG) diluted 1:5 at required concentration in blocking solution, for 2 h at room temperature.
6. Wash 6 times for 5 min each with PBS.
7. Immobilize the GSG with 2.5% glutaraldehyde for 10 min at room temperature, followed by washing in distilled water (3 changes for 5 min each).
8. Silver-enhance using freshly prepared silver-enhancement solution (Silver Enhancer Kit, Sigma Chemical Co., St Louis, MO, or any other brands available) at 15°C for 12 to 18 min, followed by washing with distilled water, treatment with 2.5% Na-thiosulfate for 2 min, and wash with distilled water.

The specimens are further processed for semithin sections as follows.

12.3.2 *Preparation of Semithin Sections from Thick Sections Treated with IGSS*

1. The pre-embedding immunogold-stained sections are counterstained with 0.3% pyronine Y in water for 15 min or longer for easy visualization with naked eyes.
2. Dehydration is carried out through an ascending ethanol series.
3. Infiltration and embedding are accomplished in a Visio-Bond (ESPE-Fabrik Pharmazeutischer Preparate, GmBH & Co, Oberbay, Germany) based resin (Gao and Peng, 1987). A "light pistol" (ESPE-Fabrik Pharmazeutischer Preparate, GmBH & Co., Oberbay, Germany) is used to produce a blue light illumination that activates the solidification of the resin, which cures in 20 to 40 sec, while activated by a blue light. Other resins (such as Epon 812, etc.) can also be used for the same purpose; however, none of them can cure as fast as Visio-Bond does.
4. The IGSS and Visio-Bond embedded thick sections are viewed under a dissecting binocular, and the selected area of interest is cut with a scalpel and then mounted with Visio-Bond, or any super glue, onto the top of a blank resin block. Semithin sections (0.35 to 1.0 µm) perpendicular (while the thick section is mounted in a narrow furrow on the top of the blank block), or oblique, or level to the original plane of thick section are cut with a diamond or sapphire knife on an ultramicrotome.
5. The resulting semithin sections are collected with a fine hair-loop from the trough and transferred onto a clean glass slide, and dried over a 60 to 80°C hot plate.
6. After drying and cooling, semithin sections are coverslip-mounted in Visio-Bond for EPM detection (such as using a Zeiss Axoscope or a Nikon microphot FX microscope equipped with IGS block). Color image is recorded by microphotography (e.g., using 35 mm Kodak Ektachrome tungsten T160 positive color slides or Fuji Reala 100 color print negative film).

12.3.3 *Simultaneous Visualization of Antigenic Sites and Tissue Profiles*

A pre-embedding IGSS of 10 µm cryosections of 4% paraformaldehyde perfusion-fixed normal male rat liver is used as a model for localization of phosphoenolpyruvate carboxykinase, a rate limiting enzyme of carbohydrate metabolism. The immunostained section is counterstained with pyronin Y and reembedded in Visio-Bond. The semithin sections are then made and coverslip-mounted. A Nikon microphot-FXA microscope

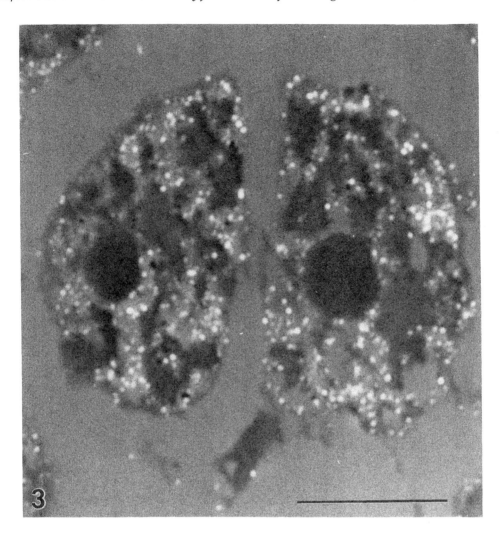

Figure 3 Hepatic phosphoenol pyruvate carboxykinase (PEPCK) antigen sites (white dots) in pyronine Y stained 0.5 μm section of 10 μm IGSS cryosection and viewed by EPM with a combination of low intensity of transmitted light. Bar = 1 μm. (Magnification ×4085.) (From Gao, et al., *Proc. 51st Annual Meeting of the Microscopy Society of America*, Bailey, G. W. and Rieder, C. L. [Eds.], San Francisco Press, 1993. Permission granted by Microscopy Society of America.)

equipped with an IGS polarizing filter block in an EPI-FL3 house (Nikon Corporation, Tokyo, Japan) is used as EPM. The polarized light is produced by an HBO-100 W mercury lamp. A Fluor X100/1.3 oil Ph 4D immersion objective lens with an adjustable iris (0.5 to 1.3) is used for high magnification. The images are recorded on Kodak Ecktachrome 160 T positive color slides with a Nikon FX-35DX camera, setting the control panel at S (standard). The resolution and the sensitivity of detection are satisfied for the detection of GSG in 0.5 μm semithin sections.

When epipolarized illumination is combined with transmitted light, the image of antigenic sites produced by epipolarized light is very bright green spots, and the tissue counterstaining (rose-red) is shown by transmitted light illumination. The two images are demonstrated simultaneously (Figure 3). In order to optimize the condition, the diaphragm in the 100X oil immersion objective lens has to be properly adjusted to match with different intensity settings of transmitted light. If the intensity of epipolarized light is too high, halos appear all around GSGs; whereas if the intensity of the transmitted light is adjusted higher, the signal of GSG decreases gradually, and finally disappears. By adjusting the aperture in the objective lens and the neutral density filters in the transmitted light pathway, a combination of transmitted and epipolarized lights can form. Different setting conditions are recorded on film for comparison. The best image is achieved which represents the maximum amount for antigenic sites with properly balanced tissue profile (Gao et al., 1994).

12.3.4 High Power Objective Lens with Adjustable Iris

The importance of an adjustable iris has been described in the previous section. The high intensity epi-illumination may cause a loss of sharpness of GSG due to the surrounded halos, and even secondary and complex halos (formed by the adjacent GSGs), which may be as a result of multiple reflections of the light scattered from GSG surface between the glass slide and the coverslip. In such a circumstance, the use of adjustable objective is helpful.

12.3.5 Other Options

Color filters offer the advantage of producing micrographs with better contrast. A blue filter or an orange filter in the transmission light path, or a blue filter plus a phase contrast can change the background color. Other color filters may also be inserted in the epipolarized light path to alter the color of GSG marker. Video image enhancement or pseudo color processing can also be used to improve the life image obtained from the semithin section in the EPM after IGSS (Gao and Cardell, 1994).

12.4 Conclusion

After the development and extensive practice in the past decade, the IGSS method, with its improved sensitivity and reproducibility, has become a mature approach for immunohisto-cytochemistry. The IGSS method is not restricted to antigenic site localization in tissues, but also extends to *in situ* hybridization for gene and its products. This means that a new important area related to the modern genetic engineering has opened a door to IGSS-EPM. As a consequence, it will increase the number of users of the method. Furthermore, IGSS-EPM provides a way to view the marker and the tissue profile simultaneously; fluorescence microscopy cannot obtain such information. The detection of GSG in EPM using semithin section after pre-embedding IGSS is of high resolution. The magnification of enlarged microphotograph thus obtained is about 4000× (Figure 3). The high resolution plus high sensitivity of IGSS-EPM detection bridges the gap between LM and EM. It becomes possible to compare subcellular localization of antigenic sites with both LM and EM. We hope that the IGSS-EPM method will be tried by more immunocytochemists, who will both get benefit from and make new contribution to that method. Colloidal gold today will be the IGSS-EPM tomorrow.

Acknowledgments

Thanks to Dr. Gerard Morel, Department of Neuroendocrinology at the University of Lyon, France, for mailing to me, and permitting to be used in this chapter, an unpublished photo of EPM, which shows the silver granules after *in situ* hybridization (Figure 4; for more information refer to Ouchit et al., 1993).

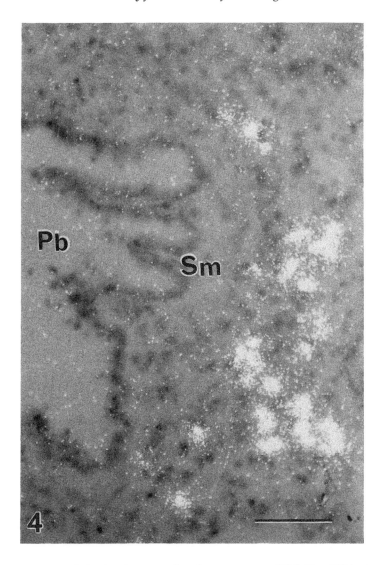

Figure 4 Regional and cellular localization of prolactin receptor (PRL-R) mRNAs in rat lung after *in situ* hybridization with ^{35}S-labeled oligoprobes and autoradiography and viewed with EPM. The intensity of silver granules of PRL-Rs mRNA was highest in cells that surround smooth muscle cells (Sm) of pulmonary bronchus (Pb). Bar = 1 μm. (Magnification ×200.) (Courtesy of Dr. Gerard Morel, Neuroendocrinology, University of Lyon I. Cedex France.)

References

Danscher, G. and Nörgaard, J. O. R. 1983. Light microscopic visualization colloidal gold on resin-embedded tissue. *J. Histochem. Cytochem.* 31: 1394.

Delarue, F., Virone, A., Hagege, J., Lacave, R., Peraldi, M. N., Adida, C., Rondeau, E., Feunteun, J., and Sraer, J. D. 1991. Stable cell line of T-SV 40 immortalized human glomerular visceral epithelial cells. *Kidney Int.* 40: 906.

De Waele, M., De Mey, J., Renmans, W., Labeur, C., Jochmans, K., and Van Camp, B. 1986. Potential of immunogold-silver staining for the study of leukocyte subpopulations as defined by monoclonal antibodies. *J. Histochem. Cytochem.* 34: 1257.

De Waele, M., Renmans, W., Sergers, E., Jochmans, K., and Van Camp, B. 1988. Sensitive detection of immunogold-silver staining with dark field and epipolarization microscopy. *J. Histochem. Cytochem.* 36: 679.

De Waele, M., Renmans, W., Segers, E., De Valck, V., Jochmans, K., and Van Camp, B. 1989. An immunogold-silver staining method for detection of cell surface antigens in cell smear. *J. Histochem. Cytochem.* 37: 1855.

Faulk, W. P. and Taylor, G. M. 1971. An immunocolloid method for the electron microscope. *Immunochemistry* 8: 1081.

Frens, G., 1973. Controlled nucleation for the regulation of the particle size in nomodisperse gold suspensions. *Nature (London) Phys. Sci.* 241: 20.

Gao, K. X. and Huang, L. 1987. Preparation of colloidal gold-labeled agarose-gelatin microspherules for electron microscopic studies of phagocytosis in cultured cells. *J. Histochem. Cytochem.* 35: 163.

Gao, K. X. and Peng, B. 1987. Visio-Bond as an embedding medium for light and electron microscopy. *J. Cell Biol.* 105: 226a (1278).

Gao, K. X. and Godkin, J. D. 1991. A new method for transfer of polyethylene glycol-embedded tissue sections to silanated slides for immunocytochemistry. *J. Histochem. Cytochem.* 39: 537.

Gao, K. X. and Cardell, E. L. 1994. Pseudocolor image processing of PEPCK subcellular distribution in rat hepatocytes shown with IGSS and epipolarization microscopy. *J. Histochem. Cytochem.* 42: 1651.

Gao, K. X., Liu, K. H., and Godkin, J. D. 1991a. Immunohistochemical localization of bovine placental retinol-binding protein. *Int. J. Dev. Biol.* 35: 485.

Gao, K. X., Smith, S. E., and Godkin, J. D. 1991b. Phagocytosis of protein coated colloidal-gold-agarose-gelatin microbeads by cultured uterine glandular epithelial and stromal cells. *Biotechnic Histochem.* 1: 1.

Gao, K. X., Morris, R. E., Giffin, B. F., Cardell, E. L., and Cardell, R. R. 1993a. Immunogold-silver staining and epipolarized microscopic detection of phosphoenolpyruvate carboxykinase and glycogen phosphorylase in rat liver. *Histochemistry* 99: 341.

Gao, K. X., Morris, R. E., Giffin, B. F., Cardell, E. L., and Cardell, R. R. 1993b. Immunogold electron microscopic localization of phosphoenolpyruvate carboxykinase in rat liver. In: *Proc. 51st Annual Meeting of the Microscopy Society of America.* Bailey, G. W. and Rieder, C. L. (Eds.), San Francisco Press.

Gao, K. X., Giffin, B. F., Morris, R. E., Cardell, E. L., and Cardell, R. R. 1994. Optimized condition for epipolarization microscopic detection of immunogold silver staining of PEPCK in rat liver. *J. Histochem. Cytochem.* 42: 823.

Giffin, B. F., Gao, K. X., Morris, R. E., and Cardell, R. R. 1993. Enhancement of antigenic site detection with gold labeled secondary and tertiary antibodies using the immunogold-silver staining method. *Biotech. Histochem.* 68: 309.

Geoghegan, W. D., Scillian, J. J., and Ackerman, G. A. 1978. The detection of human B-lymphocytes by both light and electron microscopy utilizing colloidal gold-labeled anti-immuno-globulin. *Immunol. Commun.* 7: 1.

Gupta, J., Gendelman, H. E., Naghashfar, Z., Gupta, P., Rosenshein, N., Sawad, E., Woodruff, D., and Shah, K. 1985. Specific identification of human papillomavirus type in cell smears and paraffin sections by in situ hybridisation with radioactive probes: a preliminary communication. *Int. J. Gynecol. Pathol.* 4: 311.

Hamilton, R. L., Wang, J. S., Guo, L. S. S., Krisans, S., and Havel, R. J. 1990. Apolipoprotein E localization in rat hepatocytes by immunogold labeling of cryothin sections. *J. Lipid. Res.* 31: 1589.

Handley, D. A. 1989. The development and application of colloidal gold as a microscopic probe. In: *Colloidal Gold: Principles Methods, and Applications.* pp. 1–12. Hayat, M. A. (Ed.), Academic Press, San Diego.

Hagège, J., Peraldi, M. N., Rondeau, E., Adida, A., Delarue, F., Medcalf, R., Schleuning, D. W., and Srear, J. D. 1992. Plasminogen activator inhibitor-1 deposition in the extracellular matrix of cultured human mesangial cells. *Am. J. Pathol.* 141: 117.

Holgate, C. S., Jackson, P., Cowen, P. N., and Bird, C. C. 1983. Immunogold-silver Staining: new method for immunostaining with enhanced sensitivity. *J. Histochem. Cytochem.* 37: 938.

Horisberger, M. 1979. Evaluation of colloidal gold as a cytochemical marker for transmission and scanning electron microscopy. *Biol. Cell.* 36: 253.

Jackson, P., Lewis, F. A., and Wells, M. 1989. *In situ* hybridization technique using an immunogold silver stain system. *Histochem. J.* 21: 425.

Mahdihassan, S. 1984. Tan, cinnabar, as drug of longevity prior to alchemy. *Am. J. Chin. Med.* 12: 50.

Ouchit, A., Morel, G., and Kelly, P. A. 1993. Visualization of gene expression of short and long forms of prolectin receptor in rat. *Endocrinology* 113: 135.

Roth, J. 1982. Applications of immunocolloids in light microscopy. *J. Histochem. Cytochem.* 30: 691.

Shires, M., Goode, N. P., Crellin, D. M., and Davison, A. M. 1990. Immunogold-silver staining of mesangial antigen in Lowicryl K4M- and LR gold-embedded renal tissue using epipolarization microscopy. *J. Histochem. Cytochem.* 38: 287.

Weiser, H. B. 1949. *A Textbook of Colloid Chemistry.* pp. 1–444. Wiley and Sons, New York.

Chapter 13

Application of Gold Toning to Immunogold-Silver Staining

Ryohachi Arai and Ikuko Nagatsu

Contents

13.1 Introduction

The immunogold silver staining method employs immunoglobulin adsorbed to colloidal gold as a secondary antibody (Holgate et al., 1983). The gold particles that are then localized at antigenic sites are revealed by silver staining. This method has the following advantages: (1) it is very sensitive, (2) there is no endogenous peroxidase interference, and (3) handling of reagents is safe because no hazardous reagents are involved. The most notable successes of the immunogold silver staining method in studies of the central nervous system have been in the localization of large molecules, such as tyrosine hydroxylase (van den Pol, 1985, 1986; Chan et al., 1990), choline acetyltransferase (Chan-Palay, 1988), somatostatin (Chan-Palay, 1987), galanin (Chan-Palay, 1988), and parvalbumin (Chang and Kita, 1992).

0-8493-2449-1/95/$0.00+$.50

Gold toning causes the replacement of metallic silver precipitated in tissue by metallic gold (Feigin and Naoumenko, 1976). The gold toning procedure has been previously used (1) in ultrastructural studies of silver-impregnated neurons to transform the silver deposits to fine gold particles (Fairén et al., 1977), (2) in retrograde tracing and immunohistochemical studies at electron microscopic levels to stabilize the deposits of silver staining of peroxidase-diaminobenzidine reaction products (Liposits et al., 1982, 1984), and (3) in retrograde tracing with colloidal-gold-conjugated tracers which is revealed by silver staining, to protect against loss of the silver precipitates during the osmication that is required for electron microscopy (Basbaum and Menétrey, 1987).

We have recently applied the gold toning procedure to the immunogold-silver staining method and detected a small molecule, dopamine, in the rat brain (Arai et al., 1992). The purpose of the present study is to examine the influence of gold toning on immunogold-silver staining at light and electron microscopic levels. Here we detected vasopressin immunoreactivity in the supraoptic nucleus of the rat, using two methods; (1) the single immunogold-silver staining and (2) the combination of immunogold silver staining with gold toning, and then morphologically compared labels of the two methods. The animals were injected with colchicine to increase the content of the neurosecretory material in the cell bodies (Norström et al., 1971; Parish et al., 1981).

13.2 Methods

Male Sprague-Dawley rats (250 to 300 g) were used. In order to increase the content of vasopressin in cells of the supraoptic nucleus, under anesthesia with sodium pentobarbital (60 mg/kg, intraperitoneally), 100 µg of colchicine (Sigma) was injected in the lateral ventricle. The animals were allowed to survive for 48 h. Immunogold-silver staining and gold toning procedures were performed based on the previous studies (Fairén et al., 1976; Arai et al., 1992).

13.2.1 Tissue Preparation

Under anesthesia, the animals were perfused through the ascending aorta with 50 ml of 0.01 M phosphate-buffered saline (PBS) (pH 7.4, room temperature), followed by 300 ml of a fixative containing 4% paraformaldehyde, 0.3% glutaraldehyde, and 0.2% picric acid in 0.1 M phosphate buffer (pH 7.4, 4°C). Brains were dissected out and placed in a fixative containing 4% paraformaldehyde and 0.2% picric acid in 0.1 M phosphate buffer for 24 h at 4°C, and then immersed in 0.1 M phosphate buffer containing 15% sucrose for 24 h at 4°C. Sections of the hypothalamus including the supraoptic nucleus were cut on a cryostat in a frontal plane at 40 µm and collected in 0.1 M PBS (pH 7.4).

13.2.2 Single Immunogold-Silver Staining

For penetration of immunoreagents, free-floating sections were first permeabilized with 0.1% trypsin in PBS for 5 min at room temperature. The sections were then incubated with: (1) 5% normal goat serum in PBS for 1 h at room temperature, (2) guinea pig-antivasopressin antiserum (GAS8103, Peninsula Lab.; diluted 1:30,000 in PBS) for 48 h at 4°C, and (3) 1-nm colloidal gold bound to goat-antiguinea pig immunoglobulin G (GAG1, BioCell Lab., diluted 1:100 in PBS) for 2 h at room temperature. After each step, the sections were rinsed with PBS. The sections were then incubated in a silver staining solution (Intense M, Amersham) for 20 min at 20°C. Before and after this step, the sections were rinsed with distilled water. Some of the stained sections were photographed under an Olympus BH-2 light microscope (Figure 1A), and others were further osmicated.

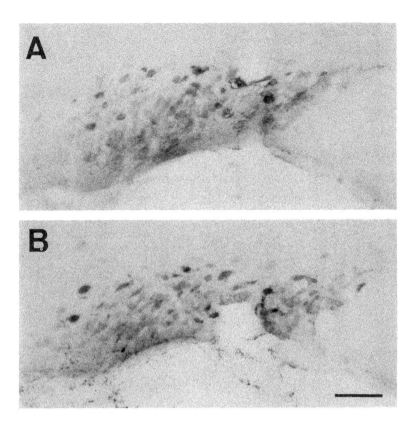

Figure 1 Photomicrograph showing vasopressin-immunoreactive neurons in the supraoptic nucleus. (A) The section was processed by the single immunogold-silver staining. (B) The section was processed by the combination of immunogold-silver staining with gold toning. Bar = 100 µm.

13.2.3 Combination of Immunogold-Silver Staining with Gold Toning

Sections were first processed by the immunogold-silver staining as described above. For gold toning, the sections were then treated as follows: (1) 0.05% gold chloride for 20 min at 4°C, (2) 0.05% oxalic acid for 2 min at 4°C, and (3) 1% sodium thiosulfate for 1 h at room temperature. After each step, the sections were rinsed in distilled water. Some of the stained sections were photographed (Figure 1B), and others were then osmicated.

13.2.4 Osmication

Sections stained by the two methods were further osmicated (1% osmium tetroxide in 0.1 M phosphate buffer, pH 7.4) for 1 h at 4°C, dehydrated in graded alcohols, and flat-embedded in Epon. These sections were photographed (Figure 2) and processed for electron microscopy.

13.2.5 Electron Microscopy

Small areas containing the supraoptic nucleus were trimmed from the osmicated and Epon-embedded sections, and cut on a Reichert ultramicrotome. Thin sections were collected on grids, stained with uranyl acetate, and examined under a Philips EM 300 electron microscope (Figures 3 and 4).

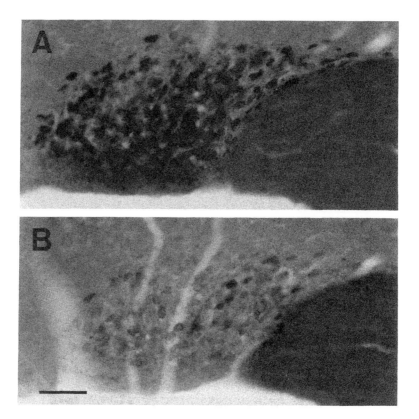

Figure 2   Photomicrograph showing vasopressin-immunoreactive neurons in the supraoptic nucleus. (A) The section was processed by the single immunogold-silver staining, and osmicated. (B) The section was processed by the combination of immunogold-silver staining with gold toning, and osmicated. Bar = 100 µm.

13.3 Results

13.3.1 Light Microscopy

With the single immunogold-silver staining, the immunolabeled vasopressin-containing cells appeared brown in color (Figure 1A). The background was clear. By using the combination of immunogold-silver staining with gold toning, the immunoreactive cells showed blue color (Figure 1B). The background was still clear. Labels of the latter method were somewhat more transparent than those of the former. No significant differences in the staining intensity were detected between the two methods. The distribution of vaso-pressin-containing cells in the supraoptic nucleus revealed by the two methods was consistent with previous studies (Rhodes et al., 1981; Hou-Yu et al., 1986).

After osmication, labels of the single immunogold-silver staining appeared to be darkened (Figures 1A and 2A). For labels of the combined method, significant difference was not observed in color and staining intensity before or after osmication (Figures 1B and 2B).

13.3.2 Electron Microscopy

With the single immunogold-silver staining, the labels appeared as highly electron-dense particles and were easily detected at a low-power magnification (Figure 3A). They were

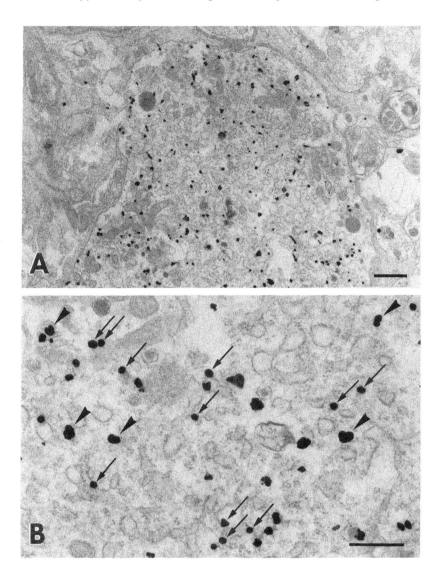

Figure 3   Electron micrograph showing a vasopressin-immunoreactive dendrite in the supraoptic nucleus. The section was processed by the single immunogold-silver staining, and osmicated. (A) Low magnification. The labels were distributed throughout the cytoplasm of the dendrite. (B) Higher magnification. The particles were generally round in shape and commonly 60 to 80 nm in diameter (arrows). A number of particles appeared to adhere to one another (arrowheads). Bars = 1 μm (A) and 0.5 μm (B).

distributed throughout the cytoplasm of the perikaryon and dendrite. The particles were generally round in shape and commonly 60 to 80 nm in diameter (Figure 3B). A number of particles appeared to adhere to one another.

By the combined method, the labels were also highly electron-dense and easily detected at a low-power magnification (Figure 4A). They were distributed throughout the cytoplasm of the perikaryon and dendrite. At a high-power magnification, the label observed at the low magnification came in view to be an aggregation of fine grains (Figure 4B). The grains were less than 15 nm in diameter. The aggregations of grains were commonly 80 to 130 nm long and 60 to 80 nm wide.

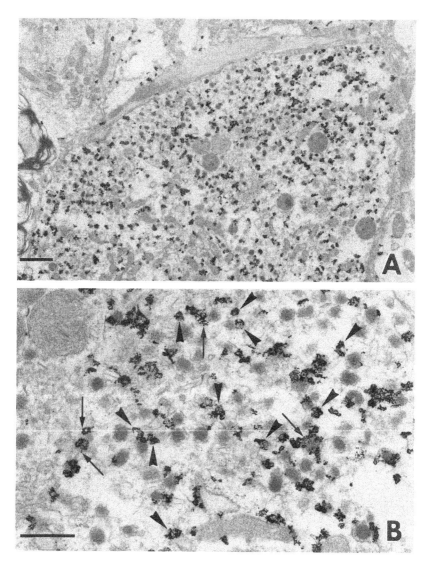

Figure 4 Electron micrograph showing a vasopressin-immunoreactive perikaryon in the supraoptic nucleus. The section was processed by the combination of immunogold-silver staining with gold toning, and osmicated. (A) Low magnification. The labels were distributed throughout the cytoplasm of the perikaryon. (B) Higher magnification. The labels were constituted by aggregations (arrowheads) of fine grains (arrows). The grains were less than 15 nm in diameter. The aggregations of grains were commonly 80 to 130 nm long and 60 to 80 nm wide. Bars = 1 µm (A) and 0.5 µm (B).

13.4 Discussion

13.4.1 Effect of Gold Toning on Immunogold-Silver Staining: Light Microscopy

The effect of gold toning on immunogold-silver staining is estimated by comparing the labels of the two methods: the single immunogold-silver staining and the combination of immunogold-silver staining with gold toning. The conspicuous effect of gold toning is to turn the color of the silver products blue. Label of the gold toning would be distinct from a brown product of the peroxidase-diaminobenzidine reaction.

The immunogold-silver staining has been previously used for double labeling with the immunoperoxidase method (van den Pol, 1985, 1986; Chan-Palay, 1987, 1988; Chan et al., 1990; Chang and Kita, 1992). Color of labels of the immunogold silver staining appears to depend on the amount of silver deposits as well as on quantity of the antigen and immunoreactivity of the antibody. Short incubation times in the silver staining solution resulted in a brown product, whereas long incubation times provide a black product (Chan et al., 1990). In the present study, silver staining for 20 min resulted in a brown product. However, longer incubation would cause self-nucleation of silver precipitate and result in an unacceptable background (product note of IntenSE M, Amersham). Therefore, gold toning would be useful when one needs to change the color of labels of the immunogold-silver staining so that it is distinct from the peroxidase-diaminobenzidine product.

Osmication has reduced, in a previous study, the number of labeled neurons that were revealed by silver staining of the gold particles in retrograde tracers with a physical developer. The loss is due to oxidation of the silver precipitate to silver salt which is lost in subsequent washing steps (Basbaum and Menétrey, 1987; Basbaum, 1989). However, we used IntenSE M as a silver staining solution; the labels of the silver staining were not faded by osmication but intensified. Chan et al. (1990) also applied the IntenSE M and detected no notable loss of the silver product after osmication. It is likely that loss of the silver product after osmication may depend on the kind of silver staining solution used. As regarding labels of the combination of immunogold-silver staining with gold toning, we did not detect any change in color and intensity after osmication. As Basbaum and Menétrey (1987) described, the gold product is stable to osmication.

13.4.2 Effect of Gold Toning on Immunogold-Silver Staining: Electron Microscopy

Labels of the single immunogold-silver staining are particles that are generally round in shape. It seems that one particle reflects one antigenic site. In contrast, the combined method provides the aggregation of fine grains as its label. Although the aggregation of the combined method is somewhat larger than the particle of the single method, it is our impression that the aggregation corresponds to the particle. It would be possible that gold toning might transform the particle of the immunogold-silver staining to the aggregation of fine grains.

The combination of immunogold-silver staining with gold toning provides the label that is ultrastructurally distinct from the label of the single immunogold-silver staining. For ultrastructural double-labeling, future studies will examine the applicability of this combined method to the single immunogold-silver staining or to the immunoperoxidase method.

13.5 Summary

We examined the effect of gold toning on immunogold-silver staining by comparing labels of the two methods: the single immunogold-silver staining and the combination of immunogold-silver staining with gold toning. At the light microscopic level, the gold toning changes the color of the silver product. At the electron microscopic level, the gold toning might transform the particle of immunogold-silver staining to the aggregation of fine grains.

Acknowledgment

We thank Professor Toshihiro Maeda for his constructive suggestions.

References

Arai, R., Geffard, M., and Calas, A. 1992. Intensification of labelings of the immunogold silver staining method by gold toning. *Brain Res. Bull.* 28: 343.

Basbaum, A. I. 1989. A rapid and simple silver enhancement procedure for ultrastructural localization of the retrograde tracer WGAapoHRP-Au and its use in double-label studies with post-embedding immunocytochemistry. *J. Histochem. Cytochem.* 37: 1811.

Basbaum, A. I. and Menétrey, D. 1987. Wheat germ agglutinin-apoHRP gold: a new retrograde tracer for light- and electron-microscopic single- and double-label studies. *J. Comp. Neurol.* 261: 306.

Chan, J., Aoki, C., and Pickel, V. M. 1990. Optimization of differential immunogold-silver and peroxidase labeling with maintenance of ultrastructure in brain sections before plastic embedding. *J. Neurosci. Method.* 33: 113.

Chan-Palay, V. 1987. Somatostatin immunoreactive neurons in the human hippocampus and cortex shown by immunogold/silver intensification on vibratome sections: coexistence with neuropeptide Y neurons, and effects in Alzheimer-type dementia. *J. Comp. Neurol.* 260: 201.

Chan-Palay, V. 1988. Galanin hyperinnervates surviving neurons of the human basal nucleus of Meynert in dementias of Alzheimer's and Parkinson's disease: a hypothesis for the role of galanin in accentuating cholinergic dysfunction in dementia. *J. Comp. Neurol.* 273: 543.

Chang, H. T. and Kita, H. 1992. Interneurons in the rat striatum: relationships between parvalbumin neurons and cholinergic neurons. *Brain Res.* 574: 307.

Fairén, A., Peters, A., and Saldanha, J. 1977. A new procedure for examining Golgi impregnated neurons by light and electron microscopy. *J. Neurocytol.* 6: 311.

Feigin, I. and Naoumenko, J. 1976. Some chemical principles applicable to some silver and gold staining methods for neuropathological studies. *J. Neuropathol. Exp. Neurol.* 35: 495.

Holgate, C. S., Jackson, P., Cowen, P. N., and Bird, C. C. 1983. Immunogold-silver staining: new method of immunostaining with enhanced sensitivity. *J. Histochem. Cytochem.* 31: 938.

Hou-Yu, A., Lamme, A. T., Zimmerman, E. A., and Silverman, A.-J., 1986. Comparative distribution of vasopressin and oxytocin neurons in the rat brain using a double-label procedure. *Neuroendocrinology* 44: 235.

Liposits, Z., Görcs, T., Gallyas, F., Kosaras, B., and Sétáló. G. 1982. Improvement of the electron microscopic detection of peroxidase activity by means of the silver intensification of the diaminobenzidine reaction in the rat nervous system. *Neurosci. Lett.* 31: 7.

Liposits, Z., Sétáló, G., and Flerkó, B. 1984. Application of the silver-gold intensified 3,3'-diaminobenzidine chromogen to the light and electron microscopic detection of the luteinizing hormone-releasing hormone system of the rat brain. *Neuroscience* 13: 513.

Norström, A., Hansson, H.-A., and Sjöstrand, J. 1971. Effects of colchicine on axonal transport and ultrastructure of the hypothalamo-neurohypophyseal system of the rat. *Z. Zellforsch.* 113: 271.

Parish, D. C., Rodriguez, E. M., Birkett, S. D., and Pickering, B. T. 1981. Effects of small doses of colchicine on the components of the hypothalamo-neurohypophyseal system of the rat. *Cell Tiss. Res.* 220: 809.

Rhodes, C. H., Morrell, J. L., and Pfaff, D. W. 1981. Immunohistochemical analysis of magnocellular elements in rat hypothalamus: distribution and numbers of neurophysin, oxytocin and vasopressin containing cells. *J. Comp. Neurol.* 198: 45.

van den Pol, A. N. 1985. Silver-intensified gold and peroxidase as dual ultrastructural immunolabels for pre- and postsynaptic neurotransmitters. *Science* 228: 332.

van den Pol, A. N. 1986. Tyrosine hydroxylase immunoreactive neurons throughout the hypothalamus receive glutamate decarboxylase immunoreactive synapses: a double pre-embedding immunocytochemical study with particulate silver and HRP. *J. Neurosci.* 6: 877.

Chapter 14

Pre-Embedding Immunocytochemistry with Silver-Enhanced Small Gold Particles

Richard W. Burry

Contents

14.1 Introduction

Pre-embedding electron microscope (EM) immunocytochemistry requires a system of labeling that will take advantage of a high level of resolution, while also maintaining morphological details of cellular organelles. The use of horseradish peroxidase as a label for EM immunocytochemistry (Nakane and Pierce, 1967) has been popular, but this method has not given discrete labeling that takes advantage of the high resolution in the EM. The use of colloidal gold particles is a major advance in EM immunocytochemistry because these particles are small, discrete, and highly electron dense (Faulk and Taylor,

1971). The use of colloidal gold particles in pre-embedding preparations is most important for identifying cell surface proteins in the TEM and the SEM (Romano et al., 1974; Horisberger and Rosset, 1977). In addition, colloidal gold has been used as an extracellular tracer since without detergent, it is too large to penetrate cell membranes.

For pre-embedding immunocytochemistry, the large size of colloidal gold particles prevents their penetration through membrane barriers into tissue sections. The smallest colloidal gold particles are ~5 nm in diameter (Horisberger and Rosset, 1977). Consequently, they only penetrate cells when harsh detergent treatments are employed, which alter the morphology of plasma membrane and membranes of the cellular organelles. Sections from tissues treated with detergents show that the morphology in the labeled portion of the tissue is of poor quality, and, even though label is present, it is difficult if not impossible to determine which structures are labeled.

Optimal pre-embedding EM immunocytochemistry requires electron dense particulate labeling that can be produced in the tissue with a minimal disruption of tissue morphology. The advent of the 1-nm colloidal gold particles by Jan Leunissen (Leunissen and de Mey, 1989) and subsequently 1.4-nm gold compounds (NanoGold) by Hainfeld (1987, 1988) and Hainfeld and Furuya (1992, in this volume) provided markers that could penetrate into tissue when mild detergent treatment is employed. While the small gold particles solved the problem of penetration, they raised a new problem, the particles were too small to be seen in the resulting epoxy thin sections by TEM.

Silver intensification (silver-enhancement) technique is required in order to visualize small gold particles in thin sections. Silver-enhancement of heavy metal ions had been used previously as a physical development technique for EM autoradiography to obtain small silver grains of uniform shape (Caro and van Tubergen, 1962). In this case, sodium sulfite was used as a buffer to maintain a low pH, p-phenylenediamine was used as the reducing agent, and silver nitrate was used as the source of silver ions. The first silver-enhancement solutions that were used for colloidal gold had similar components to those of the physical developers. One of the first silver-enhancement solutions for colloidal gold is that of Danscher (Danscher, 1981, 1983; Danscher and Norgaard, 1983; Hacker et al., 1988), which is designed for use with sections prepared for light microscopy. In the Danscher solution, sodium citrate is the buffer, hydroquinone is the reducing agent, silver lactate is the source of silver ions, and gum arabic controls the reaction rate. The Danscher silver-enhancement solution is very effective for small gold particles (Lah et al., 1990). However, we determined that the low pH (3.5) of the sodium citrate buffer resulted in damage to the cellular structure such that most membranes were lost (Figure 1a). In experiments with cultured neurons and a primary antibody to synaptophysin, the Danscher solution was used to silver-enhance small gold particles over the synaptic vesicles in the synaptic terminals (Figure 1a, arrows), but it was not possible to identify most of the cellular organelles. Our experiments showed that by using HEPES buffer and increasing the pH to about 6.0, the quality of the morphology was greatly increased which allowed for good localization of gold particles in tissue sections and cell cultures (Lah et al., 1990). Sections prepared from cultures treated with this buffer and silver-enhancement solution showed better morphology (Figure 1b) than cultures prepared with the Danscher silver-enhancement solution (Figure 1a). With the HEPES-buffered silver-enhancement solution, not only are the plasma membranes intact, but the clear details of the rough endoplasmic reticulum are seen (Figure 1b, arrowheads).

We present here new procedures which increase reproducibility and allow far more uniform silver-enhancement. With the newly developed 1.4-nm NanoGold probes (NanoProbes, Inc.) and the silver-enhancement procedure described here, pre-embedding EM immunocytochemistry is easily reproducible (Gilerovitch et al., 1995).

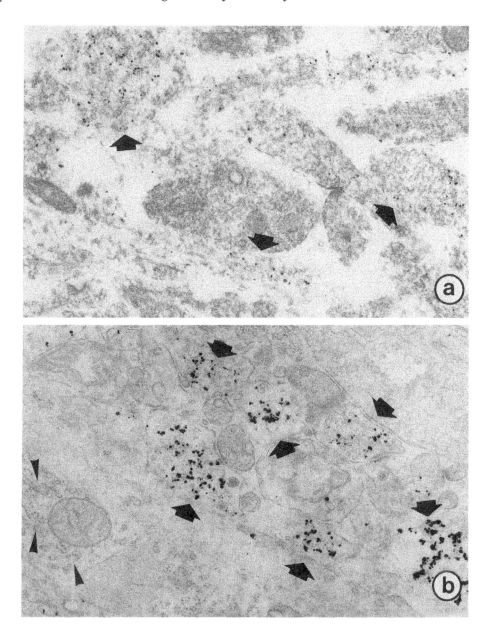

Figure 1   (a) Cultured neurons processed for silver-enhancement with the Danscher solution at pH 3.5, showing the loss of plasma membranes. Cerebellar cultures fixed with 4% paraformaldehyde, then incubated with a primary antibody to synaptophysin (a synaptic vesicle protein), goat-antimouse AuroProbe-One, and silver enhanced with the Danscher solution at pH 3.5. The solid arrows indicate presynaptic terminals with synaptic vesicle labeling with silver-enhanced gold particles. (Magnification ×30,000.) (b) Silver-enhancement with the HEPES-buffered silver-enhancement solution with hydroquinone clearly show the plasma membrane and the improved morphological preservation. Cerebellar cultures fixed with 4% paraformaldehyde, incubated with primary antibody to synaptophysin (a synaptic vesicle protein), goat-antimouse AuroProbe-One, and silver enhanced with HEPES-buffered silver-enhancement solution at a higher pH. Solid arrows indicate presynaptic terminals containing synaptic vesicles labeled with silver-enhanced gold particles. The increased quality of morphological preservation is seen in the detail of the rough endoplasmic reticulum at the arrowheads. (Magnification ×30,000.)

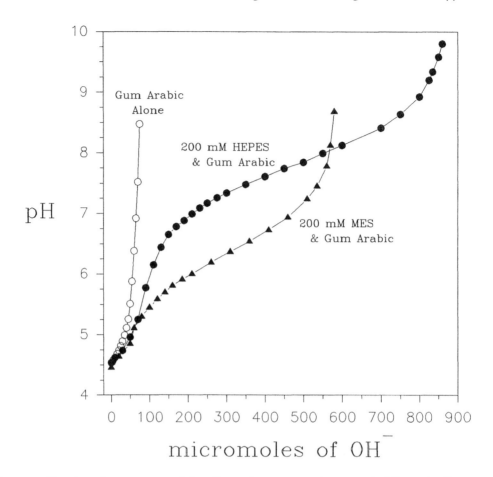

Figure 2 Titration of components of the silver-enhancement solution with different buffer systems shows that MES has greater buffering capacity than HEPES in the pH range of 5.5 to 6.5 range. Titration of a 50% solution of gum arabic showed that it has only slight buffering capacity at pH below 5.0. The 200 mM HEPES with 50% gum arabic solution buffered best from pH 6.8 to 8.2 with an apparent pK of ~7.8. The 200 mM MES with 50% gum arabic solution buffered best from pH 5.5 to 6.8 with an apparent pK of ~6.4. (Reproduced with permission from Gilerovitch et al., *J. Histochem. Cytochem.* 43, 337, 1995.)

14.2 Silver-Enhancement Method

14.2.1 Buffering the Enhancement Solution

The Danscher procedure for silver-enhancement used citrate buffer at pH 3.5 for an extended enhancement time of 30 min. Lah et al. (1990) increased the pH by using organic HEPES buffer (20 mM), which shortened the enhancement time to between 5 and 10 min and led to greatly improved preservation of cellular morphology. However, we have recently determined that the actual pH of the final silver-enhancement solution mixed with 20 mM HEPES (pH 6.8), has a pH of 4.0 (Burry et al., 1992a).

   We titrated components of the silver-enhancement solution to investigate the effect of the different components of the solution on the final pH. In the silver-enhancement solution, the gum arabic is added as a protective colloid to slow the autocatalytic reaction of the silver ions and the reducing agent (Gallyas, 1979; Namork and Heier, 1989; Stierhof et al., 1991). Titration of the gum arabic solution showed that it is an important buffer with a pK below 4.0 (Figure 2, open circles). To bring the pH of the solution close to a final pH of 5.8, and to better regulate the pH of the silver-enhancement solution, 200 mM HEPES (pH 6.8) was added (Burry et al., 1992a). However, titration curve of the gum arabic and HEPES buffer showed that over the pH range of 5.5 to 6.5, HEPES did not have good buffering capacity (Figure 2, solid circles). The variable nature of HEPES buffering capacity at pH 6.0 suggested that other buffers should be examined. We determined that another organic buffer, 200 mM MES, mixed at its pK of 6.15 gave more stable buffering at pH 6.0 (Figure 2, solid triangles). Our new method uses the 200 mM MES, and gives a final pH of 6.0 for the silver-enhancement solution.

14.2.2 Enhancement Reducing Agents

The increase in pH of the silver-enhancement solution, changed the nature of the silver-enhancement and made the technique more variable with respect to time of enhancement and size of the particles. The shape and size of the silver-enhanced particles from the higher pH solution (Figure 1b) are larger and more irregular than those from a lower pH solution (Figure 1a). In an attempt to find a better reducing agent, we examined other compounds with aromatic rings similar to hydroquinone but with different charged side groups. The use of p-phenylenediamine, at a pH above 5.0 even at low concentrations (10% hydroquinone), gives rapid reaction that is difficult to control. N-propyl-gallate (NPG) has a similar molecular structure to hydroquinone but has an additional hydroxyl group. When tested, NPG gives more uniform growth of silver-enhanced gold particles than does hydroquinone (Burry et al., 1992a). We routinely use NPG for silver-enhancement of 1.4-nm gold particles and find it to be more reproducible than the other enhancing reagents we have utilized.

14.2.3 N-Propyl-Gallate (NPG) Silver-Enhancement Method

An important component of the NPG silver-enhancement method is gum arabic. We have shown (Burry et al., 1992a) that without this component silver lactate will be completely converted to metallic silver by the reducing agent NPG in a few seconds, thereby, leaving no silver ions available to be deposited on the small gold particles (Figure 3). The role of the gum arabic has been previously suggested to minimize autocatylic reaction (Stierhof et al., 1991). Our results suggest that gum arabic sequesters some of the reagents and then slowly releases them thus permitting the enhancement of small gold particles to proceed at a more controlled rate.

Because silver-enhancement proceeds via oxidation-reduction reactions, we investigated the possibility that dissolved oxygen in the gum arabic could affect the rate of the reaction. Degassing the gum arabic in a vacuum leads to better consistency of the reaction time. To prepare this solution, gum arabic was dissolved slowly over 2 days without stirring. Contrary to previous reports, the powder gum arabic available (e.g., Sigma) no longer requires filteration through gauze. Next, the gum arabic solution was placed in a large vacuum flask and a mechanical vacuum pump was applied slowly to control the foaming of the gum arabic. The degassed gum arabic was stored in a freezer at –20°C in tightly capped tubes containing enough gum arabic for one use.

To prepare the complete silver-enhancement solution, four components must be mixed immediately before use. We have attempted to premix different combinations of these four components before use and store them at –20°C. None of the possible combinations of these components proved to be stable. The most obvious combination of mixing a solution of gum arabic and NPG gives no reactivity after 2 weeks storage at –20°C, but when additional NPG was added again to this stored mixture, normal silver-enhancement was seen. These results imply that the reducing agent, NPG, was not stable in gum arabic, even at low temperatures.

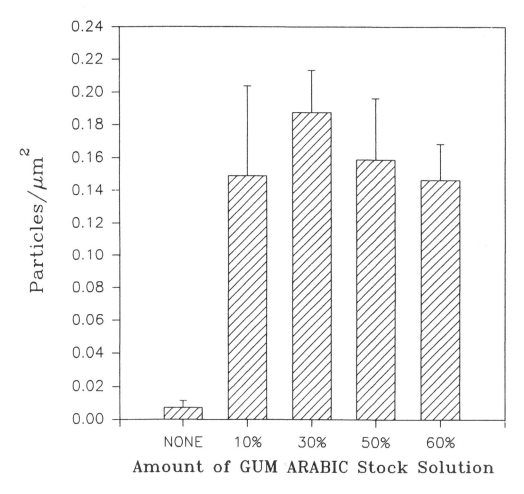

Figure 3 Particle density differences seen with different amounts of gum arabic stock solution in the silver-enhancement solution. With 10 to 60% gum arabic, the silver-enhancement solution gave similar particle densities. Samples were 50-μm vibratome sections of agar mixed with AuroProbe One, and silver enhanced for 15 min with the HEPES solution. Particle density was determined from TEM negatives, with an image analysis computer. (Reproduced with permission from Burry et al., *J. Histochem. Cytochem.*, 40, 1849, 1992a.)

It is useful to test the NPG silver-enhancement solution when new batches of gum arabic solution are made or when the solution is used to ensure that the time for enhancement is correct. Different batches of gum arabic may, for example, contain different amounts of chloride ions that will change the rate of silver-enhancement. To test the silver-enhancement solution we used a strip of nitrocellulose filter paper with 1.4-nm gold dots made from solutions of different dilutions that can be incubated just as tissue sections or cell cultures. To prepare a test strip, antibody labeled with 1.4-nm gold particles are diluted to 1:10, 1:50, 1:100, 1:500, and then a small drop (~0.5 μl) of each are dotted on a nitrocellulose filter. The test strip is run at the same time as the tissue to help determine the correct time for the reaction. The dots of antibody labeled with 1.4-nm gold will turn from faint brown to dark brown during the incubation. The time selected and the density of selected dots on nitrocellulose should be compared to the particle size seen on sections in the TEM. To get 15- to 20-nm particles in the TEM, we found that the best silver-enhancement time in tissue sections will be the time just before the reaction is seen in the section in the light microscope, when the test strip shows a medium brown spot at the 1:50 dilution.

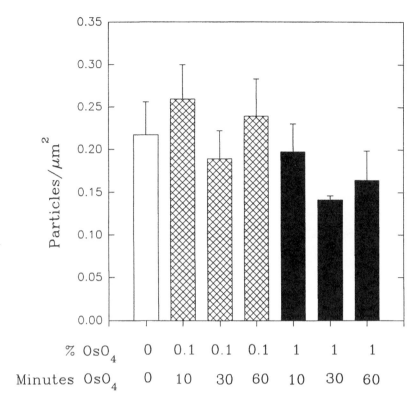

Figure 4 Particle density changes were found after incubation with OsO_4. Agar sections with silver-enhanced gold particles that were treated with 0.1% OsO_4 (cross-hatched bars), or treated with 1% OsO_4 (solid bars) show a slight but not significant reduction in particle density from untreated controls (open bar). The time of OsO_4 treatment did not affect the particle density. Agar sections with AuroProbe One were enhanced with HEPES and NPG for 10 min before OsO_4 treatment. (Reproduced with permission from Burry et al., *J. Histochem. Cytochem.*, 40, 1849, 1992a.)

14.3 Postfixation Use of OsO_4

Following silver-enhancement, cells or tissue sections are postfixed with OsO_4. The use of OsO_4 following silver-enhancement has been questioned, since light microscopic loss of labeling has been reported (Danscher and Norgaard, 1983). Several reports have showed that OsO_4 has not affected silver-enhanced gold particles (van den Pol, 1986; Chan et al., 1990; Burry et al., 1991). We have investigated the use of osmium postfixation using agar hardened with 1.4-nm gold particles that were subsequently silver-enhanced and examined in the TEM (Burry et al., 1992a). Quantitative results showed that there was a slight reduction in the size of silver-enhanced particles following treatment with 1% OsO_4, indicating that this very active oxidizing compound can affect silver-enhanced particles. In test sections treated with 0.1% OsO_4, no reduction in size of silver-enhanced particles was observed (Figure 4). These results suggest that 0.1% OsO_4 should be used in place of 1.0% for postfixation following silver-enhancement. We have not observed any loss of membrane staining due to reduced osmium concentration.

14.4 Technique Differences for Tissue Sections and Cell Cultures

The NPG silver-enhancement solution was initially tested in neuronal cell cultures and gave the previously reported optimal results. However, we observed poor labeling when we began using this procedure applied to vibratome sections from brain tissue. This result

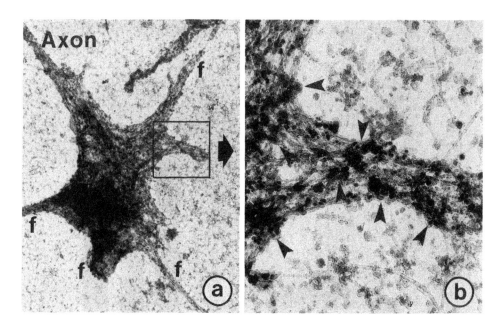

Figure 5 (a) Whole mount of a neuronal growth cone prepared by pre-embedding immunocy-tochemistry with anti-GAP-43, GAM AuroProbe One, and HEPES-buffered hydroquinone silver-enhancement solution. The cells were grown on a Formvar coated grid processed for immunocy-tochemistry, and critical-point dried before examination in a TEM. The axon connecting growth cone to the cell body enters from the top of the micrograph. (f = Individual filopodia.) (b) This micrograph is a higher magnification micrograph of the indicated filopodia in (a). Numerous individual silver-enhanced gold particles are seen with clusters of particles shown by arrowheads. Labeling for the protein GAP-43 is not seen when the detergent is omitted, which indicates that GAP-43 is present on the cytoplasmic side of the plasma membrane and not on the extracellular surface. (Reproduced with permission from Burry et al., *J. Neurocytol.*, 21, 413, 1992b.)

indicated that different conditions were needed for these tissue sections as opposed to cell cultures. The following are recommendations for processing of cell cultures or vibratome sections to be used in pre-embedding EM immunocytochemistry.

14.5 Cell Cultures

The following procedure has been used for cultures of neuronal cell that were labeled either for specific proteins found in synaptic vesicles (synaptophysin) or for proteins found on the cytoplasmic side of the plasma membrane of growth cones (GAP-43). Analysis of thin sections of silver-enhanced cultures has been published (Burry et al., 1992a). In addition, we have used this procedure for whole mount cultures that were critical point dried instead of epoxy embedded (Burry et al., 1992b). In the latter case, stereo-pair analysis of whole mounted growth cones (Figure 5a) showed the distribution of a specific protein, GAP-43, on the cytoplasmic side of the plasma membrane of filopodia (Figure 5a,b,f). The silver-enhanced gold particles are clearly seen in whole mounts in the TEM at 120 kV as individual particles and small clusters (Figure 5b, arrowheads).

14.5.1 Paraformaldehyde Fixation

Paraformaldehyde at 4% is the standard fixative used in most EM immunocytochemistry studies. The addition of glutaraldehyde has been found to reduce the specific labeling by at least two possible mechanisms. First, glutaraldehyde could denature the protein that is to be bound by the primary antibody and prevent antibody binding. This may not be important because of the low concentrations of glutaraldehyde (0.05 to 0.5%) used in EM immunocytochemistry. The second mechanism involves the cross-linking of proteins of the tissue with glutaraldehyde, creating a cross-linked network of chemical polymers that will reduce penetration of antibodies into the tissue (Hopwood, 1972; Hayat, 1989). An additional problem with glutaraldehyde is that many reactive groups are not bound to tissue proteins when fixation is complete, and these reactive groups (free aldehydes) may lead to nonspecific binding of antibodies resulting in high background labeling. This type of background can be eliminated by including reagents in buffers that bind or block aldehyde groups, such as those discussed in the section on blocking buffer in this chapter.

In cell cultures, we have found that the best fixative is 4% paraformaldehyde because it gives good cellular morphology while allowing for specific labeling. Glutaraldehyde was not used because of the increased nonspecific labeling seen at the exposed edges of cultured cells. Most cells in cell culture are directly exposed to the incubation solutions and thus are similar to the edges of a tissue section with high levels of nonspecific labeling. The conditions in cell culture must be optimized for antibody binding of proteins within the first 5 or 10 µm at the surface of the cultured cells. We have examined the distribution of the synaptic vesicle protein, synaptophysin, in cultured neurons with silver-enhanced particles localized over synaptic terminals (Figure 6a, solid arrow).

14.5.2 Saponin Permeabilization

Our experiments have confirmed that for cell cultures the detergent of choice for maintaining cell morphology is saponin (Goldenthal et al., 1985). Our attempts to use Triton-X100 in cell cultures showed that although antibodies did penetrate the cytoplasm of cells, most plasma membranes were lost. On the other hand, saponin leaves the membranes intact by reversibly inserting into the plasma at cholesterol groups (Assa et al., 1973), and allowing antibodies to penetrate into the cytoplasm. Saponin introduces holes in plasma membranes without damaging the morphology of the membranes seen in the TEM (Seeman, 1967; Bohn, 1978; Willingham et al., 1980). One caveat in the use of saponin is that it must be present continuously in all solutions, because it will be washed out of membranes in buffers without saponin. To allow antibody penetration, saponin must be present in the blocking buffer, incubation buffer, and in all rinse buffers. Thus, saponin at 0.1% allows penetration of antibodies and 1.4-nm gold particles, while maintaining good membrane morphology.

14.6 Tissue Sections

Electron microscope immunocytochemistry of tissue sections presents greater problems for penetration of antibodies than do cell cultures. In our samples, the use of 4% paraformaldehyde combined with saponin, gives little if any labeling in tissue sections, and gives morphology that is less than acceptable. The following technique was designed to give the highest possible morphological preservation combined with good penetration of antibodies. We have used this method with only a few antibodies in the brain; however, we suspect that with proper testing this method or its modification will be useful in other experimental tissues.

14.6.1 Glutaraldehyde/Metabisulfite Fixation

Experiments were conducted with vibratome sections from adult mouse and rat brain perfused in 4% paraformaldehyde combined with glutaraldehyde (0 to 3%). Although the

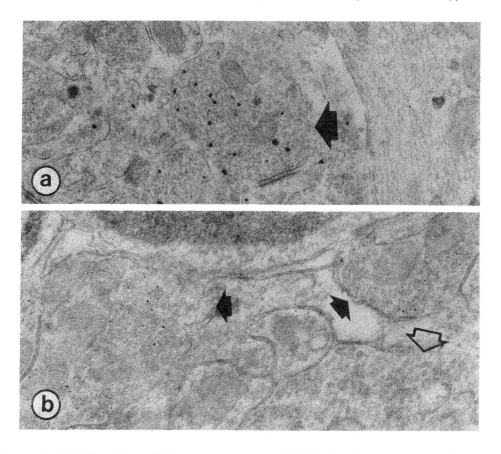

Figure 6 (a) Cell culture of the nervous system labeled for the synaptic vesicle protein synaptophysin. A labeled synaptic terminal is indicated by the solid arrow. In this section, the size of the silver-enhanced particles is between 35 and 50 nm. Note the excellent morphology of the synaptic terminals and cells. Cultures were fixed with 4% paraformaldehyde, labeled with GAM 1.4-nm gold, and silver-enhanced with the NPG solution buffered with MES. (b) Tissue section from a mouse brain labeled for an enzyme GAD. The small synaptic terminals indicated by the solid arrows are positive for GAD, while the larger synaptic terminal shown by the open arrows is not labeled for GAD. The size of the silver-enhanced particles in this section is between 15 and 20 nm. The animal was perfused with glutaraldehyde/metabisulfite, incubated with a monoclonal antibody to GAD, labeled with GAM 1.4-nm gold, and silver-enhanced with the NPG solution buffered with MES. Note the excellent quality of cellular morphology in the synaptic terminals. (Magnification ×30,000.)

quality of the morphology increased dramatically, the penetration of antibodies, as seen by the depth of silver-enhanced particles, decreased. In an attempt to reduce the problems associated with high concentrations of glutaraldehyde, we used a glutaraldehyde/ metabisulfite procedure of Oleskevich et al. (1991). Animals were perfused with 3% glutaraldehyde and 0.2% sodium metabisulfite in buffer, and the resulting vibratome sections were prepared for labeling with goat-antimouse (GAM) Fab' 1.4-nm gold and subsequently silver-enhanced. The penetration of antibodies into these sections was the best of any other procedures we tested. With an antibody to the enzyme glutamic acid decarboxylase (GAD) that synthesizes the neurotransmitter gamma amino butyric acid (GABA), we determined that GAD was distributed over one type of presynaptic terminals (Figure 6b, solid arrows) and not over other types of terminals (Figure 6b, open arrow; Gilerovitch et al., 1995).

14.6.2 *Permeabilization with Triton X-100*

In the experiments testing different fixatives, we consistently used 0.3% Triton-X100 as the detergent. As mentioned above, saponin tested at concentrations from 0.1 to 1% did not give notable penetration in tissue sections. In the experiments with glutaraldehyde/metabisulfite there was little apparent loss of morphology in the tissue permeabilized with 0.3% Triton-X100. In brain tissue, long stretches of plasma membrane remained intact, and the numerous synaptic vesicles clustered in synaptic terminals (Figure 6b). Evidently the use of Triton-X100 following glutaraldehyde/metabisulfite fixation does not remove large amounts of the plasma membrane but produces small breaks in the plasma membrane. While it is important to open the plasma membrane to allow penetration of reagents into the cytoplasm, it is also important to preserve the internal membranes to allow identification of labeled compartments.

14.7 *Blocking Buffer*

An important aspect of any immunocytochemistry method is the solution used to block nonspecific binding of antibodies to the cells or tissues. In our procedure, the same buffer solution is used as a blocking buffer, an incubation buffer, and a rinse buffer. The blocking buffer is used immediately after fixation, the incubation buffer is used to dilute the antibodies, and the rinse buffer is used after antibody incubations. Phosphate-buffered saline + (PBS+) at pH 7.3 is composed of: PBS, 1% normal goat serum, 1% BSA, 0.1% saponin (or 0.3% Triton-X100), 0.1% gelatin, 50 mM glycine, 0.02% NaN$_3$. In the PBS we have included normal goat serum, bovine serum albumin (BSA), fish scale gelatin (Amersham), and glycine, which reduce background by binding to reactive groups in the cells that could nonspecifically bind the antibodies used in incubations. The sodium azide is included to prevent bacterial growth during overnight incubations. When we use saponin as a permeabilization agent, incubations are done at room temperature and not at 4°C. Experiments where incubations were done at 4°C did not allow good penetration of antibodies into cells. We suspect that the change in the phase of membrane lipids caused by the low temperature alters the ability of saponin to bind to membrane cholesterol or the ability of saponin to generate holes in the plasma membrane for penetration of antibodies. All of the incubations with either saponin or Triton-X100 were done at room temperature with sodium azide.

14.8 *Procedures*

14.8.1 *Procedure for Use of NPG Silver-Enhancement*

This procedure begins after the primary antibody incubation and rinsing in PBS+.

1. Incubate cultures or tissue in GAM Fab' 1.4-nm gold in PBS+ at room temperature and rinse 3 times with PBS+ and 2 times with PBS each for 5 min. PBS+ (1% normal goat serum, 1% BSA, 0.1% saponin, 0.1% gelatin, 50 mM glycine, 0.02% NaN$_3$).
2. Fix in 1.6% glutaraldehyde in PBS for 15 min.
3. Rinse 3 times in 50 mM MES buffer with 200 mM sucrose (pH 5.8) for 15 min (check the pH each day of use).
4. Incubate tissue sections or cultures in the NPG silver-enhancement solution in a darkroom. This procedure can be safely done under a sodium vapor safelight.
5. Immediately rinse the dish 3 times in Neutral Fixer Solution for 5 min.
6. Rinse 3 times in PBS for 15 min or until gum arabic is gone.
7. Incubate for 30 min in 0.1% OsO$_4$ in rinse buffer.

8. Rinse 3 times in PBS for 15 min.
9. Dehydrate and embed.
10. Thin sections can be stained with heavy metals: 1% uranyl acetate for 5 to 10 min and lead citrate for 3 min.

14.8.2 Procedure for Mixing the NPG Silver-Enhancement Solution

Gum arabic stock (final 50%)	5.0 ml
MES 0.5 M stock pH 6.15 (final 200 mM)	2.0 ml
NPG stock (final 0.3 mg/ml)	1.5 ml
Mix for 5 min and add the next solution in darkroom with a sodium vapor safelight	
Silver lactate solution (final 1.1 mg/ml)	1.5 ml
Total volume	10.0 ml
Mix in a large plastic disposable tube	

14.8.3 Stock Solutions

MES Stock
0.5 M MES (pH 6.15); adjust pH with 1.0 M NaOH. **Do not use HCl.**
MES Buffer
50 mM MES, 200 mM sucrose pH = 5.8; adjust pH with 1.0 M NaOH. **Do not use HCl.**
Gum Arabic Stock

Gum Arabic powder (Sigma G-9752)	50.0 g
Double distilled water	100.0 ml

Combine ingredients in a container of at least 3 times the volume of water to be added. Allow 2 to 3 days for the gum arabic to dissolve with gentle agitation. Do not use stir plate. When the gum arabic is dissolved, remove the dissolved gases (degas) with a high vacuum pump. The gum arabic stock can be stored directly in small plastic tubes in a –20°C freezer; once thawed, the remaining solution should be discarded.
NPG Stock

NPG	10.0 mg
Ethyl alcohol	0.25 ml
Mix and dissolve the NPG	
Double distilled water	4.75 ml
Total volume	5.0 ml

Silver Lactate Stock
Silver lactate (7.3 mg/ml) should be made the day of use. Store in a light tight box, but mixing can be done in dim room lights.
Neutral pH Fixer Solution
250 mM sodium thiosulfate, 20 mM MES (or HEPES) (pH 6.8)

14.9 Summary

We have reviewed methods for use of silver-enhanced small gold particles in pre-embedding EM immunocytochemistry. Silver-enhancement procedures that are carried out near neutral pH give excellent morphology and allow the identification of labeled cellular compartments. The best buffer system for the silver-enhancement solution is MES mixed at its pK of 6.15 and used in the presence of the protective colloid, gum arabic. The best reducing agent for a neutral pH silver-enhancement solution is NPG because the reaction is easily controlled. Silver lactate is used as a source of silver ions. For postfixation, 0.1% OsO_4 is used to stain membranes without decreasing the diameter or loss of silver-enhanced particles. The use of these methods for either cell cultures or tissue sections requires adjustment of the aldehydes in the fixative and detergents used to permeabilize the cells. The labeling of primary antibodies with secondary antibodies bound to 1.4-nm gold particles and the silver-enhancement of these particles was found to provide the highest resolution for localizing antigens in cells.

Acknowledgments

The author is indebted to Dr. Helen G. Gilerovitch and Dr. Dale D. Vandré for assistance with the development of this new procedure. The author gives thanks to Drs. Georgia Bishop, James King, and John Robinson for their support and advice during this work. The help of Dr. James Hainfeld over the last several years is gratefully acknowledged. Funding was from the National Science Foundation BNS-8909835 (RWB).

References

Assa, Y., Shany, S., Gestetner, B., Tencer, Y., Birk, Y., and Bondi, A. 1973. Interaction of alfalfa saponins with components of the erythrocyte membrane in hemolysis. *Biochim. Biophys. Acta* 307: 83.

Bohn, W. 1978. A fixation method for improved antibody penetration in electron microscopical immunoperoxidase studies. *J. Histochem. Cytochem.* 26: 293.

Burry, R. W., Lah, J. J., and Hayes, D. M. 1991. Redistribution of GAP-43 during growth cone development in vitro; Immunocytochemical studies. *J. Neurocytol.* 20: 133.

Burry, R. W., Vandre, D. D., and Hayes, D. M. 1992a. Silver enhancement of gold conjugated antibody probes for pre-embedding electron microscopic immunocytochemistry. *J. Histochem. Cytochem.* 40: 1849.

Burry, R. W., Lah, J. J., and Hayes, D. M. 1992b. GAP-43 distribution is correlated with development of growth cones and presynaptic terminals. *J. Neurocytol.* 21: 413.

Caro, L. G. and van Tubergen, R. P. 1962. High-resolution autoradiography. I. Methods. *J. Cell Biol.* 15: 173.

Chan, J., Aoki, C., and Pickel, V. M. 1990. Optimization of differential immunogold-silver and peroxidase labeling with maintenance of ultrastructure in brain sections before plastic embedding. *J. Neurosci. Methods* 33: 113.

Danscher, G. 1981. Localization of gold in biological tissue. *Histochemistry* 71: 81.

Danscher, G. 1983. A silver method for counterstaining plastic embedded tissue. *Stain Tech.* 58: 365.

Danscher, G. and Norgaard, J. O. R. 1983. Light microscopic visualization of colloidal gold on resin-embedded tissue. *J. Histochem. Cytochem.* 31: 1394.

Faulk, W. P. and Taylor, G. M. 1971. An immunocolloid method for the electron microscope. *Immunochem.* 8: 1081.

Gallyas, F. 1979. Light insensitive physical developers. *Stain Tech.* 54: 173.

Gilerovitch, H. G., Bishop, G. A., King, J. S., and Burry, R. W. 1995. GAD-immunocytoreactive terminals in the cerebellar nuclei labeled with silver enhanced small gold particles for electron microscopy. *J. Histochem. Cytochem.* 43: 337.

Goldenthal, K. L., Hedman, K., Chen, J. W., August, J. T., and Willingham, M. C. 1985. Postfixation detergent treatment for immunofluorescence suppresses localization of some integral membrane proteins. *J. Histochem. Cytochem.* 33: 813.

Hacker, G. W., Grimelius, L., Danscher, G., Beratzky, G., Muss, W., Adam, H., and Thurner, J. 1988. Silver acetate autometallography: an alternative technique for immunogold-silver staining (IGSS) and silver amplification of gold, silver, mercury and zinc in tissues. *J. Histotechnol.* 11: 213.

Hainfeld, J. F. 1988. Gold cluster-labeling antibodies. *Nature* 333: 281.

Hainfeld, J. F. 1987. A small gold-conjugated antibody label: improved resolution for electron microscopy. *Science* 236: 450.

Hainfeld, J. F., and Furuya, F. R. 1992. A 1.4-nm gold cluster covalently attached to antibodies improves immunolabeling. *J. Histochem. Cytochem.* 40: 177.

Hayat, M. A. 1989. *Principles and Techniques of Electron Microscopy: Biological Applications.* 3rd ed. pp. 24–27. CRC Press, Boca Raton, FL.

Hopwood, D. 1972. Theoretical and practical aspects of glutataldehyde fixation. *Histochem. J.* 4: 267.

Horisberger, M. and Rosset, J. 1977. Colloidal gold, a useful marker for transmission and scanning electron microscopy. *J. Histochem. Cytochem.* 25: 295.

Lah, J. J., Hayes, D. M., and Burry, R. W. 1990. A neutral pH silver development method for the visualization of 1 nanometer gold particles in pre-embedding electron microscopic immunocytochemistry. *J. Histochem. Cytochem.* 38: 503.

Leunissen, J. L. M., and De Mey, J. R. 1989. Preparation of gold probes. In: *Immuno-gold Labeling in Cell Biology.* pp. 3–16. Vrekleij, A. J. and Leunissen, J. M. (Eds.), CRC Press, Boca Raton, FL.

Nakane, P. K. and Pierce, G. B. 1967. Enzyme-labeled antibodies: preparation and application for the localization of antigens. *J. Histochem. Cytochem.* 14: 929.

Namork, E. and Heier, H. E. 1989. Silver enhancement of gold probes (5–40 nm): Single and double labeling of antigenic sites on cell surfaces imaged with backscattered electrons. *J. Elect. Microsc. Tech.* 11: 102.

Oleskevich, S., Descarries, L., Watkins, K. C., Seguela, P., and Daszuta, A. 1991. Ultrastructural features of the serotonin innervation in adult rat hippocampus: an immunocytochemical description in single and serial thin sections. *Neuroscience* 42: 777.

Romano, E. L., Stolinski, C., and Hughes-Jones, N. C. 1974. An antiglobulin reagent labeled with colloidal gold for use in electron microscopy. *Immunochemistry.* 11: 521.

Seeman, P. 1967. Transient holes in the erythrocyte membrane during hypotonic hemolysis and stable holes in the membrane after lysis by saponin and lysolecithin. *J. Cell Biol.* 32: 55.

Stierhof, Y. D., Humbel, B. M., and Schwarz, H. 1991. Suitability of different silver enhancement methods applied to 1nm colloidal gold particles: An immunoelectron microscopic study. *J. Elect. Microsc. Tech.* 17: 336.

van den Pol, A. N. 1986. Tyrosine hydroxylase immunoreactive neurons throughout the hypothalamus receive glutamate decarboxylase immunoreactive synapses: A double pre-embedding immunocytochemical study with particulate silver and HRP. *J. Neurosci.* 6: 877.

Willingham, M. C. 1980. Electron microscopic immunocytochemical localization of intracellular antigens in cultured cells: the EGS and ferritin bridge procedures. *Histochem. J.* 12: 419.

Chapter 15

Visualization of Microtubules With Immunogold-Silver Staining and Backscattered Scanning Electron Microscopy

Dennis Goode

Contents

15.1 Introduction

Gold-labeled secondary or tertiary antibodies can be combined with the high specificity and sensitivity of monoclonal primary antibodies to provide precise localization of many antigens by both light and electron microscopy (Horisberger, 1981; DeMey, 1984). Gold has a high atomic number (79), so even small gold particles scatter electrons sufficiently to provide high contrast in both transmission electron microscopy and backscatter electron imaging (BEI). Immunogold-labeled antigens can be detected with light microscopy by their red-brown color, but the contrast is low. Silver-enhancement catalyzed by the gold particles provides a high-contrast image for both bright-field light microscopy and for electron microscopy (Goode and Maugel, 1987; and Chapters 1 and 2.)

Dramatic progress in describing and characterizing the cytoskeletal components of cells has been made in the past 20 years. Immunofluorescence techniques, confocal microscopy, and high-voltage transmission electron microscopy on whole mounts of cells provide an appreciation for the patterns and interactions of cytoskeletal elements that were not obvious in thin sections of cells (Lazarides and Weber, 1974; Brinkley et al., 1975;

Buckley and Porter, 1975; Goode, 1975; Schollmeyer et al., 1976; White et al., 1987). Scanning electron microscopy (SEM) of detergent-extracted cells (Bell, 1981) provides a unique pseudo three-dimensional view of the spatial distribution of the cytoskeleton, but individual filament types are not easily identified or followed for long distances in these secondary electron images of metal-coated specimens.

The development of polyclonal (Brinkley et al., 1975) and monoclonal (Kilmartin et al., 1982) antibodies against tubulin has led to the clarification of the structure, functions, and dynamic changes in the microtubule networks in animal (Osborn and Weber, 1977; DeMey et al., 1981), plant (DeMey et al., 1982), and protist (Goode and Cachon, 1985) cells. Goode and Maugel (1987) developed a method that uses monoclonal antitubulin, 5-nm colloidal gold-labeled tertiary antibody, and silver-enhancement to identify and visualize the microtubular components of the cytoskeleton in detergent-extracted animal cells grown in monolayer cultures. This method provides both high-contrast views of microtubule patterns by light microscopy and higher resolution images of microtubules that can be viewed by backscattered electron microscopy on whole mounts of lysed cells.

15.2 Backscattered Electron Imaging

Backscattered electron imaging (DeNee and Abraham, 1976; Robinson, 1980) of heavy metal atoms bound to cell structures promises to be a very useful localization method. Both surface (de Harven et al., 1984) and subsurface (Becker and Sogard, 1979; Small et al., 1980; Tellez et al., 1982; Thiebant et al., 1984; Goode and Maugel, 1987) components can be studied in the backscattered mode, and their location can be compared to surface structures identified by secondary electron imaging (SEI). Backscattered electrons are those electrons from the incident beam that are scattered at angles of greater than 90° by interaction with the nuclei of atoms within a specimen. Therefore, backscatter is a sensitive function of atomic weight, proportional to Z^2 (DeNee and Abraham, 1976). Since backscattered electrons can retain most of their original energy, they can escape from much deeper layers in a specimen and can be detected separately from the lower energy secondary electrons emitted from orbits in the atoms of the specimen.

15.3 Fixation and Permeabilization

A number of different techniques for cell lysis and fixation has been developed to preserve and detect microtubules in whole mounts of cultured cells by immunocytochemical techniques. Six of these were evaluated for the quality of microtubule preservation and visualization by light microscopy and BEI scanning electron microscopy. PtK$_2$ cells (American Type Culture Collection, CCL 56) were removed from culture flasks with 0.2% trypsin (Worthington) and grown at medium density for 1 d on sterile, 12-mm round, # 1 glass coverslips (Bellco Glass, Inc.) in multiwell plates with Earle's minimal essential medium plus 10% fetal calf serum (both GIBCO). Coverslips were rinsed twice for 1 min each in Earle's calcium-magnesium free balanced salt solution (GIBCO) to remove serum and divalent cations before lysis and fixation by one of the procedures listed below. Steps shorter than 1 min were done by repeatedly dipping the coverslips in a small beaker of the solution; steps of 1 min or longer were done by floating the coverslips, cell-side down, on large drops of solution on Parafilm. All steps should be at room temperature except where noted. The lysis and fixation protocols are listed below in approximately the order of their effectiveness for microtubule preservation and visualization.

1. *A modification of the Osborn et al. (1978) and Osborn and Weber (1977) procedures.* In this procedure, coverslip cultures are rinsed in two 30-sec changes of microtubule

stabilizing buffer (MTSB: 4% (w/v) polyethylene glycol 8000, 1 mM EGTA, 0.5 mM MgCl$_2$ in 0.1 M PIPES buffer pH 6.8) and lysed 4 min in MTSB plus 0.2% (v/v) Triton X-100. After two 30-sec rinses in MTSB, cells are fixed for 10 min with 1% glutaraldehyde in MTSB, rinsed twice in MTSB and twice in Dulbecco's phosphate-buffered saline (PBS) solution and reduced with two 4-min changes of 0.5 mg/ml NaBH$_4$ in PBS. After 30-sec, 5-min, and 30-sec rinses in PBS, coverslips are processed for immunocytochemistry.

2. *A modification of the DeMey et al. (1981, 1982) procedure.* Coverslip cultures are fixed in 0.5% glutaraldehyde plus 0.1% Triton X-100 in 0.1 M phosphate buffer (pH 6.9) for 5 min and then in 1% glutaraldehyde in phosphate buffer for 10 min. After a rinse in PBS, coverslips are floated on 0.5% Triton X-100 in PBS for 10 min, rinsed in PBS, and reduced with two 4-min treatments of NaBH$_4$ freshly prepared in PBS. After 30-sec, 5-min, and 30-sec rinses in PBS, coverslips are processed for immunocytochemistry.

3. *A modification of the Cande et al. (1981) lysis and fixation procedure for mitotic cells.* Coverslip cultures are rinsed twice for 30 sec each in medium A (0.1% Brij 58 [Sigma], 2.25 mM MgSO$_4$, 1.0 mM EGTA, and 1.5 × 10^{-4} CaCl$_2$ in 85 mM PIPES buffer: pH 6.94) at 30°C and lysed for 5 min at 30°C in medium B (medium A plus 2.5% [w/v] polyethylene glycol 20M). After a brief rinse in medium B, cells are fixed in medium B plus 1% glutaraldehyde for 10 min, rinsed twice in 0.1 M PIPES buffer and twice in PBS, and reduced with two 4-min changes of NaBH$_4$ in PBS. After three rinses in PBS, coverslips are processed for immunocytochemistry.

4. *A modification of the Goode and Sarma (1986) lysis procedure for tubulin incorporation.* Coverslip cultures are rinsed in 2 M glycerol in reassembly buffer (50 mM PIPES, 1 mM EGTA, 0.5 mM MgCl$_2$, pH 6.8), lysed with 0.1% Nonidet P-40 in 2 M glycerol-containing reassembly buffer for 4 min, rinsed in 2 M glycerol reassembly buffer, and fixed in 1% glutaraldehyde in 2 M glycerol reassembly buffer for 10 min. After two rinses in reassembly buffer plus 2 M glycerol and two rinses in PBS, free aldehydes are reduced with NaBH$_4$ in PBS for two 4-min periods. After three rinses in PBS, coverslips are processed for immunocytochemistry.

5. *A variation on the Osborn et al. (1978) procedure using glutaraldehyde-paraformaldehyde fixation.* This procedure is identical to Procedure 1; only 2% by weight of paraformaldehyde was added to the 1% glutaraldehyde fixative.

6. *A modification of the Walsh (1984) fixation procedure for immunocytochemistry on protozoan microtubules.* After rinsing in calcium and magnesium-free BSS, coverslip cultures are fixed and lysed in 0.9% formaldehyde, 0.1% Nonidet P-40, and 0.125 M sucrose in 50 mM phosphate buffer (pH 7.2) for 10 min, rinsed in four changes of PBS for a total of 20 min, and processed for immunocytochemistry.

15.4 Immunocytochemistry

To block nonspecific immunoglobulin binding sites, fixed cultures are incubated for 20 min to 1 h at 37°C with 1% normal goat serum plus 0.1% (w/v) bovine serum albumin in PBS with 20 mM sodium azide. This solution is removed with bibulous paper, and cultures are incubated at 37°C for 1 h in a water-saturated atmosphere on 40 μl drops of rat monoclonal antitubulin 34 (Kilmartin et al., 1982), diluted 1:40 with PBS containing 1% normal goat serum (NGS), 0.1% bovine serum albumin (BSA), and 20 mM sodium azide. After rinsing once with PBS and twice with PBS plus NGS and BSA for a total of 20 min, cultures are incubated for 1 h at 37°C on 40 μl drops of polyclonal rabbit antirat IgG (Sigma), diluted 1:40 in PBS-NGS-BSA. They are then rinsed once in BSA and twice for a total of 20 min in a solution of 0.1% BSA, 0.9% NaCl, and 20 mM Tris buffer at pH 8.2 (BSA-Tris), and incubated for 2 h at 37°C in a humid chamber on 40 μl drops of polyclonal goat-antirabbit IgG linked to 5-nm mean diameter colloidal gold particles (Janssen Pharmaceutica), diluted 1:40 in BSA-Tris solution. The preparations are rinsed three times for a total of 30 min in BSA-Tris and twice for 20 min in PBS. Cells are then treated for 30 min with 1% glutaraldehyde in the same buffer used for the initial fixation to fix the antibodies in place on the microtubules. This is followed by two rinses in the fixative buffer, three rinses in distilled water, and two rinses with 0.2 M citrate buffer (23.5 mg/ml trisodium citrate dihydrate and 25.5 mg/ml citric acid monohydrate).

The silver-enhancer solution is prepared under low light conditions just before use. Hydroquinone (95 mg) is dissolved in 1.5 ml of deionized, distilled water in a foil-wrapped container. One ml of freshly prepared 2 M citrate buffer is mixed with 6 ml of deionized, distilled water and added to the hydroquinone solution. Silver lactate (11 mg, Janssen Pharmaceutica) is dissolved in 1.5 ml of deionized, distilled water, mixed with the hydroquinone solution, and used immediately. Coverslips are dipped in the silver enhancer under a red safelight, and incubated for 5 min at room temperature or 7 min at 10°C on 100 μl drops of enhancer in the dark. Enhancement is stopped by transferring the preparation to 10% (v/v) fixing solution (Janssen Pharmaceutica) for 3 min after which they are washed in many changes of distilled water. At this point, coverslips can be mounted on slides for light microscopic examination or prepared for electron microscopy. Other silver-enhancement protocols are available that are less light sensitive.

Controls for each immune reagent are done by substituting the same concentration of pre-immune goat, rabbit, or rat serum in PBS from the specific reagent. A control for the gold specificity of the silver-enhancement procedure is performed by using unlabeled goat-antirabbit IgG rather than gold-labeled goat-antirabbit IgG. Nonenhanced immunogold-labeled cultures are also examined by light and electron microscopy.

15.4.1 *Preparation for and Examination by Backscatter Scanning Electron Microscopy*

To freeze dry the preparations, coverslips rinsed 10 times with distilled, deionized water and left slightly moist are frozen in liquid nitrogen, placed in a brass coverslip holder, and dried for 48 h at –80°C on the cold stage of a Pierce-Edwards tissue dryer containing molecular sieve desiccant to absorb water molecules. Dehydrated coverslip cultures are mounted with silver paint on aluminum stubs and coated with aluminum to provide conductivity with a low atomic weight coating that is relatively transparent to the incident and backscattered electrons.

For critical point drying, coverslips are dehydrated through a graded ethanol series, transferred to isoamyl acetate, and dried from liquid CO_2. The dried coverslips are mounted on aluminum stubs with silver paint and coated with approximately 40 nm of aluminum, as described above.

Specimens are examined in an AMRay 1000A scanning electron microscope equipped with both a secondary electron detector and a prototype quartz scintillation-type backscattered electron detector (M. E. Taylor Engineering, Inc., Kensington, MD). Specimens are tilted at a 36 or 45° angle and examined and photographed at a 10 to 12 mm working distance with a 20 or 30 KV electron beam.

15.4.2 *Comparison of Techniques*

The quality of microtubule preservation and visualization varies greatly with the method of fixation and permeabilization. The methods using formaldehyde fixation, including the Walsh (1984) procedure, that provide excellent preservation of microtubules in many

Figures 1–4 are light micrographs of PtK_2 cell cultures permeabilized and fixed by two of the different procedures tested. All were incubated with rat monoclonal antitubulin, rabbit-antirat IgG, 5-nm gold-labeled goat-antirabbit IgG, and silver lactate enhancing solution (except controls, where one substitution was made on each). Bar = 10 μm.

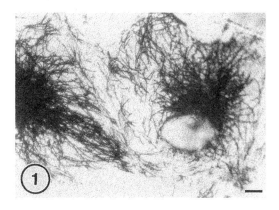

Figure 1 DeMey et al. procedure (glutaraldehyde and Triton X-100). Microtubules are well preserved in general, but some fragmentation is seen.

protists for light microscopy (Goode, unpublished observations), do not preserve PtK_2 microtubules for this procedure. The combination of glutaraldehyde and paraformaldehyde (Procedure 5) provides better preservation than formaldehyde alone, but does not preserve microtubules as well as does glutaraldehyde alone (Procedures 1 to 3). Lysis with Nonidet P-40 in microtubule reassembly buffer plus 2 M glycerol (Goode and Sarma, 1986) preserves microtubules well but retains a high background level of stainable material, perhaps soluble tubulin. The lysis procedure of Cande et al. (1981) followed by glutaraldehyde fixation (Procedure 3) preserves microtubules well, but they are labeled in only about 60% of the cells. The other 40% of the cells apparently remain intact and thus impermeable to the antibodies used. The procedures of De Mey et al. (1981, 1982) (Figure 1) and Osborn and Weber (1977) both preserve microtubules well, but the latter (Procedure 1) gives the lowest level of background labeling (Figure 2).

Controls in which normal sera are substituted for the primary (Figure 3), secondary, or gold-labeled antibody (Figure 4) and then silver enhanced show no microtubule staining, but a small amount of nonspecific silver often is seen around the nuclear envelope, plasma membrane, or mitotic chromosomes (Figures 3, 4, 15, and 16).

The microtubule patterns of individual PtK_2 cells are clearly resolved by SEM using a detector for backscattered electrons when cells are lysed and fixed by the Osborn et al. (1978) method, labeled with monoclonal antitubulin and immunogold, silver-enhanced, and critical point dried (Figures 5 and 6). Microtubules appear as 100 to 200 nm bright rods.

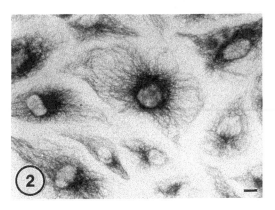

Figure 2 Osborn et al. procedure (polyethylene glycol, Triton X-100 and glutaraldehyde). Microtubules are well preserved, and background staining is low.

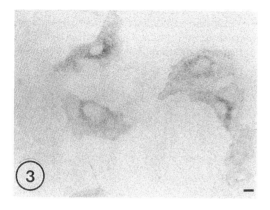

Figure 3 Control cells: normal serum was substituted for the primary monoclonal antitubulin, but other procedures were as Figure 2. No labeled microtubules are seen.

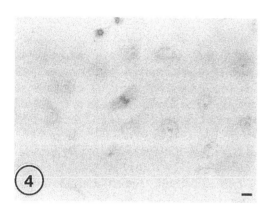

Figure 4 Control cells: normal goat serum was substituted for gold-labeled goat-antirabbit, but other procedures were as in Figure 2. No labeled microtubules are seen.

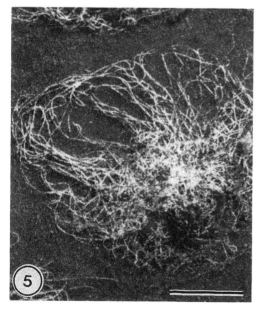

Figure 5 Backscattered electron image (BEI) of a PtK$_2$ cell lysed and fixed by the Osborn et al. method, incubated with monoclonal antitubulin and 5-nm gold-labeled anti-IgG, enhanced with silver, and critical-point dried. The microtubule network of the entire cell is displayed. Bar = 10 μm.

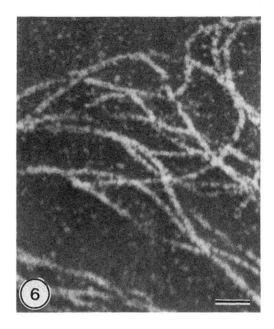

Figure 6 BEI at higher magnification of the microtubules in the upper central sector of the cell in Figure 5. Microtubules appear as 120-nm bright rods after labeling. Bar = 1 µm.

The increase in diameter from the 26-nm diameter of sectioned microtubules is expected due to the addition of successive layers of rat monoclonal antitubulin, rabbit-antirat IgG, 5-nm gold conjugated to goat-antirat IgG, and silver metal deposited around the gold particles. In addition to amplifying the signal, these steps increase microtubule diameters up to the resolving power of the light microscope.

When the detector is changed to pick up the secondary electron image, other, unlabeled cytoskeletal components can be seen in cells whose upper plasma membrane was removed by the detergent treatment (Figure 7). Fine (microfilaments?) and coarser (intermediate filaments?) unlabeled filaments are seen between the labeled microtubules.

At higher magnifications, the labeling along microtubules appears granular or patchy in critical point dried specimens (Figures 6 and 8 [BEI]), and the higher resolution of the

Figure 7 Secondary electron image of microtubules and other filaments in a gold-labeled, silver-enhanced, and critical-point dried PtK$_2$ cell cytoskeleton. Microtubules are coated with large silver granules. Unlabeled filaments with large (intermediate filaments?) and small (microfilaments?) diameters are visible. The beading of the label seen here appears to be an artifact of dehydration in nonaqueous solvents (compare to Figure 12, which was freeze dried). Bar = 1 µm.

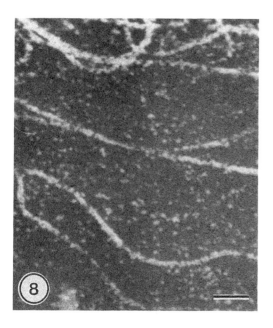

Figure 8 BEI of gold- and silver-labeled, critical-point dried microtubules in another cell. Note that the label appears patchy or beaded in places (compare to Figure 10). Bar = 1 μm.

secondary signal makes the granularity more obvious (Figure 7). In some places, the appearance of a smaller filament or tubule can be seen in the gaps (Figure 8).

In an effort to identify the source of this apparent artifact, silver-enhanced cells are observed by light microscopy as the dehydrating alcohol solutions are passed under the coverslip prior to critical point drying. The silver-enhanced microtubules appear initially as dark lines but become progressively more granular as they are dehydrated through the higher alcohols. Air drying also produces granular-appearing silver on microtubules. To avoid this apparent dehydration artifact, silver-enhanced coverslip cultures are freeze dried from distilled water. The silver-enhanced microtubules in these freeze-dried cytoskeletons appear smoother and more uniformly decorated (Figures 9 and 10) than critical-point dried microtubules. At higher magnification with 30 KV electrons, these

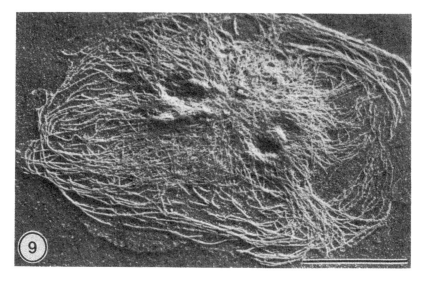

Figure 9 The microtubule network of a binucleate PtK$_2$ cell is viewed by BEI. These cells were prepared by the Osborn et al. procedure (1978) and freeze dried. Bar = 10 μm.

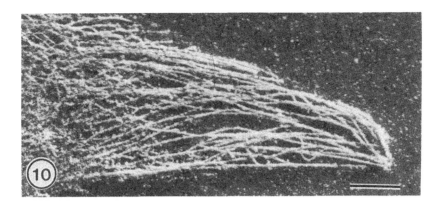

Figure 10 BEI view of microtubules in a cell projection of a freeze-dried PtK$_2$ cell. Note that the microtubules appear more smooth walled than after critical-point drying (compare Figure 8). Bar = 10 μm.

labeled microtubules measure 180 to 200 nm in diameter with a 60 to 70 nm core of lower signal intensity (Figure 11). The core probably represents the original microtubule plus the unlabeled primary and secondary antibodies. This unlabeled core is much less obvious in secondary electron images of microtubules prepared in the same way (Figure 12).

Cultures lysed and fixed by methods 2 (De May et al., 1982) and 3 (Cande et al., 1981) followed by monoclonal antitubulin, immunogold, and silver-enhancement exhibit similar microtubule patterns to the Osborn et al. (1978) lysis procedure. However, the microtubules are less clear (e.g., Figures 13 and 14), and occasional cells are completely unlabeled after the Cande lysis procedure.

Control cells incubated with normal serum in place of the antitubulin contain no labeled fibers in either secondary or backscattered electron images (Figures 15 and 16), although some unlabeled fibrous components can be seen in the secondary image (Figure 15). Some nonspecific silver accumulation can be seen around the nucleus in Figure 16, as is seen by light microscopy (Figure 3). Cells from control experiments in which the gold-labeled goat-antirabbit IgG is replaced by unlabeled normal goal serum or in which the silver-enhancement step is omitted did not exhibit microtubules by BEI.

Specific visualization of microtubules in cultured cells by this technique requires: (1) an initial rinse to remove serum proteins and Ca^{2+} ions, (2) cell lysis in a medium that makes the plasma membrane permeable to soluble proteins (especially soluble tubulin and

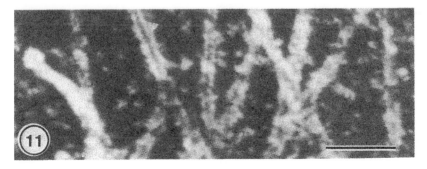

Figure 11 A high magnification BEI view of labeled microtubules. A 60-nm unlabeled central core is seen in each microtubule, probably the original microtubule plus the unlabeled primary and secondary antibodies used before a gold-labeled tertiary antibody and silver enhancement were added and the cell was freeze dried. Bar = 1 μm.

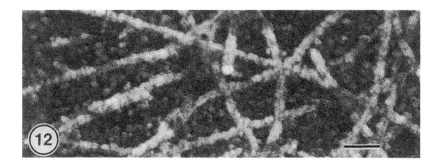

Figure 12   A secondary electron image of PtK$_2$ microtubules that were labeled, coated with aluminum and freeze dried as in Figure 11. Note that labeled, freeze-dried microtubules are not as beaded as Figure 7. Bar = 1 µm.

added antibodies) while preserving microtubule stability, (3) fixation to preserve cytoskeletal components through the many preparative steps (glutaraldehyde in microtubule-stabilizing buffer is most effective), (4) blocking or reducing unreacted aldehyde groups on the bound fixative molecules, (5) binding specific antibodies to the microtubules, (6) marking the bound antibodies with 5-nm gold-conjugated secondary or tertiary antibodies against the IgG chains of the bound antibodies, (7) enhancing the signal with gold-catalyzed silver deposition, (8) dehydration by freeze drying (preferable) or critical point drying, (9) light coating with a conductor of low atomic number such as aluminum to reduce charging and (10) examination with a scanning electron microscope equipped with a backscattered electron detector. The use of larger gold particles may allow the silver-enhancement step to be omitted, but the faster penetration and higher labeling efficiency of antibodies labeled

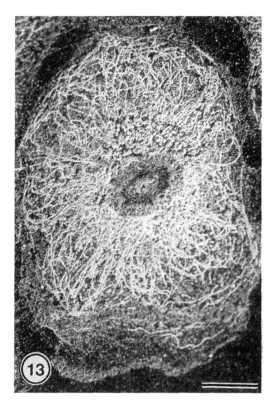

Figure 13   Secondary electron image of PtK$_2$ cell prepared by the Cande et al. method (1981), labeled as in Figure 5, and freeze dried. Bar = 10 µm.

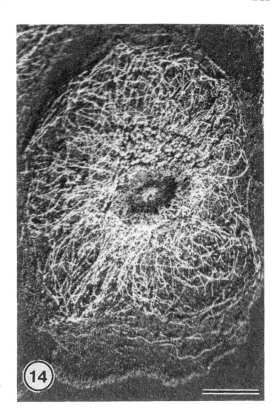

Figure 14 BEI of the same PtK$_2$ cell as Figure 13. Bar = 10 μm.

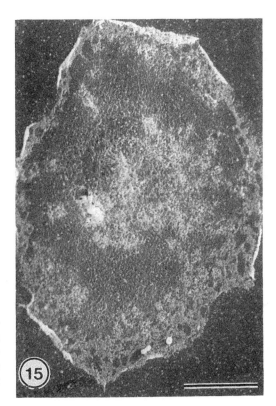

Figure 15 Secondary electron image of a control PtK$_2$ cell incubated with normal serum in place of antitubulin, then with gold-labeled anti-IgG. After silver enhancement, the culture was freeze dried. Cytoskeletal elements are present in the cytoplasm. Bar = 10 μm.

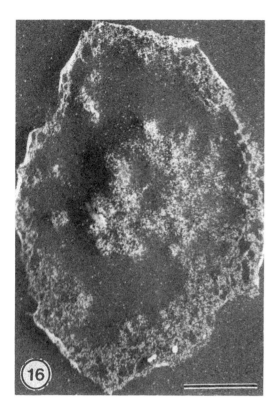

Figure 16 BEI of the same cell as Figure 15. Some nonspecific silver is seen around the nucleus and cell periphery, but the cytoskeletal filaments seen in Figure 14 are unlabeled when antitubulin is replaced with normal serum. Bar = 10 μm.

with small gold particles is an advantage (DeMey, 1984; de Harven et al., 1984; de Harven and Soligo, 1989; Namork, 1991).

Of the lysis and fixation methods used, the methods of Osborn et al. (1987) and DeMey et al. (1981, 1982) give the best results in this procedure. The Osborn et al. (1978) method gives the lowest background and most consistent results, probably because the cells are lysed in a microtubule-stabilizing buffer before fixation, which removes soluble tubulin while retaining microtubules. The DeMey et al. (1981, 1982) procedure involves simultaneous detergent lysis and glutaraldehyde fixation. It does, however, seem to result in more intensely labeled microtubules. The Cande et al. (1981) lysis and fixation procedure is a more gentle extraction and works well except that some cells remain completely unlabeled, apparently because they are unlysed until after at least the first antibody step. The Goode and Sarma (1986) procedure gives a higher background, and other procedures involving formaldehyde in the fixative solution inadequately preserve microtubules through the process.

15.5 Advantages and Limitations

Backscattered electron imaging has several advantages as a method to study cytoskeletal elements in cultured cells, and these advantages may apply to many other intracellular structures as well. The entire cell or cytoskeleton can be viewed at once. A pseudo three-dimensional image of the cytoskeleton can be obtained (e.g., Figure 10), although the greater depth of interactions with atoms that produce backscattered electrons can produce some topographic ambiguity. A principal limitation of SEI is the difficulty of seeing structures that lie under other components in cells or extracted cytoskeletons; BEI can provide images of metal-labeled components that lie under other cellular or extracellular components. This extends SEM to the study of intracellular components in ways that were not feasible previously. The major advantage of this approach is the ability to specifically mark and examine one set of structures (microtubules in this case) and be able to relate them to the positions of other cell components in a secondary SEM view of the same cell or cytoskeleton. The resolution is better than light microscopy, but because of the reduced efficiency of the backscattered electron detector, resolution is less than in the standard SEI mode and poorer than TEM of whole-mounted cytoskeletons. Our backscatter detector does not resolve 5-nm gold particles, but the 60-nm core (distance between silver layers on either side of the microtubule) of silver-enhanced microtubules is clearly resolved (Figure 11). Both 20- and 45-nm gold particles can be detected as cell surface markers by BEI, although particles in contact may not be resolved as two entities (de Harven, 1984).

This technique could be applied to many other cell structures. Clearly other cytoskeletal polymers composed of actin, myosin, keratin, desmin, vinculin, etc. could be visualized in detergent-extracted whole mounts with this technique. Also, monoclonal antitubulins, immunogold, and silver-enhancement could be used to study the patterns of microtubule-containing organelles in diverse protists, which are of major phylogenetic significance (Lynn and Small, 1981; Goode, 1981). Using BEI, Small et al. (1980) have studied the subsurface fibers forming the kinetids of ciliates after silver staining with the Protargol silver albuminose method. Individual kinetosomes or basal bodies are clearly resolved under the cell membrane by this method. The Fernandez-Galiano silver impregnation method stains the ciliate kinetodesmal fibers as well as basal bodies (Tellez et al., 1982). In theory, any intracellular component with discrete localization sites and an available, highly specific antibody or metal-binding ability could be studied by BEI.

15.6 Summary

Silver-enhanced immunogold labeling and BEI bridges the gap between fluorescence microscopy and transmission electron microscopy. This method can obtain a high-contrast, medium resolution view of the entire microtubule complex of each cell. For best results, monolayer cultures of animal cells are lysed with Triton X-100 in a microtubule stabilizing buffer, fixed with 1% glutaraldehyde, reduced with $NaBH_4$, incubated with monoclonal antitubulin and 5-nm gold-labeled anti-IgG, silver-enhanced, freeze dried, lightly coated with aluminum and examined in an SEM equipped with a backscattered electron detector. Silver-enhanced microtubules in freeze-dried preparations have relatively smooth surfaces, whereas those in critical-point dried preparations are more irregular or beaded. At high magnifications, an unstained inner core of each microtubule can be resolved. Backscattered electron imaging appears to be a promising technique for localizing cytoskeletal proteins and other intracellular antigens that can be labeled with immunogold and enhanced with silver.

Acknowledgments

The author thanks Tim Maugel for his advice and help on the research on which much of this review is based, M. E. Taylor for providing the prototype backscatter detector, John Kilmartin for the monoclonal antitubulin, and John Johnson and John Wiley & Sons, Inc. for permission to publish some material that appeared in the *Journal of Electron Microscopy Technique*.

References

Becker, R. P. and Sogard, M. 1979. Visualization of subsurface structures in cells and tissues by backscattered electron imaging. *Scanning Electron Microsc.* II: 836.

Bell, P. B. 1981. The application of scanning electron microscopy to the study of the cytoskeleton of cells in culture. *Scanning Electron Microsc.* II: 139.

Brinkley, B. R., Fuller, G. M., and Highfield, D. P. 1975. Cytoplasmic microtubules in normal and transformed cells in culture: Analysis by tubulin antibody immunofluorescence. *Proc. Natl. Acad. Sci. U.S.A.* 72: 4981.

Buckley, I. K. and Porter, K. R. 1985. Electron microscopy of critical point dried whole cultured cells. *J. Microsc.* 104: 107.

Cande, W. Z., McDonald, K., and Meeusen, R. L. 1981. A permeabilized cell model for studying cell division: A comparison of anaphase chromosome movement and cleavage furrow constriction in lysed PtK$_1$ cells. *J. Cell Biol.* 88: 618.

DeMey, J., Moeremans, M., Geuens, G., Nuydens, R., and De Brabander, M. 1981. High resolution light and electron microscopic localization of tubulin with the IGS (Immuno-Gold Staining) method. *Cell Biol. Int. Rep.* 5: 889.

DeMey, J., Lambert, A. M., Bajer, A. S., Moeremans, M., and De Brabander, M. 1982. Visualization of microtubules in interphase and mitotic plant cells of *Haemanthus* endosperm with the immuno-gold staining method. *Proc. Natl. Acad. Sci. U.S.A.* 79: 1898.

DeMey, J. 1984. Colloidal gold as marker and tracer in light and electron microscopy. *Electr. Micr. Soc. Amer. Bull.* 14: 54.

DeNee, P. B. and Abraham, J. L. 1976. Backscattered electron imaging. Application of atomic number contrast. In: *Principles and Techniques of Scanning Electron Microscopy.* Vol. 5. pp. 144–180. chapter 8. Hayat, M. A. (Ed.), Van Nostrand Reinhold, New York.

de Harven, E., Leung, R., and Christensen, H. 1984. Backscattered electron imaging of the colloidal gold surface marker. *Proc. Electron Microsc. Soc. Amer.* 42: 254.

de Harven, E. and Soligo, D. 1989. Backscattered electron imaging of the colloidal gold marker on cell surfaces. In: *Colloidal Gold: Principles, Methods, and Applications,* Vol. 1. pp. 230–251. Hayat, M. A. (Ed.), Academic Press, San Diego.

Goode, D. 1975. Mitosis of embryonic heart cells *in vitro:* an immunofluorescence and ultrastructural study. *Cytobiologie* 11: 203.

Goode, D. 1981. Microtubule turnover as a mechanism of mitosis and its possible evolution. *BioSystems* 14: 271.

Goode, D. and Cachon, M. 1985. Immunofluorescence and ultrastructural studies of spindle microtubules during mitosis in the colonial radiolarian *Collozoum pelagicum. Biol. Cell* 53: 41.

Goode, D. and Maugel, T. K. 1987. Backscattered electron imaging of immunogold-labeled and silver-enhanced microtubules in cultured mammalian cells. *J. Electron Microsc. Tech.* 5: 263.

Goode, D. and Sarma, V. 1986. Incorporation and turnover of labeled exogenous tubulin in the mitotic spindles of *Chaetopterus* oocytes and HeLa cells. *Cell Motil. Cytoskel.* 6: 114.

Horisberger, M. 1981. Colloidal gold: A cytochemical marker for light and fluorescence microscopy and for transmission and scanning electron microscopy. *Scanning Electron Microsc.* II: 9.

Kilmartin, J. V., Wright, B., and Milstein, C. 1982. Rat monoclonal antitubulin antibodies derived by using a new nonsecreting rat cell line. *J. Cell Biol.* 93: 576.

Lazarides, E. and Weber, K. 1974. Actin antibody: the specific visualization of actin filaments in non-muscle cells. *Proc. Natl. Acad. Sci. U.S.A.* 71: 2268.

Lynn, D. H. and Small, E. B. 1981. Protist kinetids: structural conservatism, kinetid structure, and ancestral states. *BioSystems* 14: 377.

Namork, E. 1991. Double labeling of antigenic sites on cell surfaces imaged with backscattered electrons. In: *Colloidal Gold: Principles, Methods, and Applications,* Vol. 3. pp. 188–209. Hayat, M. A. (Ed.), Academic Press, San Diego.

Osborn, M. and Weber, K. 1977. The display of microtubules in transformed cells. *Cell* 12: 561.

Osborn, M., Webster, R. E., and Weber, K. 1978. Individual microtubules viewed by immunofluorescence and electron microscopy in the same PtK$_2$ cell. *J. Cell Biol.* 77: R24.

Robinson, V. N. E. 1980. Imaging with backscattered electrons in a scanning electron microscope. *Scanning* 3: 15.

Schollmeyer, J. E., Furcht, J. E., Goll, D. E., Robeson, R. M., and Stromer, M. M. 1976. Localization of contractile proteins in smooth muscle cells and in normal and transformed fibroblasts. *Cold Spring Harbor Conf. Cell Proliferation* 3: 361.

Small, E. B., Wetzel, B., Maugel, T. K., and Meola, P. 1980. Preliminary study of ciliates stained with silver albuminose (Protargol) examined in the backscattered electron mode with the SEM. *Scanning Electron Micros.* III: 543.

Tellez, C., Small, E. B., Corliss, J. O., and Maugel, T. K. 1982. The ultrastructure of specimens of *Paramecium multimicronucleatum* impregnated with silver by the Fernández-Galiano method. *J. Protozool.* 29: 627.

Thiebant, F., Rigant, J. P., Feren, K., and Reith, A. 1984. The application of the nuclear organizer region silver staining (AgNOR) to backscattered electron imaging. *Biol. Cell* 52: 103.

Walsh, C. 1984. Synthesis and assembly of the cytoskeleton of *Naegleria gruberi* flagellates. *J. Cell Biol.* 98: 449.

White, J. G., Amos, W. B., and Fordham, M. 1987. An evaluation of confocal microscopy versus conventional imaging of biological structures by fluorescence light microscopy. *J. Cell Biol.* 105: 41.

Chapter 16

Pre-Embedding Immunogold-Silver Staining of Plasma Membrane-Associated Antigens

Gian Carlo Manara, Corrado Ferrari, Lucilla Badiali-De Giorgi, and Gianandrea Pasquinelli

Contents

16.1 Introduction

Colloidal gold was originally introduced as an electron dense marker for immunoelectron microscopy (IEM) by Faulk and Taylor (1971). Colloidal gold markers have the advantage of low unspecific background when compared to ferritin- and peroxidase-coupled markers. On the other hand, one of their major drawbacks is a concomitant loss of sensitivity (Courtoy et al., 1983; Kerjaschki et al., 1986; Singer et al., 1987), especially when large-sized colloidal gold particles are utilized (Horisberger and Tacchini-Vonlanthen, 1983; Slot and Geuze, 1983). In order to increase the sensitivity of the method, an immunogold-silver staining (IGSS) procedure was introduced by Holgate et al. (1983). The silver intensification of small colloidal gold particles yields a higher sensitivity as compared to the use of large gold particles without silver-enhancement (Van den Pol, 1989).

    In the IGSS method, immunogold-labeled specimens are incubated in a physical developer which deposits concentric layers of metallic silver around the gold particles. The colloidal gold particles catalyze the reduction of silver ions into metallic silver, which is deposited on the surface of the particles (Holgate et al., 1983). Consequently, at the end of the developmental phase of the process, each gold granule becomes encapsulated in a coating of silver (for more details see Chapter 1). Because the IGSS procedure effectively increases the diameter of the gold particle, which increases its detectability by means of different imaging systems, it has been subsequently used for detecting cell surface antigens in cell suspensions at the light microscope (De Waele et al., 1986a,b, 1989; Otsuki et al., 1990), as well as at the transmission (Manara et al., 1989, 1991; Otsuki et al., 1990), and scanning (Scopsi et al., 1986) electron microscope levels. We will describe the detection of blood cells and Langerhans cells plasmalemma-associated antigens in both transmission electron microscopy (TEM) and scanning electron microscopy (SEM).

16.2 Methodology

16.2.1 Blood Cell Preparation

Human peripheral heparinized blood was obtained from five randomly selected healthy donors. Mononuclear cells were isolated by Ficoll-Hypaque density gradient centrifugation (Böyum, 1986) and resuspended in RPMI 1640 (Gibco, Paisley, Scotland) containing 10% fetal calf serum (FCS), and 50 µg/ml gentamycin, at a concentration of 5×10^6 cells/ml.

16.2.2 Epidermal Cell Suspension Preparation

Freshly suspended human epidermal cells were obtained from normal skin of five healthy subjects, undergoing breast or abdomen plastic surgery reduction operations. Trimmed skin was split-cut with a keratotome set at 0.4 mm. The resulting slices were incubated, dermal side down, in Hanks' balanced salt solution (Gibco) containing 0.25% trypsin for 1 h at 37°C. The epidermis was then peeled from the dermis with fine forceps and the epidermal sheets were placed in RPMI 1640 supplemented with 10% FCS, shaken and pipetted vigorously for several minutes. After filtration through sterile gauze, the cells were washed in RPMI 1640 plus 10% FCS and resuspended in the same medium, containing 50 µg gentamycin at a concentration of 5×10^6 cells/ml. Epidermal cell viability ranged between 80 and 90%, as determined by eosin exclusion.

16.2.3 Prefixation

Samples for TEM were prefixed with 0.1% glutaraldehyde in 0.1 M phosphate buffer (pH 7.2) at 4°C for 10 min, or with 4% formaldehyde in 0.1 M phosphate buffer (pH 7.2) at 4°C for 30 min. For SEM, 50 µl of cell suspensions were allowed to adhere onto glass coverslips, previously coated with 1% polylysine (77,000 mol. wt.) in 0.01 M phosphate-buffered saline (PBS) solution (pH 7.2) for 30 min at room temperature. Specimens were washed with PBS to remove unattached cells and then fixed with one of the following solutions: (1) 0.1% glutaraldehyde in 0.1 M phosphate buffer (pH 7.2) for 5 min at room temperature; (2) 2% formaldehyde-0.1% glutaraldehyde in 0.1 M phosphate buffer (pH 7.2) for 5 min at room temperature; or (3) 4% formaldehyde in 0.1 M phosphate buffer (pH 7.2) for 15 to 30 min at room temperature. The choice of the fixative depends on the antigen sensitivity to fixatives. After fixation, cells were washed three times in PBS. If a glutaraldehyde solution is chosen, rinsing buffers contain 50 mM glycine to quench aldehyde groups.

16.2.4 Immunogold Labeling

Immunogold labeling was performed at the TEM as well as at the SEM levels according to the methods previously described (De Harven et al., 1984; De Harven and Soligo, 1986;

Manara et al., 1986, 1990; Ferrari et al., 1989). After several washings with PBS, cells were incubated in PBS containing 20% normal goat serum (Amersham, Buckinghamshire, England) and 1% bovine serum albumin (BSA) for 30 min, in order to saturate Fc receptors as well as unspecific protein binding sites. Cells were washed twice in PBS-0.1% BSA and incubated with the specific monoclonal antibody (MoAb) diluted in PBS-1% BSA for 60 min. Subsequently, cells were washed in PBS-0.1% BSA and then in 0.02 M Tris-HCl buffer (pH 8.2) containing 0.02 M sodium azide and 0.1% BSA (0.1% BSA buffer).

Cells for TEM were incubated for 1 h in a goat-antimouse antibody coupled to either 5 nm- or 1 nm-sized colloidal gold particles (Amersham) diluted 1:10 in 0.02 M Tris-HCl buffer (pH 8.2) containing 0.02 M sodium azide and 1% BSA (1% BSA buffer). Cells for SEM were incubated for 30 min in a goat-antimouse antibody coupled to either 5 nm- or 15 nm-sized gold particles, diluted 1:10 in 1% BSA buffer. Samples were washed twice in 0.1% BSA buffer, and then in 0.1 M cacodylate buffer (pH 7.4). Incubations were performed at room temperature. All SEM procedures were carried out in a moist chamber. Finally, in order to minimize the loss of gold label during the silver-enhancement step, cells were fixed with 2% glutaraldehyde in cacodylate buffer for 30 min at room temperature.

Immunolabeling specificity was checked by (1) omitting the incubation with the MoAb, (2) substituting the MoAb with unreactive immunoglobulins of the same isotype (Pel Freez Biologicals, Rogers, USA), or (3) substituting the MoAb with an unrelated MoAb.

16.2.5 Silver-Enhancement Procedure

Fixed cells were washed twice in cacodylate buffer, and rinsed in distilled water. The silver-enhancement procedure was performed by incubating the specimen cells in a developer prepared by mixing equal volumes of the "initiator solution" and "enhancer solution" of the commercially available light-insensitive IntenSE M kit (Amersham). The two solutions were mixed immediately prior to use, and the enhancement procedure was performed in a water bath for 3 to 7 min at 22°C. Finally, cells were thoroughly washed in deionized water.

16.2.6 Ultrastructural Studies

After the IGSS procedure, cells for TEM were packaged in 2% Bacto-Agar (Difco Laboratories, Detroit, MI) at 45°C, postfixed in OsO_4 reduced with potassium ferricyanide (Karnovsky, 1971), dehydrated in a graded acetone series, and embedded in Durcupan ACM (Fluka, Buchs, Switzerland). Specimens were observed under a Philips EM300 TEM. Cells for SEM were dehydrated through graded ethanols, critical-point dried in liquid CO_2, mounted on a suitable stub by means of silver or colloidal-graphite glue, and coated with evaporated carbon. Specimens were observed under a Philips 505 SEM using either secondary electron imaging (SEI) or backscattered electron imaging (BEI).

16.3 Concluding Remarks

The use of small colloidal gold particles enhances the labeling degree when compared with larger-size particles (Horisberger and Tacchini-Vonlanthen, 1983; Kellenberger and Hayat, 1991). On the other hand, small-sized gold particles are moderately electron dense and not ideally suited for an easy detection. Consequently, a relatively high magnification is needed to detect small gold particles on labeled cells. Therefore, the ability to discern the overall distribution of an immunoreaction is not only highly time consuming, but also

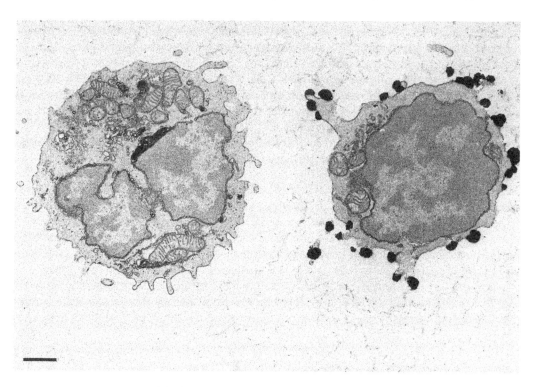

Figure 1   TEM. Immunogold-silver staining of a cell surface-localized antigen. Two peripheral blood lymphocytes, one of which shows on its surface large electron-dense deposits due to metallic silver precipitation on 5-nm colloidal gold particles. The CD4 molecule has been labeled. Ultrastructural details are well preserved. Prefixation performed with 0.1% glutaraldehyde in 0.1 M phosphate buffer (pH 7.2). Enhancement time is 5 min. MoAb used: OKT4 (Ortho Diagnostic System, Raritan, N.J.) diluted 1:10. Section not poststained. Bar = 1 µm.

considerably reduced. This is especially the case when a low percentage of labeled cells and/or low moieties of antigens are present in the sample under investigation.

The IGSS largely overcomes this pitfall and considerably facilitates the detection of cells labeled by a small-sized gold marker because of metallic silver deposition as concentric shells around colloidal gold particles (Figure 1). It has been evaluated that the gold probe can be amplified by a factor of about 100 (Hodges, 1993). On the other hand, a major drawback of the IGSS technique, when compared to immunogold methods, is represented by the impossibility to perform semiquantitative investigations (Manara, 1991). Further, unwanted metallic silver formation can also occur which gives rise to some background. In fact, chemical groups in the tissue may reduce the silver ions and lead to silver granule appearance (Gallyas, 1979). The silver ions and the reducing agent (hydroquinone) may also interact autocatalytically within the solution itself. This process is favored by the presence of contaminating chloride ions with the consequent generation of silver chloride (De Waele, 1989). Rinsing the specimens with deionized water before and after the silver-enhancement procedure minimizes the contamination with chloride ions (Danscher, 1981). The cells should be free of heavy metal contamination, and all contact of the enhancer solution with metal should be avoided.

The silver-enhancement step is essential for 1-nm colloidal gold particles because of low marker contrast (Stierhof et al., 1991). Due to its small size, IgG-coupled 1-nm colloidal gold yields a greater sensitivity in that a larger amount of primary antibody or antigen is

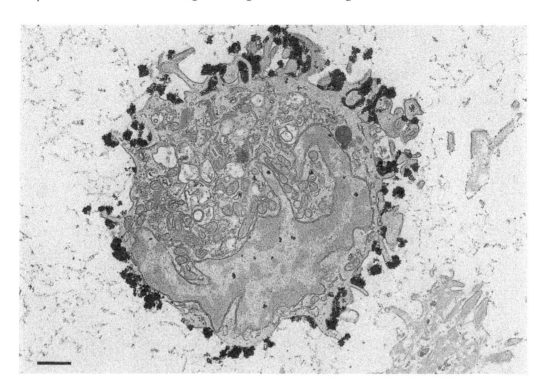

Figure 2 TEM. Immunogold-silver staining of the CD11b molecule exposed on the cell surface of a freshly suspended epidermal Langerhans cell. Large electron dense deposits, which are generated by metallic silver deposition around 1-nm gold granules, are easily detectable on the cell surface. Few unspecific silver granules are visible, mainly on the nucleus. Prefixation performed with 0.1% glutaraldehyde in 0.1 M phosphate buffer (pH 7.2). Enhancement time is 7 min. MoAb used: Dako-C3bi-R (Dakopatts, Glostrup, Denmark) diluted 1:10. Section not poststained. Bar = 1 μm.

detectable. Therefore, even limited amounts of antigenic moieties can be intensely stained and easily detected as, for example, the heterodimer CD11b/CD18 in the case of Langerhans cells (De Panfilis et al., 1989) (Figure 2). The IGSS performed using 1-nm gold granules besides being used in pre-embedding detection of cell surface-localized molecules, has been efficiently employed in postembedding labeling of intra- and extracellular antigens (Shimizu et al., 1992). Moreover, because of the extremely small size, the penetration of 1-nm gold particles into suspended cells can be achieved with very mild permeabilization procedures. The subsequent silver-enhancement procedure allows visualization of intracellular-labeled antigens (Rocchi et al., 1993).

At the SEM level, IGSS provides an accurate visualization of surface immunoreactions, even at a low magnification, by using the BEI mode (Figures 3 and 4b). Individual silver-enhanced gold particles are also clearly detectable at higher magnification by using the SEI mode (Figure 4a). The IntenSE M kit offers a simple silver-enhancement method which is characterized by a faster intensification process in comparison to the silver lactate/hydroquinone/gum arabic procedure (Stierhof et al., 1991). Furthermore, the IntenSE M kit is light insensitive, has a reduced sensitivity to contaminating ions, and does not affect ultrastructural features. The development has proven to be variable between experiments, however. The silver-intensified particles are not equal in size and the size of some of these particles could preclude precise ultrastructural identification of labeled structures. The fast reaction probably leads to an unequal and uncontrollable enhancement of single particles

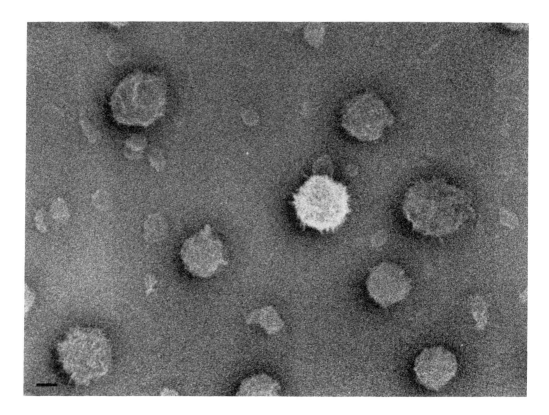

Figure 3   SEM. Immunogold-silver staining of the 72/73 kDa heat shock protein localized on the cell surface of a human peripheral blood lymphocyte. Backscattered electron mode with normal polarity (+) clearly highlights the intensity of the labeling even at low magnification. The electron dense marker is represented by silver-enhanced 15-nm gold particles. Prefixation performed with 2% paraformaldehyde-0.1% glutaraldehyde in 0.1 M phosphate buffer (pH 7.2). Enhancement time is 3 min. MoAb used: Anti-HSP70/HSC70 (Stress Gen, Victoria, Canada) diluted 1:1000. Bar = 2 μm.

by favored growth of some active sites (Danscher, 1981 and in this volume). Moreover, the reproducibility of enhancement is difficult to reach due to the variation in gold particle size of the original gold probe and to the gold size variation from batch to batch. Conversely, the embedding media used do not influence silver-enhancement (Stierhof et al., 1991).

Finally, differences in temperature and stirring conditions give rise to size variations. We suggest using the IntenSE M kit at 22°C. A higher temperature would require a shorter enhancement time. It has been recently proposed to improve the enhancement quality of the IntenSE M kit by adding protective colloid gum arabic which minimizes the autocatalytic precipitation of the intensifying solution (Stierhof et al., 1991). However, the use of slow-acting developer is more time consuming.

Despite these drawbacks, it is reasonable to conclude that the pre-embedding IGSS represents a reliable tool for a rapid and sensitive examination of suspended cells immunophenotype at the TEM and SEM levels. The mild prefixation preserves both antigenicity and ultrastructural features; the use of small colloidal gold particles allows detection of even small quantities of antigen, and the metallic silver precipitation leads to a large reaction product which permits the easy detection of labeled cells even at low magnifications.

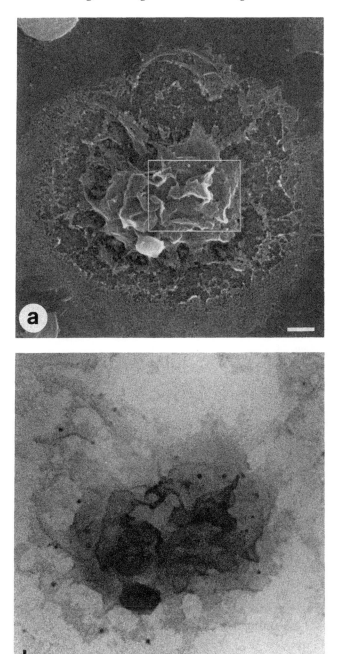

Figure 4 SEM. Immunogold-silver staining. (a) Secondary electron (+) imaging mode of a human spreaded monocyte showing prominent surface ruffles. Some silver-enhanced gold granules are visible on the cell surface. (b) The backscattered electron (–) imaging mode of the boxed area shows scattered highly dense silver-enhanced gold particles linked to the monocyte surface. The low expression of the 72/73 kDa heat shock protein on monocytes is well preserved when the prefixation procedure is performed with 4% paraformaldehyde in 0.1 M phosphate buffer (pH 7.2). On the other hand, the avoidance of glutaraldehyde in the prefixation buffer leads to loss of cell surface details to some extent. Enhancement time is 3 min. MoAb used: Anti-HSP70/HSC70 (Stress Gen, Victoria, Canada) diluted 1:1000. Bar = 1 μm.

References

Böyum, A. 1968. Isolation of mononuclear cells and granulocytes from human blood. *Scand. J. Clin. Lab. Invest.*, Suppl. 97: 77.

Courtoy, P. J., Picton, D. H., and Farquhar, M. G. 1983. Resolution and limitations of the immunoperoxidase procedure in the localization of extracellular matrix antigens. *J. Histochem. Cytochem.* 31: 1945.

Danscher, G. 1981. Histochemical demonstration of heavy metals. A revised version of the silver sulphide method suitable for both light and electron microscopy. *Histochemistry* 71: 1.

De Harven, E., Leung, R., and Christensen, H. 1984. A novel approach for scanning electron microscopy of colloidal gold-labeled surfaces. *J. Cell Biol.* 99: 53.

De Harven, E. and Soligo, D. 1986. Scanning electron microscopy of cell surface antigens labeled with colloidal gold. *Am. J. Anat.* 175: 277.

De Panfilis, G., Soligo, D., Manara, G. C., Ferrari, C., and Torresani, C. 1989. Adhesion molecules on the plasma membrane of epidermal cells. I. Human resting Langerhans cells express two members of the adherence-promoting CD11/CD18 family, namely H-Mac-1 (CD11b/CD18) and gp150,95 (CD11c/CD18). *J. Invest. Dermatol.* 93: 60.

De Waele, M., De Mey, J., Renmans, W., Labeur, C., Reynaert, P., and Van Camp, B. 1986a. An immunogold-silver staining method for the detection of cell surface antigens in light microscopy. *J. Histochem. Cytochem.* 34: 935.

De Waele, M., De Mey, J., Renmans, W., Labeur, C., Jochmans, K., and Van Camp, B. 1986b. Potential of immunogold-silver staining for the study of leukocyte subpopulations as defined by monoclonal antibodies. *J. Histochem. Cytochem.* 34: 1257.

De Waele, M. 1989. Silver-enhanced colloidal gold for the detection of leukocyte cell surface antigens in dark-field and epipolarization microscopy. In: *Colloidal Gold: Principles, Methods, and Applications.* Vol. 2. p. 443. Hayat, M. A. (Ed.), Academic Press, New York.

Faulk, W. and Taylor, G. 1971. An immunocolloid method for the electron microscope. *Immunocytochemistry* 8: 1081.

Ferrari, C., De Panfilis, G., and Manara, G. C. 1989. Preembedding immunogold staining of cell surface-associated antigens performed on suspended cells and tissue sections. In: *Colloidal Gold: Principles, Methods, and Applications.* Vol. 2. pp. 323–343. Hayat, M. A. (Ed.), Academic Press, New York.

Gallyas, F. 1979. Light insensitive physical developers. *Stain Technol.* 54: 173.

Hodges, G. M. 1993. Immunocytochemical techniques for TEM and SEM. In: *Procedures in Electron Microscopy.* p. 8:0.1. Robards, A. W. and Wilson, A. J. (Eds.), Wiley & Sons Ltd, London.

Holgate, C. S., Jackson, P., Cowen, P. N., Bird, C. C. 1983. Immunogold-silver staining: new method of immunostaining with enhanced sensitivity. *J. Histochem. Cytochem.* 31: 938.

Horisberger, M. and Tacchini-Vonlanthen, M. 1983. Stability and steric hindrance of lectin-labeled gold markers in transmission and scanning electron microscopy. In: *Lectins.* Vol. 3. p. 189. Bog-Hansen, T. C. and Spengler, G. A. (Eds.), De Gruyler, Berlin.

Karnovsky, M. J. 1971. Use of ferricyanide reduced osmium tetroxide in electron microscopy (abstract). *J. Cell Biol.* 51: 284A.

Kellenberger, E. and Hayat, M. A. 1991. Some basic concepts for the choice of methods. In: *Colloidal Gold: Principles, Methods, and Applications,* Vol. 3. pp. 1–30. Hayat, M. A. (Ed.), Academic Press, San Diego.

Kerjaschki, D., Sawada, H., and Farquhar, M. G. 1986. Immunoelectron microscopy in kidney research: Some contributions and limitations. *Kidney Int.* 30: 229.

Manara, G. C., Ferrari, C., Scandroglio, R., Rocchi, G., Pagani, L., and De Panfilis, G. 1986. Characterization of two morphologically distinct Leu-7+ cell subsets with respect to Leu-15 antigen. Evaluation of Leu-15 determinant distribution on both E rosetting and non-adherent non-E rosetting cell populations. *Scand. J. Immunol.* 23: 225.

Manara, G. C., Ferrari, C., Torresani, C., Sansoni, P., and De Panfilis, G. 1989. The immunogold-silver staining approach in the study of lymphocyte subpopulations in transmission electron microscopy. *J. Immunol. Methods* 128: 59.

Manara, G. C., Soligo, D., Lambertenghi-Deliliers, G., Ferrari, C., and De Panfilis, G. 1990. Immunogold scanning electron microscopy applied to the study of Langerhans cells immunophenotype. *Dermatologica* 180: 141.

Manara, G. C., Ferrari, C., Torresani, C., and De Panfilis, G. 1991. The immunogold-silver staining procedure in the study of freshly suspended Langerhans cells at the transmission electron microscopic level. *Dermatologica* 182: 221.

Manara, G. C. 1991. Immunogold silver staining method at the transmission electron microscopic level (letter). *J. Histochem. Cytochem.* 39: 385.

Otsuki, Y., Maxwell, L. E., Magari, S., and Kubo, H. 1990. Immunogold-silver staining method for light and electron microscopic detection of lymphocyte cell surface antigens with monoclonal antibodies. *J. Histochem. Cytochem.* 38: 1215.

Rocchi, G., Pavesi, A., Ferrari, C., Bolchi, A., and Manara, G. C. 1993. A new insight into the suggestion of a possible antigenic role of a member of the 70 kDa heat shock proteins. *Cell Biol. Int.* 17: 83.

Scopsi, L., Larsson, L.-I., Bastholm, L., and Hartvig Nielsen, M. 1986. Silver-enhanced colloidal gold probes as markers for scanning electron microscopy. *Histochemistry* 86: 35.

Shimizu, H., Ishida-Yamamoto, A., and Eady, R. A. J. 1992. The use of silver-enhanced 1-nm gold probes for light and electron microscopic localization of intra- and extracellular antigens in skin. *J. Histochem. Cytochem.* 40: 883.

Singer, S. J., Tokuyasu, K. T., Keller, G.-A., Takata, K., and Dutton, A. H. 1987. Immunoelectron microscopy and molecular structure of cells. *J. Electr. Microsc.* 36: 63.

Slot, J. and Geuze, H. J. 1983. The use of protein A-colloidal gold (pAg) complexes as immunolabels in ultrathin sections. In: *Immunohistochemistry*. p. 323. IBRO Handbook Series. Cuello, A. C. (Ed.), Wiley, Chichester, England.

Stierhof, Y.-D., Humbel, B. M., and Schwarz, H. 1991. Suitability of different silver enhancement methods applied to 1 nm colloidal gold particles: an immunoelectron microscopic study. *J. Electr. Microsc. Tech.* 17: 336.

Van den Pol, A. N. 1989. Neuronal imaging with colloidal gold. *J. Microsc.* 155: 27.

Chapter 17

Immunogold-Silver Staining of Fungal Microorganisms

J. C. Cailliez and E. Dei-Cas

Contents

17.1 Introduction

During the last decade, most advances in cell biology of pathogenic fungi have resulted from the development of topochemical methods. The localization of fungal molecules having potentially high biological significance by using highly specific probes is becoming an increasingly important strategy in mycology, as well as in the laboratory diagnosis of fungal diseases. These fungal molecules are mainly represented by structural proteins, cell wall receptors, adhesins, secreted glycoproteins, enzymes, sexual pheromones, and cytoskeletal proteins (Linehan et al., 1988; Polonelli and Morace, 1986). They are specifically detected by lectins or monospecific antibodies (affinity-purified or monoclonal antibodies).

The analytical power of conventional light and electronic microscopy cytochemical methods was improved by the development of biological ligands and immunodetection high-resolution techniques. During the Seventh International Congress of Parasitology (ICOPA, Paris, 1990), it was considered that most of the important progress in parasite cell biology was due to the popularity of these fine analytical methods for the study of parasitic organisms (De Puytorac et al., 1991). This finding is also true for parasite fungi. Similar detection techniques, using highly specific reagents, were applied to identify fungal molecules by electrophoresis.

Now, immunogold-silver staining (IGSS) represents a new step in the evolution of immunodetection high-resolution methods. Biological probes are obtained by coupling ligands such as lectins, affinity-purified or monoclonal antibodies to colloidal gold particles of different sizes. Following pre- or postembedding methods, the gold-labeled probes can be used to localize fungal cells in the tissue embedded in paraffin (Springall et al., 1984) or in cryostat or whole-mount preparations as in conventional microscopical immunodetections. They can also be applied to immunoblotting techniques and enzyme-linked immunosorbent assays (ELISA) (see Chapter 19 in this volume). Primary ligands can be unlabeled. In this case, a gold-labeled secondary probe, enabling the detection of the primary ligand, is used (indirect IGSS).

The IGSS step consists of adding silver ions and hydroquinone which simultaneously adhere to the surface of gold particles. These aspects were recently analyzed by Danscher et al. (1993) and in this volume. After releasing its electrons, the hydroquinone molecule is detached from the gold particle, allowing new reducing molecules to adhere. Electrons are conducted to positive silver ions, causing transformation of silver ions into metallic silver atoms which remain intimately connected to the crystalline colloidal particles. In fact, it is the growth of the crystalline gold that allows the process to take place and leads to the formation of "silver-encapsulated gold particles" (Danscher et al., 1993). Gold labeling is thus followed by silver intensification which forms spheres of heavy metal around gold granules (enhancement similar to photographic development). The contrast and ultrastructural definition are better, or at least similar, to that obtained by using conventional immunological methods. Moreover, IGSS is four to eight times more sensitive than the standard peroxidase-antiperoxidase (PAP) technique (Danscher et al., 1993).

Immunogold-silver staining is a powerful topochemical tool for medical or veterinary mycologists. It was applied in mycology research for localizing and characterizing fungal molecules of potential biological significance (Reiss et al., 1992; Van den Bossche et al., 1992). Thus, it became possible to study fungal proteins and glycoproteins because of their significant biological properties and involvement in the fungal cell metabolism (Cailliez et al., 1992b; Reiss et al., 1992; Van den Bossche et al., 1992).

17.2 Fixation and Embedding Procedures of Fungal Microorganisms

For light microscopy, animal tissues containing parasite yeasts are fixed in formaldehyde saline solutions or Bouin's fluids before dehydration through graded alcohols. Fixation lasts between 12 and 24 h. Samples are embedded in paraffin (Cailliez et al., 1990). Temperatures above 58°C must be avoided. Sections of 5 to 8 µm thickness are mounted on glass slides and eventually treated by 2% aminopropyl-triethoxysilane in acetone (Rentrop et al., 1986). Sections are dewaxed twice for 5 min each in toluene before being rehydrated in graded alcohols. Slides are washed in gently running tap water for 5 to 20 min, especially when Bouin's solution is used. Following this standard protocol, the presence of *Candida* yeast cells in animal tissues was obtained by using IGSS with specific antibodies (Cailliez et al., 1990).

For electron microscopy, mixtures of low concentrations of glutaraldehyde and paraformaldehyde are effective in preserving protein and glycoprotein fungal antigens. Negative effects of fixation on the immunoreactivity of some fungal antigens have been reported, however. For example, fixation with high concentrations of glutaraldehyde altered the structural conformation of a *Pichia anomala* killer toxin and decreased the binding capacity of monoclonal antibodies by killer yeast antigens (Cailliez et al., 1992a). Nonspecific labeling can occur in the IGSS method depending on the fixative used. Fixation procedures are also involved in the dissociation of fungal proteins and glycoproteins thereby affecting their accessibility to antibodies or lectins (Li and Cutler, 1991). Glutaraldehyde, a dialdehyde producing irreversible cross-links between the amine residues of proteins, reacts with the Σ-amino groups of lysine in proteins and glycoproteins and, probably, with the glucosamine of the fungal chitin (Gorman et al., 1980). The ability of live and glutaraldehyde-fixed *Candida albicans* cells to bind lectins and/or monospecific antibodies has been studied (Mleczko et al., 1989). Glutaraldehyde-fixed yeast cells show little binding to lectins and antibodies. Moreover, the fixative penetrates slowly through the cell wall and plasmalemma of fungi. Surface glycoproteins, which are readily exposed to the denaturing action of glutaraldehyde, are altered in their tertiary structure; thus, preventing the binding of lectins and antibodies (Cailliez et al., 1992a).

The relevance of the fixation-embedding procedures to the preservation of cell structure and molecular reactivity was strikingly illustrated by *Pneumocystis carinii*, a parasitic fungus-like microorganism. A large number of papers on *P. carinii* ultrastructure were published before 1990 (for review, see Palluault and Akono, 1993) and various fixation protocols were tested. However, they did not succeed to sufficiently protect *P. carinii* cells from cytoplasmic extraction. Moreover, while ultrastructural studies of *P. carinii* showed the presence of carbohydrate moieties in the cell wall (Pesanti and Shanley, 1988; Yoshikawa et al., 1988; Palluault et al., 1990) and revealed host IgG bound to the parasite surface (Blumenfeld et al., 1990), studies showing clear labeling of cytoplasmic structures (which usually were altered by fixation) were lacking. In contrast, by using high osmotic pressure (≈900 mOsm) during fixation and embedding steps, Palluault et al. (1992a,b) obtained good preservation of the *P. carinii* cell structure and antigenicity. New fixation-embedding protocols enabled these authors to clearly show lectin labeling of endoplasmic reticulum and Golgi complex, as well as marked human IgG affinity of cytosol and surface parasite structures (Palluault et al., 1992b). The improvement in the preservation of parasite structure, resulting from the high osmolarity fixation procedures, opens the way for the development of immunocytochemical investigations of *P. carinii* with IGSS.

17.3 Gold-Conjugated Ligands

17.3.1 Lectin Gold-Silver Staining

Among biological probes displaying binding properties with fungal antigens, lectins were the most popular. They have been widely used in mycological studies because of their ability to bind to specific sequences of oligosaccharides belonging to a large panel of fungi (Virtanen et al., 1981).

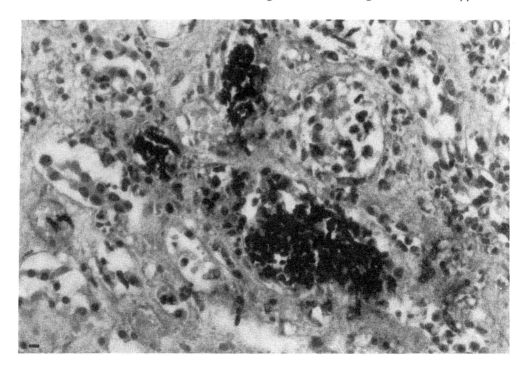

Figure 1   Concanavalin A-gold silver staining of *Candida albicans*-infected human kidney sections embedded in paraffin. Proliferation of fungal cells in kidney epithelia was strongly stained with silver intensification. Bar = 100 μm.

Lectins can be conjugated with colloidal gold particles and used both at the light and electron microscopic level to visualize the topography of polysaccharide moieties on fungal cell wall or cytoplasmic compartments. Among the most commonly used lectins, are the α-D-Glucose or α-D-Mannose binding lectins: Concanavalin A (Con A), *Pisum sativum* agglutinin (PSA), *Lens culinaris* agglutinin (LCA), and N-acetyl glucosamine specific lectins such as wheat germ agglutinin (WGA). By Concanavalin A-gold silver-staining (Con A-GSS), both yeast and hyphal cells of *C. albicans* were strongly labeled (Cailliez et al., 1990). A strong contrast was obtained with Con A-GSS which allowed to clearly localize the site of proliferation of pathogenic fungi in host tissues, even at low magnification (Figure 1). Although the silver intensification allowed strong enhancement of yeast and hyphal labeling, a gray background could be observed. This nonspecific staining was probably due to lectin binding molecules which were naturally present in host tissues. However, a good localization of fungi in infected tissues was seen because this background was very faint.

17.3.2 Monospecific Antibody Gold-Silver Staining

Antisera are produced by immunization with pathogenic fungi. Polyclonal antibodies are rendered monospecific by adsorption with closely related fungal antigens. A great variety of antibodies has been linked to colloidal gold particles (De Mey, 1986). Antibodies reacting with cell wall compounds and secreted fungal glycoproteins were used in IGSS for the labeling of tissue sections containing fungal cells (Figure 2). Intense immunostaining was also obtained by IGSS performed on isolated yeast cells (Figure 3). The inherent density of the staining was found to improve the localization of fungi in paraffin sections and to allow the use of histological counterstaining for classical identification of host tissues (Cailliez et al., 1990).

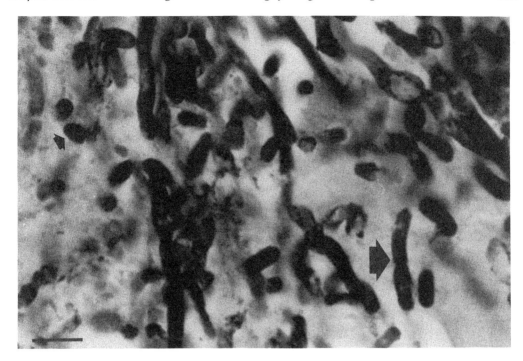

Figure 2 Indirect immunogold-silver staining reaction with whole anti-*Candida albicans* germ tube antiserum on infected human kidney section. This reaction provided a good definition of infection sites. *C. albicans* germ tubes were well stained in transverse (small arrow) and longitudinal (large arrow) sections. Bar = 100 μm.

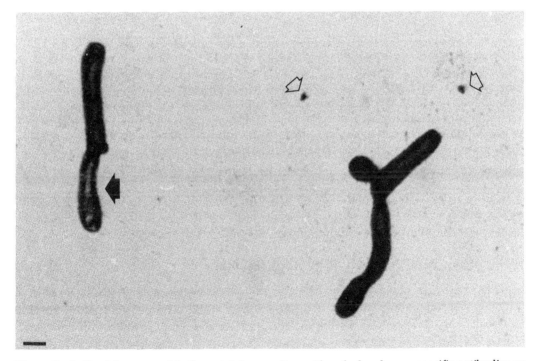

Figure 3 Indirect immunogold-silver staining reaction with polyclonal monospecific antibodies on yeast cells of *Candida pseudotropicalis*. This reaction provided a good contrast of yeast cell wall (full arrow). Some gold-conjugated antibodies were adsorbed on the grid membrane. They were enhanced by silver intensification, providing a weak background staining (opened arrows). Bar = 10 μm.

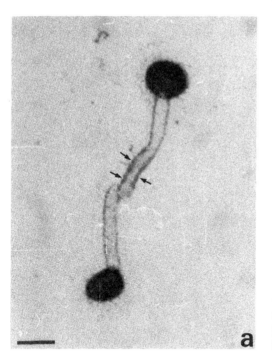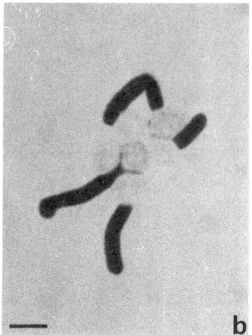

Figure 4 Two indirect immunogold-silver staining reactions with monospecific polyclonal antisera on *Candida albicans* germ tubes grown for 3 h at 37°C in RPMI 1640 medium. Each monospecific antibody was specific for *C. albicans* blastoconidia (a) or germ tubes (b). A punctuated labeling was observed at the surface of germ tubes (arrows). This particular labeling was easily visualized with the silver enhancement of gold particles. Bar = 10 μm.

Rabbit antisera were rendered monospecific after adsorption with different *Candida* species (for review, see Fukazawa, 1989). These monospecific antisera were used both in IGSS and immunofluorescence assay (IFA). By using appropriate antibody dilutions, IGSS allowed a good discrimination between yeast species (or between different parasitic stages such as blastoconidia and germ tubes). Although monospecific antibodies reacted only with mother blastoconidia of *C. albicans*, thereby, leaving germ tubes unlabeled (Figure 4a), other antibodies remained highly specific for germ tubes (Figure 4b). Young buds originating from blastoconidia were also labeled (even at low dilutions) by rabbit monospecific antisera which reacted with germ tubes of the same yeast species (Cailliez et al., 1990). Monospecific antibodies were produced against germ tubes of *C. albicans.* They were used in IGSS to localize the proliferation sites of fungi in human tissues (Figure 2). Finally, reproducible patterns of reactivity were obtained by using IGSS with monospecific antisera on blastoconidia and pseudo-hypha of different *Candida* species.

17.3.3 Monoclonal Antibody Gold-Silver Staining

In the last ten years, about 100 monoclonal antibodies (mabs) have been produced against antigenic determinants belonging to different fungal species (for a review, see Cailliez et al., 1992b; Reiss et al., 1992). Many of them were conveniently used as immunological probes to characterize proteins or glycoproteins involved in the pathogenesis of fungal infections (Cailliez and Poulain, 1988; Poulain et al., 1989). These molecules had peculiar biological and biochemical properties and were included in different molecular families (Linehan et al., 1988; Polonelli and Morace, 1986). The main part of these glycoproteins, secreted or not, were accumulated in the cell wall before being released into the culture medium (Cailliez et al., 1992b).

Most of mabs were "in house" probes which were developed and used in individual laboratories. They were directly linked to gold particles and used with different aims (often to localize the secretion sites of yeast glycoproteins or proteins carrying their corresponding epitopes) (Poulain et al., 1989). Each mab was specific for either the protein part or polysaccharidic fraction of fungal glycoproteins (Cailliez et al., 1992b). One of these mabs (named mab 5B2) was specific for *C. albicans* secreted glycoproteins. Mab 5B2 was used with success in IGSS to identify conglomerates of blastoconidia in infected human kidney sections (Cailliez et al., 1990). It was also used in cytological and biochemical studies of the antigenic composition of the cell wall of *C. albicans* strains (Cailliez and Poulain, 1988).

17.4 Indirect Gold-Silver Staining

Gold particles have became a widely employed marker for the localization of antibodies or lectins in both light, scanning, and transmission electron microscopy (Goodman et al., 1991; Hayat, 1989–1991). Indirect IGSS presents the advantage of increasing the labeling when antigenic determinants are present at low concentrations. However, because many commensal fungal species (such as *Candida*) are ubiquitous, secondary antibodies can directly react with these fungi, giving unspecific labeling (i.e., independent of the presence of primary antibody). In order to resolve this problem, primary ligands can be revealed by the avidin-biotin system. In this IGSS procedure, streptavidin is conjugated with gold particles and used as secondary reagent for the detection of biotinylated ligands. However, in some cases, the presence of endogenous biotin-like molecules in either parasite or host tissues can be responsible for background labeling. In these conditions, IGSS must also be performed in control assay by using another gold conjugated secondary ligand. This control attests to the specificity of immunolabeling. Alternatively, the second conjugated antibody can be linked with another marker (e.g., fluorescein, rhodamin, peroxidase, etc). In this case, results must be systematically compared with those of IGSS. When immunolabeling is performed on tissue sections, IGSS can be more easily compared with PAP staining because the same counterstaining can be used. When IGSS is performed on isolated fungal cells, comparative analysis can be made with IFA or PAP staining.

17.5 Choice and Control of Immunological Probes

17.5.1 Selection of Probes

In order to minimize the background staining associated with IGSS, primary ligands should be of high quality and specificity. Polyclonal monospecific or monoclonal antibodies generally yield specific labeling. Nevertheless, nonspecific binding occurs when corresponding epitopes belong to molecules largely represented in biological structures. In *C. albicans*, unspecific binding on cell wall compounds was reduced by using antibodies at their lowest effective concentration. As it was already shown by Roth et al. (1989), both the antibodies and the gold-lectin complexes should be optimally diluted before being used in IGSS.

The intensity of labeling moreover was correlated to the diameter of gold particles, irrespective of the specificity of each immunoprobe. It has been well established that using small gold particles results in higher labeling density as compared to using particles with larger size (Horisberger, 1981; Slot and Gueuze, 1983; Stierhof et al., 1986; Yokota, 1988; Hacker, 1989; Kellenberger and Hayat, 1991; Dansher et al., 1993). In our experience, the best results are obtained by using 5-nm colloidal gold particles. Antigens are probably more accessible to gold particles of this size, resulting in better penetration of gold-ligands and higher labeling density.

17.5.2 Precautionary Measures

Distilled water should be used in all IGSS steps. Glassware should be free of heavy metal contamination. Any contact of metallic objects with washing and incubation solutions must be avoided. This precautionary measure especially concerns the silver-enhancement mixtures. Temperature which can influence the reaction rate should be kept as constant as possible (37°C for incubation with each antibody, 25 to 28°C for incubation with lectins and 4°C for overnight incubations). Moreover, the stability of the gold colloid is improved by increasing the pH of the buffered saline solution which is employed for diluting immunogold reagents. The pH of the colloid must be adjusted to favor stable formation of ligand-gold complexes and to optimize long-term storage (Horisberger, 1985; Hayat, 1989). Immunogold probes are generally stored at 4°C in buffered solutions containing a 0.01% sodium azide solution as preservative for at least 1 month. The gold conjugates should be regularly checked in the electron microscope for the potential number of clustered particles. Indeed, agglutination of gold particles can be falsely interpreted as a concentration of antigens through the fungal cell structures.

17.6 Immunogold-Silver Staining in Medical and Veterinary Mycology

17.6.1 Laboratory Diagnosis of Fungal Diseases

Immunogold-silver staining methods were applied to the laboratory diagnosis of mycoses — a group of human and veterinary diseases of growing importance. The trend of various fungal species to develop parasitism in mammals seems independent of their taxonomic and phylogenetic relationships (Dei-Cas and Vernes, 1986). Pathogenic fungi are characterized by their ability to invade and develop in healthy hosts (as *Sporothrix schenkii, Coccidioides immitis, Histoplasma capsulatum,* or *Paracoccidioides brasiliensis*) or in immunocompromised patients (as *Aspergillus fumigatus, Candida* spp., and other opportunistic fungal species).

Among the fungal diseases, the opportunistic mycoses have become a major nosocomial infection. Invasive aspergillosis is frequent in neutropenic patients, second only to invasive candidiasis as the most frequent opportunistic fungal pathogen (Roberts et al., 1992). Candidiasis is now the third leading cause of death by systemic infection in American hospitals (Rinaldi, 1993). Clinical signs and symptoms of opportunistic fungal infections are generally nonspecific, which makes early diagnosis difficult. The correct diagnosis of a fungal infection requires the identification of the pathogenic fungus at the species level. The species identification of the pathogen is needed both to help clinicians determine the clinical relevance of specific species recovered from clinical specimens and to decide the appropriate antifungal treatment (Roberts et al., 1992).

The microscopic examination of biological fluid or tissue sample smears allows the detection of the fungal agent and its identification when the morphology of the parasite stage is typical. This is the case with *Cryptococcus neoformans* and some dimorphous pathogenic fungi such as *Histoplasma capsulatum.* However, in the case of commensal microorganisms such as *Candida* species, the presence of yeast in clinical samples does not necessarily mean mycosis. Moreover, the species identification is not always possible on the basis of the blastoconidium morphology. In practice, the presence of the commensal fungus in the host tissues demonstrates its involvement in the clinical picture. Now, hematoxylin-eosin or other similar routine stain does not usually stain fungal elements. Periodic Acid-Schiff reaction,

Gomori-Grocott, Orthotoluidin blue or other histochemical methods (Emmons et al., 1970) stain fungi in smears or histological preparations but have too low specificity. For instance, the identification of *Aspergillus* hyphae in tissue sections can become problematical with similar tissue morphologies displayed by a number of filamentous fungi that are emerging as significant pathogens in immunocompromised hosts (Smith et al., 1992). It is for these reasons that labeled polyclonal or monoclonal antibodies specific of fungal species are used for the microscopic identification of fungal pathogens.

Immunogold-silver staining has increased the efficiency of fungal detection and species specific identification in tissue sections. Of course, the specificity of each immunoassay was determined by the specificity of the antibodies used. From this point of view, the immunological importance of yeast cell wall mannan has been recognized. In *Candida* species, antigenic determinants of pathogenic strains were identified by using polyclonal monospecific antibodies in direct agglutination (Shinoda et al., 1981). The production of mabs to species specific antigens could lead to significant improvements of this diagnostic approach. However, mabs would be frequently too specific in so far as they do not stain neither all strains of a given fungus species nor all its parasitic stages (Hay, 1993).

The mycological culture remains the diagnostic reference. However, the recovery of deep fungal agents by culture is frequently difficult, especially in human systemic mycoses or in veterinary mycology (Buckley et al., 1992). In the veterinary laboratory, the recovery of important pathogens (Aspergilli and Zygomycotina) from clinical specimens is usually difficult or impossible as tissue samples are often contaminated or are received already fixed in formalin (Smith et al., 1992). Moreover, although marked advances in immunodiagnosis, antigen detection, and PCR/hybridization techniques were made (Buckley et al., 1992; Roberts et al., 1992), in practice, the definitive diagnosis still frequently depends on the microscopical examination of clinical specimens. For this reason, IGSS presents a great interest in medical and veterinary diagnosis.

17.6.2 Biotyping of Fungal Microorganisms

Epidemiological markers to monitor opportunistic infectious diseases, especially nocosomial infections, are a pressing need for microbiology laboratories (Morace et al., 1989). The incidence of opportunistic infections is increasing with the growing frequence of immunodeficiency states, associated with transplantation or cancer chemotherapy and with HIV infection. Various methods have been developed to differentiate bacterial, fungal, or parasitic species (or strains of the same species) for epidemiological purposes. For example, previous studies have shown that direct agglutination with monospecific antisera permitted the identification of different pathogenic *Candida* species (Shinoda et al., 1981).

With regard to pathogenic fungi, clinical isolates can be differentiated at the phenotypic level (i.e., physiological tests, serotyping, isoenzyme profiling or killer toxin sensitivity) or at the genotypic level (i.e., electrophoretic chromosomic pattern, restriction fragment length polymorphism, Southern-blotting hybridization) (Morace et al., 1989; Odds et al., 1992). Gold-labeled immune or lectin probes can be used as epidemiological markers to identify fungal agents on clinical specimens (i.e., smears, histological sections, or culture isolates). Fungal killer toxins were already used as epidemiological markers in some mycological laboratories (Polonelli et al., 1983, 1987; Morace et al., 1989). These toxins, which bind to specific receptors at the surface of sensitive fungal cells, were immunolocalized by using gold-conjugated mabs (Cailliez et al., 1994). Immunogold-silver staining methods, which until now were not applied to pathogenic fungus typing, could reveal specific probes used as epidemiological markers. By using appropriate dilutions of monospecific antibodies, IGSS has already allowed significant identification of pathogenic *Candida* species present in isolates or host tissues embedded in paraffin (Cailliez et al., 1990).

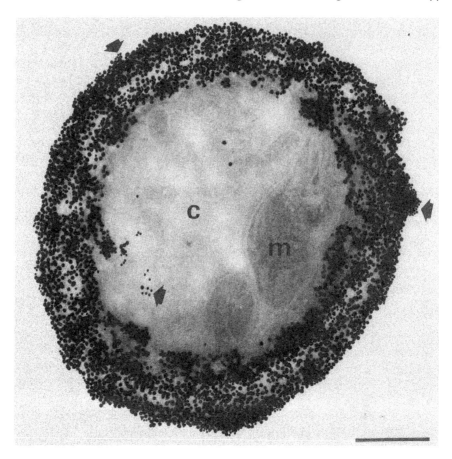

Figure 5   Ultrastructural double immunodetection with gold-conjugated Concanavalin A (14-nm size gold particles, small arrows) and gold-conjugated monoclonal antibody (mab 5B2) (40-nm size gold particles) on ultrathin section of *Candida albicans* blastoconidia grown in RPMI 1640 medium. The two labelings mainly occured in the yeast cell wall. Only few colloidal gold particles were found in the cytoplasm (c). m = mitochondria. Bar = 0.5 μm.

17.7 Immunogold-Silver Staining in Mycological Research

17.7.1 Cell Surface Distribution of Fungal Antigens

Cell surface antigens of parasite fungi have been extensively studied because of their involvement in the invasion of host tissues. They have been shown to be responsible for the development of pathogenicity and immunological responses in infected hosts.

All secreted glycoproteins accumulate in the cell wall of fungi. They were immunolocalized by using Concanavalin A-gold (14-nm size particles) and mab 5B2-gold (40-nm size particles). This double immunogold procedure was performed without silver intensification to differentiate 14- and 40-nm gold particles (Figure 5). Immunofluorescence assay was also used to localize corresponding epitopes in the yeast cell wall (Cailliez et al., 1988). However, the comparison between IFA and IEM labelings was limited because the secondary ligands used in each assay had different affinity for the primary ligand. For this reason, IGSS was considered a powerful assay to compare light and electron microscopy results. In *C. albicans* cell wall, a homogeneous immunolabeling was obtained with IFA by using mab 5B2 whose corresponding epitope is carried by high molecular weight glycoproteins. This labeling became heterogeneous by increasing the dilution of mab.

Two different heterogeneities of labeling were detected. The first one was not related to the mab concentration. In this case, a difference of staining intensity was observed on yeast daughter cells, buds, or germ tubes in comparison with mature blastoconidia. The second type of heterogeneity was directly linked to the mab concentration: some cell wall areas presented a greater concentration of mab 5B2 reacting antigens which may be due to a more intense secretion of antigenic glycoproteins (Cailliez et al., 1992b). In accord with these results, IGSS of *C. tropicalis* and *C. albicans* cells by using mab 5B2 showed a "punctuated" labeling at the surface of the cell (Cailliez et al., 1992b). A strong labeling was observed in the septum, the bud, and birth scars of yeasts.

A heterogeneous distribution of glycoproteins may be masked in the case of strong labeling. When appropriate dilutions of ligands were used, a "punctuated" labeling (especially with high dilutions of mab and/or conjugated secondary antibodies) was confirmed by IGSS and IFA, showing dashes and punctuations at the surface of labeled yeast cells. These results, which were first observed in IFA, were correlated with those obtained by IGSS in light microscopy and IEM. With the last assay, colloidal gold particles were found aggregated on selected areas through the yeast cell wall (Cailliez et al., 1992b).

In these experiments the same gold-conjugated antibodies were used both in IGSS and IEM. Thus, the ligand-receptor affinity was probably unaffected by procedures involved in the two immunocytochemical assays.

17.7.2 Antigenic Variability of Fungal Glycoproteins

None of the immunological studies reported to date have shown absolute specificity of mabs when a great number of fungal strains of the same species, or only a single strain which was cultivated under different conditions, were tested (Fruit et al., 1990). Mabs directed against surface compounds (for example, *C. albicans* cell wall glycoproteins) are excellent probes to confirm the existence of antigenic variability in fungal microorganisms (Chaffin et al., 1988; Poulain et al., 1988). The study of the antigenic variability is a mean to investigate relations between phenotypic variability of fungi and their adaptation to environmental conditions. In this respect, mab 5B2 was used to detect the presence of corresponding epitope in various strains of *C. albicans* and *Saccharomyces cerevisiae* (Cailliez and Poulain, 1988). However, the absence of mab binding at the surface of yeast does not necessarily mean that reactive proteins or glycoproteins are not synthesized (Poulain et al., 1988; Cailliez et al., 1989). It is not easy to distinguish between lack of antigenic determinants and incomplete labeling. For example, in *C. albicans* mutant and *S. cerevisiae* strains, a cytoplasmic labeling was detected by using mab 5B2 in IEM assay, even when the cell surface was unlabeled by IGSS or IFA (Cailliez and Poulain, 1988).

Regardless of the potential value of mabs for diagnosis in mycology, no definitive assessment can be made without fundamental studies on the distribution of their corresponding antigenic determinants. Thus, it is important to study their mode of expression under different culture conditions and in different phases of the fungal cell cycle before applying mabs to immunocytological diagnosis.

17.8 The Same Immunolabeling for Light and Electron Microscopy

The major advantage of IGSS is that the same gold-conjugated ligand can be used in light and electron microscopy as well as in different biochemical assays (immunoblotting and immunodotting). Before performing IEM, it is often necessary to titrate primary antibodies (especially polyclonal ones). For example, nonspecific binding of antibodies directly increases with the concentration of primary antisera. Nonspecific reactions involve a lot of time, which can be avoided by first performing the same immunodetection in light microscopy. Thus, optimal concentrations of primary and secondary antibodies can be assessed by making a suitable range of probes dilution. Titrations are easy to perform on glass slides with isolated yeast cells or paraffin-embedded sections.

Moreover, IGSS is a relatively fast procedure since results are obtained within an hour in the case of primary ligands directly bound to gold particles. In the case of isolated yeast cells, such as with *Candida* or *Saccharomyces* strains, a first IGSS assay can be made with unfixed cells simply dried on glass slides (yeast cells and germ tubes are directly deposed on glass slides without prior fixation). This preliminary immunostaining can permit the detection of antigenic determinants which are denatured by the fixation steps. Such antigens are changed or masked by the fixative which can modify their spatial conformation. A good preliminary study consists in performing IGSS on both fixed and unfixed isolated yeast cells.

17.9 Disadvantages of Immunogold-Silver Staining

17.9.1 Strong Labeling of Fungal Cells

When high concentrations of ligands are used, IGSS presents the disadvantage of thickening the labeled cell structures (especially the yeast cell wall layers containing the main part of antigens). It then becomes difficult to clearly evaluate the relationships between antigen concentration and labeling density. Similar observations were also made by using conventional immunoperoxidase methods or IFA (Cailliez et al., 1992a). By performing these methods, a heterogeneous distribution of antigens may be masked by a too strong labeling which is often due to high antigen density. For this reason, IGSS assays should be performed with a significant range of probes dilutions.

17.9.2 Nonspecific Labeling

Background labeling is caused by experimental conditions (specificity of primary and secondary ligands, stability of gold colloid, pH of buffered solutions, etc). Many modifications can be attempted to reduce this background. For example, nonspecific binding of primary and/or secondary ligand can be reduced by using detergents (Triton X-100, Tween 20, or Tween 80), or nonspecific proteins (e.g., gelatine, bovine serum albumin or goat normal serum) which are added in almost all solutions. Moreover, the use of alkaline buffer for washing sections between incubations with primary and secondary antibodies gives lower background staining. When immunostaining cannot be performed without nonspecific binding of primary ligands (more often in the case of lectins and antibodies without strict specificity), further controls should be included in order to determine the level of background staining.

17.10 Practical Advantages of Immunogold-Silver Staining

17.10.1 Sensitivity

An advantage of the IGSS procedure is its sensitivity coupled with good definition of cell structures. A strong contrast in light microscopy, comparable to the Gram or Cotton blue staining, can be obtained with a low concentration of gold-conjugated antibodies. Since it is possible to identify subcellular structures such as the periplasmic space of yeast blastoconidia or cell wall aggregates of secreted glycoproteins, IGSS remains effective even at high dilutions of primary and secondary markers (Holgate et al., 1983; Cailliez et al., 1990).

17.10.2 Specificity

The specificity of IGSS is classically assessed by omission of the primary antibody. This specificity can be demonstrated by a good correlation between the distribution of biological ligands (antibodies or lectins) and that obtained with other immunoassays such as PAP and IFA. An observation of the same distribution of binding sites attests to the absence of nonspecific reactions. Such comparative immunoassays were performed, revealing differential surface staining of yeast cells, depending on the biological probes used (monospecific and monoclonal antibodies) (Cailliez et al., 1990).

When IGSS is performed with fungal microorganisms, specificity failure can be revealed in two major cases. First, a large number of antigenic molecules can be produced by fungi, resulting in widespread cross-reactions between related and unrelated species (Buckley et al., 1992). Second, many polyclonal antisera (including commercial gold-conjugated antibodies) are produced in animals harboring opportunistic fungal species in their digestive tracts (especially *C. albicans*). For this reason, specificity controls are needed to prevent binding of such antibodies to fungal cells. Any new batch of gold conjugate should be tested with a known positive control, such as isolated yeast cells on glass slides.

Another advantage of IGSS, compared to methodologies using probes coupled to enzymatic markers, is that the specificity of immunolabeling is not impaired (especially on unfixed fungal cells) by endogenous enzymes (e.g., peroxidase). The specificity of IGSS has been described in biotyping experiments. Immunogold-silver staining, performed on isolated yeast cells dried on glass slides or present in infected tissue sections, showed that it was possible, by using discriminant dilutions of antibodies labeled by gold particles, to identify species of fungal microorganisms (Cailliez et al., 1990).

17.10.3 Contrast and Definition of Fungal Cell Structures

Immunohistochemical methods have been used to localize fungal antigens on paraffin-embedded tissue sections with polyclonal antisera and mabs (Cailliez et al., 1990). Comparative studies showed that IGSS gave better results than those obtained with peroxidase methods. Immunogold-silver staining performed on fungi in isolates or histopathological samples gave contrast and definition of cell structures comparable to those obtained by Gomori-Grocott silver impregnation staining. Moreover, IGSS provided higher contrast than Gram or Cotton blue staining. The very dark gold-silver labeling and sharp contrast also allowed the use of conventional counterstains such as haematoxylin or any other appropriate histological stains.

17.10.4 Preservation of Immunolabeling

In comparison with IFA, IGSS provides a long preservation of immunoreactivity without decreasing the staining intensity. Each reaction can be observed a long time after immunolabeling and the labeling can be compared with further studies (comparative studies can be carried out with immunolabelings performing at different times). The satisfactory preservation of tissue sections or fungal cells labeled by IGSS does not need particular conditions of storage. Permanent preparations can be stored in the laboratory collection, and can be reexamined in the future.

17.10.5 Avoidance of Carcinogenic Reagents

In comparison with conventional enzymatic methods, IGSS presents the advantage of avoiding carcinogenic products. Indeed, in PAP staining, a careful handling is recommended because of the use of substances having carcinogenic potential (Kobayashi et al.,

1988). In routine mycology laboratories, this aspect is particularly important because of the great number of immunoassays which are usually performed each day.

17.10.6 The Low Cost of Immunogold-Silver Staining Reactions

Immunogold-silver staining generally needs high antibody concentration (primary and secondary antibodies). Because of the large number of immunoassays which are performed for diagnosis, and the relatively wide surface of each paraffin section, the cost of immunogold-silver staining in light microscopy can become prohibitive (especially when regarding the price of the primary antibody). Immunogold-silver staining methods require a markedly lower quantity of primary antibody or gold-conjugated secondary antibodies to be bound, since labeling of gold particles is enhanced by the silver development step which is inexpensive. Even if gold-conjugated antibodies are expensive probes, the high sensitivity of IGSS methods enables the use of high dilutions of primary and secondary ligands. In many laboratories, it is possible to bind colloidal gold particles of different size to purified ligands. Direct gold-silver staining can be performed by using less expensive specific probes. This approach is being used in some laboratories, which use inexpensive, available colloidal gold solutions and housemade probes.

17.11 Summary and Conclusions

Immunogold-silver staining (IGSS) is a powerful topochemical tool for medical or veterinary mycologists. This staining procedure was evaluated for microscopical detection of fungal species such as *Candida* spp., *Cryptococcus neoformans*, *Aspergillus* spp., or *Histoplasma capsulatum*. Immunogold-silver staining assays were performed in both clinical studies (mycological diagnosis) and fundamental medical mycology research. Biological probes were obtained by coupling lectins, affinity-purified or monoclonal antibodies to colloidal gold particles of different size. Gold labeling was followed by silver intensification which formed spheres of heavy metal around gold granules. A strong enhancement, similar to photographic development, facilitated screening of histological sections even at low magnifications. Immunogold-silver staining has higher efficiency than other methods to detect and identify pathogenic fungi in tissue sections. Structural details of host tissues and fungal cells were well preserved without significant background staining. Contrast and ultrastructural definition were better, or at least similar, to that obtained by using conventional immunochemical methods.

Immunogold-silver staining was used to detect and localize fungal molecules of biological signification. It can be applied to fungal smears as well as paraffin-embedded tissue, cryostat, or electron microscopy preparations. Routine fixation and embedding procedures are usually acceptable to prepare fungal samples for IGSS in light microscopy. Mild fixation with glutaraldehyde-paraformaldehyde mixtures resulted in good preservation of host tissues and fungal cell structures, even if in some cases the antigenicity of glycoproteins was lost. However, antigenic determinants shared by the polysaccharidic moiety of fungal glycoproteins were shown to maintain their antigenicity even after strong fixation. Tested fixation methods are needed to prepare fungal samples for IGSS electron microscopical studies. Thus, low concentrations of glutaraldehyde must be used to obtain a good preservation of protein antigens of fungi. Otherwise, the fungus-like microorganism, *Pneumocystis carinii*, must be fixed and embedded under high osmolarity conditions to obtain an adequate preservation of its molecular and cell structures. Immunogold-silver staining can also be applied to immunoblotting techniques and enzyme-linked immunosorbent assays.

Biological probes were directly labeled with colloidal gold or revealed by using gold-conjugated specific reagents such as secondary antibodies or streptavidin. In the latter case, endogenous biotin-like molecules could be responsible for background labeling. In all cases, careful specificity control reactions should be performed. A disadvantage of IGSS is that excessively intense labeling may mask important heterogeneities of labeling, which can be related to fungal cell structures or functions. Another disadvantage is the background labeling, which is sometimes excessive. In this case, detergents, proteins (e.g., gelatin or albumin), additional alkaline washing, or silver reduction steps of tissue sections could reduce background staining.

In mycological research, IGSS is especially useful for studying the distribution of antigens in the fungal cell, as well as for the antigenic variability of fungus glycoproteins. Although this technique is not yet applied to epidemiological studies, it can become useful to the typing of fungus organisms in clinical samples or after their culture recovery. Results already obtained with biopsy sections and isolated fungal cells showed that IGSS could represent an important and complementary step in the diagnosis of mycoses (especially on histopathological sections).

The advantages of IGSS are the efficiency of the immunodetection, the good sensitivity and definition of fungal cell structures, the ease of screening and the relative rapidity of experimentations. Above all, the major advantage of IGSS is the possibility of performing the same immunolabeling both in light and electron microscopy. This represents a significant improvement in relation to previous immunocytochemical methods. Another advantage of IGSS is the good specificity, as attested by the positive correlation with other specific immunoassays. Furthermore, IGSS-processed tissue sections or smears do not need specific conditions of storage and can be stored for later study. In addition, the increased sensitivity linked to silver intensification enables the use of low concentrations of primary and/or secondary ligands, which are frequently expensive. Results already obtained clearly show that IGSS has a fine future in fundamental research and clinical diagnosis.

Acknowledgments

The authors would like to express their gratitude to Dr. Daniel Poulain, Abdelkrim Boudrissa, and Gilbert Lepage, Unité 42 INSERM, Villeneuve d'Ascq, France, for their assistance in the realization of this work.

References

Blumenfeld, W., Mandrell, R. E., Jarvis, G. A., and Griffiss, J. M. 1990. Localization of host immunoglobulin G to the surface of *Pneumocystis carinii*–antibody surface Ig LBA. *Infect. Immun.* 58: 456.

Breter, H. and Erdmann, B. 1993. Suitability of different protein A-gold markers for immunogold-silver staining in paraffin sections. *Biotech. Histochem.* 68: 206.

Buckley, H. R., Richardson, M. D., Evans, E. G. V., and Wheat, L. J. 1992. Immunodiagnosis of invasive fungal infection. *J. Med. Vet. Mycol.* 30 (Suppl. 1): 249.

Cailliez, J. C. and Poulain, D. 1988. Analyse cytologique de l'expression d'un épitope porté par les glycoprotéines excrétées par *Candida albicans*. *Ann. Microbiol. Inst. Pasteur* 139: 171.

Cailliez, J. C., Boudrissa, A., Mackenzie, D. W. R., and Poulain, D. 1990. Evaluation of a gold-silver staining method for detection and identification of *Candida* species by light microscopy. *Eur. J. Clin. Microbiol.* 9: 886.

Cailliez, J. C., Gerloni, M., Morace, G., Conti, S., Cantelli, C., and Polonelli, L. 1992a. Ultrastructural immunodetection of a *Pichia anomala* killer toxin: a preliminary study. *Biol. Cell* 75: 19.

Cailliez, J. C., Poulain, D., Mackenzie, D. W. R., and Polonelli, L. 1992b. Cytological immunodetection of yeast glycoprotein secretion. *Eur. J. Epidemiol.* 8: 452.

Cailliez, J. C., Cantelli, C., Séguy, N., Conti, S., Gerloni, M., Morace, G., and Polonelli, L. 1994. Killer toxin secretion through the cell wall of the yeast *Pichia anomala*. *Mycopathology* 126: 173.

Chaffin, L., Sludarek, J., and Morrow, J. 1988. Variable expression of a surface determinant during proliferation of *Candida albicans*. *Infect. Immun.* 56: 302.

Chamlian, A., Benkoël, L., Ikoli, J. F., Brisse, J., and Jacob, T. 1993. Nuclear immunostaining of hepatitis C infected hepatocytes with monoclonal antibodies to C100-3 nonstructural protein. Comparison of immunogold silver staining with other immunohistochemical methods. *Cell. Mol. Biol.* 39: 243.

Dansher, G., Hacker, G. W., Grimelius, L., and Norgaard, J. O. R. 1993. Autometallographic silver amplification of colloidal gold. *J. Histotechnol.* 16: 201.

Dei-Cas, E. and Vernes A. 1986. Parasitic adaptation of pathogenic fungi to mammalian hosts. *CRC Critical Reviews in Microbiology*, 13: 173.

Dei-Cas, E., Cailliez, J. C., Palluault, F., Aliouat, E. M., Mazars, E., Soulez, B., Suppin, J., and Camus, D. 1992. Is *Pneumocystis carinii* a deep mycosis-like agent? *Eur. J. Epidemiol.* 8: 460.

De Mey, J. 1986. The preparation and use of gold probes. In: *Immunochemistry Practical Application in Pathology and Biology*. p. 115. Pollack, J. M. and Van Noorden, J. (Eds.), Wright, London.

De Puytorac, P., Andrews, N., Barrett, J., Bayne, C. J., Carlier, Y., Dei-Cas, E., Ouaissi, A., Prensier, G., and Piras, R. 1991. Apports récents de la Biologie cellulaire en Parasitologie. *Annales des Sciences Naturelles, Zoologie*, 12: 89.

Emmons, C. W., Binford, C. H., and Utz, J. P. 1970. *Medical Mycology*. Lea & Febiger, Philadelphia.

Fruit, J., Cailliez, J. C., Odds, F. C., and Poulain, D. 1990. Expression of an epitope by surface glycoproteins of *Candida albicans*. Variability among species, strains and yeast cells of the genus *Candida*. *J. Med. Vet. Mycol.* 28: 241.

Fukazawa, Y. 1989. Antigenic structure of *Candida albicans*: Immunochemical basis of the serologic specificity of the mannans in yeasts. In: *Immunology of Fungal Diseases*. p. 37. Kurstak, E. (Ed.), Marcel Dekker, New York.

Goodman, S. L., Park, K., and Albrecht, R. M. 1991. A correlative approach to colloidal gold labeling with video-enhanced light microscopy, low voltage scanning electron microscopy and high voltage electron microscopy. In: *Colloidal Gold: Principles, Methods and Application*. Vol. 3. pp. 370–409. Hayat, M. A. (Ed.), Academic Press, San Diego.

Gorman, S. P., Scott, E. M., and Russel, A. D. 1980. Antimicrobial activity, uses and mechanism of action of glutaraldehyde. *J. Appl. Bacteriol.* 48: 161.

Hacker, G. W., Springall, D. R., Van Noorden, S., Bishop, A. E., Grimelius, L., and Polak, J. M. 1985. The immunogold-silver staining method. A powerful tool in histopathology. *Virchows Arch/ Pathol. Anat.* 406: 449.

Hacker, G. W. 1989. Silver-enhancement colloidal gold for light microscopy. In: *Colloidal Gold: Principles, Methods, and Applications*. Vol. 1. pp. 297. Hayat, M. A. (Ed.), Academic Press, London.

Hay, R. J. 1993. On fungal infections. In: *Medical Congress*. Vol. 169. p. 3.

Hayat, M. A. 1989. *Principles and Techniques of Electron Microscopy*, 3rd ed. Macmillan Press, London, and CRC Press, Boca Raton, FL.

Hayat, M. A. 1992. Quantification of immunogold labeling. *Micron Microsc. Acta* 23: 1.

Hayat, M. A. 1982–1991. Editor. *Colloidal Gold: Principles, Methods, and Applications*. Vols. 1–3. Academic Press, San Diego.

Holgate, C. S., Jackson, P., Cowen, P. N., and Bird, C. C. 1983. Immunogold-silver staining: new method of immunostaining with enhanced sensitivity. *J. Histochem. Cytochem.* 31: 938.

Horisberger, M. 1981. Colloidal gold: a cytochemical marker for light and fluorescent microscopy and for transmission and scanning electron microscopy. In: *Scanning Electron Microscopy*. Vol. 2. Johari, O. (Ed.), SEM Inc., Chicago.

Horisberger, M. 1985. Labeling of colloidal gold with protein A: a quantitative study. *Histochem.* 82: 219.

Kellenberger, E. and Hayat, M. A. 1991. Some basic concepts for the choice of methods. In: *Colloidal Gold Principles, Methods, and Application*. Vol. 3. pp. 1–31. Hayat, M. A. (Ed.), Academic Press, San Diego.

Kobayashi, K., Hayama, M., and Hotchi, M. 1988. The application of immunoperoxidase staining for the detection of causative fungi in tissue specimens of mycosis. *Mycopathology* 102: 107.

Li, R. K. and Cutler, J. E. 1991. A cell surface/plasma membrane antigen of *Candida albicans. J. Gen. Microbiol.* 137: 455.

Linehan, L., Wadsworth, E., and Calderone, R. 1988. *Candida albicans* C3d receptor, isolated by using a monoclonal antibody. *Infect. Immun.* 56: 1981.

Manara, G. C., Ferrari, C., Torresani, C., Sansoni, P., and De Panfilis, G. 1990. The immunogold-silver staining approach in the study of lymphocyte subpopulations in transmission electron microscopy. *J. Immunol. Method.* 128: 59.

Mleczko, J., Litke, L. L., Larsen, H. S., and LaJean Chaffin, W. 1989. Effect of glutaraldehyde fixation on cell surface binding capacity of *Candida albicans. Infect. Immun.* 57: 3247.

Morace, G., Manzara, S., Dettori, G., Fanti, F., Conti, S., Campani, L., Polonelli, L., and Chezzi, C. 1989. Biotyping of bacterial isolates using the yeast killer system. *Eur. J. Epidemiol.* 5: 303.

Odds, F. C., Brawner, D. L., Staudinger, J., Magee, P. T., and Soll, D. R. 1992. Typing of *Candida albicans* strains. *J. Med. Vet. Mycol.* 30 (Suppl. 1): 87.

Palluault, F. and Akono, Z. 1993. *Pneumocystis carinii* Delanoë et Delanoë, 1912 agent d'une maladie en extension: la pneumocystose. I. Aspects cellulaires. *Année Biologique* 33: 55.

Palluault, F., Dei-Cas, E., Slomianny, C., Soulez, B., and Camus, D. 1990. Golgi complex and lysosomes in rabbit derived *Pneumocystis carinii. Biol. Cell* 70: 73.

Palluault, F., Slomianny, C., Soulez, B., Dei-Cas, E., and Camus, D. 1992a. High osmotic pressure enables fine ultrastructural and cytochemical studies on *Pneumocystis carinii.* I. Epon embedding. *Parasitol. Res.* 78: 437.

Palluault, F., Soulez, B., Slomianny, C., Dei-Cas, E., Cesbron, J. Y., and Camus, D. 1992b. High osmotic pressure for *Pneumocystis carinii* London Resin White embedding enables fine immuno-cytochemistry studies: I. Golgi complex and cell-wall synthesis. *Parasitol. Res.* 78: 482.

Pesanti, E. and Shanley, J. 1988. Glycoproteins of *Pneumocystis carinii* characterization by electrophoresis and microscopy. *J. Infect. Dis.* 158: 1353.

Polonelli, L., Archibusacci, C., Sestito, M., and Morace, G. 1983. Killer system: a simple method for differentiating *Candida albicans* strains. *J. Clin. Microbiol.* 17: 774.

Polonelli, L. and Morace, G. 1986. Reevaluation of the yeast killer phenomenon. *J. Clin. Microbiol.* 24: 866.

Polonelli, L., Dettori, G., Cattel, C., and Morace, G. 1987. Biotyping of mycelial fungus cultures by the killer system. *Eur. J. Epidemiol.* 3: 237.

Poulain, D., Cailliez, J. C., Boudrissa, A., and De Mey, J. 1987. Utilisation des marqueurs à l'or colloïdal en mycologie médicale. Application en microscopie électronique, en microscopie photonique et en immunochimie. *Bull. Soc. Fr. Mycol. Med.* 26: 235.

Poulain, D., Cailliez, J. C., Fruit, J., Faille, C., and Camus, D. 1988. *Candida albicans:* Antigenic variability of cell wall surface and cellular excretion of glycoproteins. *Proceedings of the Xth Congress of the ISHAM.* p. 173. Torres-Rodriguez, J. M. (Ed.), Barcelona, Spain.

Poulain, D., Cailliez, J. C., and Dubremetz, J. F. 1989. Excretion of glycoproteins through the cell wall of *Candida albicans. Eur. J. Cell Biol.* 50: 94.

Reiss, E., Hearn, V. M., Poulain, D., and Shepherd, M.G. 1992. Structure and function of the fungal cell wall. *J. Med. Vet. Mycol.* 30 (Suppl. 1): 143.

Rentrop, M., Knapp, B., Winter, H., and Schweizer, J. 1986. Aminoalkylsilane-treated glass slides as support for in situ hybridization of keratin cDNAs to frozen tissue sections. *Histochem. J.* 18: 271.

Rinaldi, M. 1993. On fungal infections. In: *Medical Congress,* 169: p. 2.

Roberts, G. D., Pfaller, M. A., Gueho, E., Rogers, T. R., De Vroey, C., and Merz, W. G. 1992. Developments in the diagnostic mycology laboratory. *J. Med. Vet. Mycol.* 30 (Suppl. 1): 241.

Roth, J., Taatjes, D. J., and Warhool, M. J. 1989. Prevention of nonspecific interactions of gold-labeled reagents on tissue sections. *Histochem.* 92: 47.

Shinoda, T., Kaufmann, L., and Padhye, A. 1981. Comparative evaluation of the Iatron serological *Candida* Check Kit and the API 20C kit for identification of medically important *Candida* species. *J. Clin. Microbiol.* 5: 420.

Slot, J. and Gueuze, H. J. 1983. The use of protein A-colloidal gold (pAg) complexes as immunolabels in ultrathin sections. In: *Immunohistochemistry.* IBRO Handbook Series. Cuello, A. C. (Ed.), Wiley, Chichester, England.

Smith, J. M. B., Aho, R., Mattson, R., and Pier, A. C. 1992. Progress in veterinary mycology. *J. Med. Vet. Mycol.* 30 (Suppl. 1): 307.

Springall, D. R., Hacker, G. W., Grimelius, L., and Polak, J. M. 1984. The potential of the immunogold-silver staining method for paraffin sections. *Histochem.* 81: 603.

Stierhof, Y. D., Humbel, B. M., and Schwarz, H. 1991. Suitability of different silver enhancement methods applied to 1 nm colloidal gold particles: an immunoelectron microscopic study. *J. Electron Microsc. Tech.* 17: 336.

Stirling, J. W. 1993. Use of tannic acid and silver enhancer to improve staining for electron microscopy and immunogold labeling. *J. Histochem. Cytochem.* 41: 643.

Vanden Bossche, H., Kobayashi, G. S., Edman, J. C., Keath, E. J., Maresca, B., and Soll, D. R. 1992. Molecular determinants of fungal dimorphism. *J. Med. Vet. Mycol.* 30 (Suppl. 1): 73.

Virtanen, I., Lehto, V. P., and Aula, P. 1981. Lectins as probes for membranes asymetry and compartimentalization of saccharide moieties in cells. In: *Lectins, Biology, Biochemistry, Clinical Biochemistry.* pp. 215. Bog-Hansen, T. C. (Ed.), Walter de Gruyter, Berlin, New York.

Yokota, S. 1988. Effect of particle size on labeling density for catalase in protein A-gold immunocytochemistry. *J. Histochem. Cytochem.* 36: 107.

Yoshikawa, H., Morioka, H., and Yoshida, Y. 1988. Ultrastructure detection of carbohydrates in the pellicle of *Pneumocystis carinii. Parasitol. Res.* 74: 537.

Chapter 18

Immunogold-Silver Staining of Viral Antigens in Infected Cells

Antonio Marchetti and Generoso Bevilacqua

Contents

18.1 Introduction

Introduced by Holgate et al. (1983), the immunogold-silver staining (IGSS) method has been considered one of the most sensitive and accurate techniques in light microscopy immunocytochemistry (Hacker et al., 1985a; Hoefsmit et al., 1986). The sensitivity of IGSS appears to be much better than that of other immunoperoxidase-based procedures, allowing the detection of antigens present in very small quantities (Springall et al., 1984; Hacker et al., 1985a,b; Scopsi and Larsson, 1985; De Mey et al., 1986). The accuracy of localization (resolution) of IGSS is also particularly high, especially when short developing times are used, in that during the silver-enhancement single silver grains grow on individual gold particles and no diffusion artefacts occur (Romasco et al., 1985; De Waele et al., 1986; Hoefsmit et al., 1986). These features make the IGSS suitable for the immunodetection of viruses. In fact, viral antigens are sometimes expressed in infected cells at levels so low that they can be detected only by extremely sensitive methods. Moreover, due to virus size, a high-resolution immunolabeling is needed for an accurate localization of structural viral proteins.

Although primarily conceived to visualize the gold marker at the light microscopic level, IGSS can also be useful to increase the size of very small gold probes for applications in both transmission (Lackie et al., 1985; van den Pol, 1985; Danscher and Rytter Norgaard, 1985; Bienz et al., 1986; Marchetti et al., 1987) and scanning (Scopsi et al., 1986; Cohn, 1987; Stump et al., 1988) electron microscopy. The use of such probes which provide high labeling efficiency (van Bergen en Henegouwen and Leunissen, 1986; Yokota, 1988) can be advantageous for the immunoelectron microscopic localization of viruses with low antigenic masses (Oram and Crooks, 1974). Another relevant feature of the IGSS method, potentially useful in virology, is the possibility to visualize the same immunoreaction both in light and in electron microscopy (Lackie et al., 1985; Van den Pol, 1985). Cells expressing viral antigens can be first identified by light microscopy on resin sections and subsequently studied at the ultrastructural level on the same preparation.

In this chapter the main applications of the IGSS for the detection of viral antigens in infected cells under light and electron microscopy are discussed.

18.2 Applications

18.2.1 Light Microscopy

The IGSS method has been used to detect several viral antigens, both intracellular and cell membrane-associated, in frozen (Dankner and Spector, 1989) or paraffin-embedded tissue sections (Cleator et al., 1986, 1987; Mabruk et al., 1988, 1989; Dankner and Spector, 1989; Ziegler et al., 1989; Smyth et al., 1990; Magar and Larochelle, 1992; Vella et al., 1992; Larochelle and Magar, 1993; Chamlian et al., 1993). The gold-silver procedure has also been implemented for the immunolocalization of viral proteins in cell suspensions, in cultured cells, and in diagnostic cytology (Springal, 1986; Dankner and Spector, 1989; Levkutova et al., 1992; Chen et al., 1993).

Most studies were carried out as described in the original, indirect technique (Holgate et al., 1983; Springall et al., 1984) based on the use of a primary antibody (monoclonal or polyclonal) followed by a colloidal gold-labeled secondary antibody. After the immuno-reaction, gold particles are visualized at the light microscopic level by silver-enhancement. Some authors have applied variations of the standard technique in which protein A, protein A-G or streptoavidin-gold complexes are used in place of the gold-labeled anti-body (Dankner and Spector, 1989; Magar and Larochelle, 1992; Larochelle and Magar, 1993; Chen et al., 1993). The general aspects of the method and detailed protocols have been reported in several reviews (Scopsi, 1989; Hacker, 1989) as well as in other parts of this book (Chapters 1 and 2). The following paragraphs discuss some general concepts and analyze the essential steps of the IGSS procedure when used for the detection of viral antigens; in fact, the immunodetection of such a heterogeneous group of antigens cannot be performed by using a common protocol.

18.2.1.1 Fixation

The first step in the IGSS, as in any other immunocytochemical technique, generally involves the use of fixatives. The choice of the fixative is critical for the success of the immunolabeling, and depends mainly on the antigen that has to be detected. Some viral antigens are altered by fixatives commonly used in light microscopy (formalin and Bouin's fluid) in combination with paraffin embedding (Dankner and Spector, 1989), while others retain their antigenicity even after long periods of fixation with high concentrations of aldehydes (Cleator et al., 1987).

Antigens that do not survive standard treatments of fixation and embedding have been successfully immunostained by using mild tissue processing protocols. Using a series of 10 different monoclonal antibodies directed against human cytomegalovirus (HCMV) antigens, Dankner and Spector (1989) observed that formalin fixation resulted in a negative IGSS with almost all the antibodies tested. On the contrary, using an extremely mild tissue fixative technique (AMeX: fixation in acetone at –20°C overnight, dehydration in acetone at 4°C and then room temperature for 15 min each; clearing in methyl benzoate for 30 min and in xylene for 30 min) (Sato et al., 1986), the antigenicity of the HCMV proteins was retained in all cases. Unfortunately, the AMeX procedure leads to significant dehydration of the tissue with consequent distortion of the histology.

For antigens that are affected even by mild treatments of fixation and embedding, the immunoreaction can be performed on cryotome sections of unfixed or lightly fixed tissues

(Dankner and Spector, 1989). When working with cytologic material or cultured cells, one of the best fixatives for maintaining antigenicity is acetone at 4°C for 10 min. (Dankner and Spector, 1989). Cell-surface viral antigens intolerant to fixatives have been detected on cell suspensions. They perform the immunoreaction on live cells before the fixation step (Levkutova et al., 1992).

18.2.1.2 Unmasking of Antigenic Sites

The antigen of interest may be surrounded — and thus masked — by unreactive precursors or other proteins, which result in weak or absent immunolabeling. In these cases an attempt can be made to unmask the antigen by treating the samples with proteolytic enzymes which digest proteins surrounding epitopes. Larochelle et al. (1993) reported that in formalin-fixed, paraffin-embedded sections of intestinal tissue, an enzymatic digestion with 0.1% protease XIV for 10 min at 37°C was necessary to unmask transmissible gastroenteritis virus (TGEV) antigens. Proteinase K at a concentration of 10 μg/ml was successfully used to retrieve the antigenicity of two HCMV antigens after formaline fixation (Dankner and Spector, 1989).

18.2.1.3 Washing and Dilution Buffers

Among the different buffer systems described in the literature, the phosphate-buffered saline (PBS) and the tris-buffered saline (TBS) systems have been most frequently used to dilute antibodies and to wash specimens. Adjustment of the pH of the washing buffer to the same pH of the gold buffer before the immunogold or protein G-gold step can reduce the unspecific background staining (Springall et al., 1984; Larochelle and Magar, 1993). Detergents such as Tween 20 and Triton X-100 have been added to the washing buffer before the incubation with the primary antibody to improve the detection of intracellular viral antigens (Larochelle and Magar, 1993). Bovine serum albumin (BSA) at a concentration of 0.1 to 2% has been used in dilution and washing buffers to prevent aggregation of gold particles and to reduce the overall background staining (Springall et al., 1984; Dankner and Spector, 1989; Chamlian et al., 1993; Larochelle and Magar, 1993). The use of BSA should be avoided when working with protein G-gold (PGG) complexes, since PGG is sensitive to competition from proteins like BSA (Ghitescu et al., 1991; Larochelle and Magar, 1993). To localize bovine leukemia virus (BLV) antigens on the surface of infected peripheral blood lymphocytes, Levkutova et al. (1993) performed the immunoreaction before the fixation step. In this case, rinsing the cells with Eagle's minimal essential medium resulted in lymphocytes both more uniformly shaped and with smoother surfaces then those rinsed with PBS.

18.2.1.4 Primary Antibodies

Several viral antigens have been successfully localized with the IGSS method by using polyclonal or monoclonal primary antibodies as first reagents. Polyclonal antibodies raised in rabbit have been used to detect herpes virus simplex (HVS) antigens in human brain biopsies (Cleator et al., 1986) and rat brain (Cleator et al., 1987). They have also detected porcine rotavirus (PRV) proteins in pig's small intestine (Magar and Larochelle, 1992), and semliki forest virus (SLV) antigens in the dermis and musculoskeletal system of foetal mice (Mabruk et al., 1988, 1989) and in the mouse central nervous system (Smith et al., 1990). Polyclonal monkey, pig, and human antibodies were used for the immunodetection of Coxackievirus B4 (CVB4) antigens in mouse pancreas (Vella et al., 1992), TGEV antigens in small intestine of infected pigs (Larochelle and Magar, 1993), and dengue virus antigens in human endothelial cells (Chen et al., 1993), respectively. Specific monoclonal antibodies have found application for the immunolocalization of HCMV proteins in infected tissues of AIDS patients (Dankner and Spector, 1989), human T-cell leukemia virus (HTLV-1) antigens in sinovial joints (Ziegler et al. 1989), the p24 protein of BLV in peripheral blood lymphocytes (Levkutova et al., 1992), and the C100-3 nonstructural protein of hepatitis C virus (HCV) in liver biopsies (Chamlian et al., 1993).

The optimal concentration of the primary antibody which ensures a strong immunolabeling with a negligible background staining has to be determined by carrying out incubations with a series of dilutions of the immunoreagents. As a general rule, concentrations similar to or lower than those used with immunoperoxidase-based techniques should represent a good starting point. The majority of the studies has been conducted by using primary antibody incubation times of 60 to 120 min either at room temperature or at 37°C. For diagnostic purposes, the incubation may be shortened to 30 min or less (Chen et al., 1993). When the expression of the antigen of interest is very low, or the antigenicity has been considerably reduced during fixation and embedding, the incubation may be performed overnight at 4°C or even at room temperature (Larochelle et al., 1993).

18.2.1.5 Gold Labeling and Silver-Enhancement

In the standard, indirect technique, a secondary antibody directed against the immunoglobulin species of the primary antibody is conjugated with colloidal gold (Holgate et al., 1983). Protein A, a 42,000-Da protein from the cell wall of the *Staphylococcus aureus*, and protein A-G, a genetically engineered protein, have been successfully used as secondary (link) reagents, in place of immunoglobulins, for detecting TGEV, PRV, and dengue virus antigens (Magar and Larochelle, 1992; Larochelle and Magar, 1993; Chen et al., 1993). A modification of the standard technique in which a biotinilated secondary layer is followed by streptoavidin molecules has been applied by Danker and Spector (1989), for the immunolocalization of HCMV proteins.

Gold particles conjugated with different immunoreagents are commercially available in a wide range of sizes (from less than 1 nm to 40 nm), but they can also be easily prepared. Gold probes with a diameter of 1 nm or less are usually referred to as "ultrasmall". Small gold markers (4 to 5 nm) have been used for detecting viral antigens on tissue sections since they ensure a good penetration of the specimens (Cleator et al., 1987; Danker and Spector, 1989; Vella et al., 1992; Larochelle and Magar, 1993). An improvement in labeling density may be achieved by using ultrasmall gold probes, as discussed in Chapter 4. Larger gold particles (15 nm) in combination with silver-enhancement have been employed to detect surface-expressed viral proteins in cell suspensions (Levkutova et al., 1992).

Silver-enhancement for light microscopy is presently performed in most laboratories by commercially available kits such as the IntenSE BL (Janssen Biothech, Olen, Belgium) or AURION R-GENT (AURION, Wageningen, The Netherlands). It has been reported that the commercial kits work fine for silver amplification of gold probes with a diameter of more than 3 nm, but that they do not give reproducible results when applied to ultrasmall gold particles (Nielsen and Bastholm, 1993). Silver-enhancement of gold probes with particle diameters as small as 1 nm is discussed in detail in Chapters 5 and 11.

18.2.1.6 Comparison of IGSS with Other Immunocytochemical Techniques

Viral proteins can be expressed in infected cells at very low levels. The immunolocalization of such poorly concentrated proteins is strongly dependent on the sensitivity of the method used. The term "sensitivity" in immunocytochemistry has been defined by Petrusz et al. (1975) as the lowest amount or concentration of tissue antigen that can be distinguished from background. An accurate determination of the sensitivity of a given immunocytochemical method can be performed only on artificial model systems (nitrocellulose membranes spotted with serial dilutions of antigens; Scopsi 1989). However, the sensitivity of different techniques can be roughly compared on parallel sections of tissue by evaluating the intensity of the specific staining at the dilution of the primary antibody at which the maximal demonstration of the antigen occurs. In some of the studies on the detection of viral antigens in infected cells, the IGSS method has been compared with other immu-nocytochemical procedures. Cleator et al. (1987) reported that the number of virus antigen-positive cells detected by IGSS was much higher than that observed by the peroxidase-antiperoxidase (PAP) technique. Ziegler et al. (1989) found the IGSS method to be far superior to the avidin-biotin-complex (ABC) technique for detecting HTLV-1 antigens because of the higher signal-to-noise ratio. More recently, Chamlian et al. (1993), compar-ing IGSS with PAP, alkaline phosphatase-antialcaline phosphatase (APAAP) biotin-streptavidin-peroxidase (B-SA) techniques for the immunolocalization of HCV antigens, reported that the strongest specific immunostaining with no background was observed by using the IGSS method. IGSS also showed higher sensitivity when applied to localize viral antigens expressed on the surface of cultured cells (Chen et al., 1993).

18.2.2 Electron Microscopy

The IGSS method was introduced in immunoelectron microscopy to enhance the visibility of small gold particles too small to be detected at low primary magnification (Danscher and Norgaard, 1985; van den Pol, 1985; Bastholm et al., 1986; Marchetti et al., 1987) or to be resolved by either conventional scanning (Scopsi et al., 1986) or transmission electron microscopy (Nielsen and Bastholm, 1989). The use of IGSS in electron microscopy offers several advantages in virus diagnosis and research: (1) very small gold probes that allow high resolution and high efficiency labeling can be used: these probes can be particularly useful if viruses with a low antigenic mass are being examined and the immunolabeling is expected to be low; (2) viral antigens expressed in a small number of cells within a large cell population can be more readily detected by screening wide fields of the specimen at low magnification; and (3) simultaneous immunolabeling of two or more antigens with colloidal gold probes of the same size (i.e., the same efficiency) becomes possible, as first reported by Bienz et al. (1986) and discussed later.

A limited number of studies have been published concerning the use of IGSS in transmission and scanning electron microscopy for the detection of viral antigens. These applications are reviewed below. Sample processing, immunoreactions, and silver-en-hancement are described case-by-case.

18.2.2.1 Transmission Electron Microscopy

18.2.2.1.1 High Efficiency Labeling of Viral Antigens by IGSS. Several authors have reported that immunogold electron microscopy labeling efficiency, expressed as the amount of gold grains per unit antigen, increases with decreasing particle size (van Bergen en Henegouwen and Leunissen, 1986; Yokota, 1988). The denser labeling obtainable with small gold probes is related mainly to the reduced steric hindrance (Horisberger, 1981, 1989; Hodges et al., 1987) and possibly to the greater number of gold particles per immunoreagent (Slot and Geuze, 1984). Working with structural viral antigens makes it easy to assess immunolabeling efficiency, since the same amount of antigen (usually low) is expressed in each viral particle. Therefore, the number of gold grains per sectioned virion (gold/virion ratio) may be considered a reliable index of labeling efficiency. Using large (20 nm) and small (5 nm) protein A-gold complexes in pre-embedding immunolabeling experiments for the electron microscopic localization of mouse mammary tumor virus (MMTV) antigens expressed at the surface of cultured cells, we observed that labeling density was inversely proportional to the size of gold particles (Marchetti et al., 1987). The gold/virion ratio, calculated from at least 10 electron micrographs including 100 or more sectioned MMTV virions, was 5 ± 1 and 20 ± 3 with 20- and 5-nm gold probes, respectively (Figure 1a,b). These results confirm that small gold markers have higher labeling efficiency than larger gold probes. Furthermore, small gold particles ensure better penetration of the

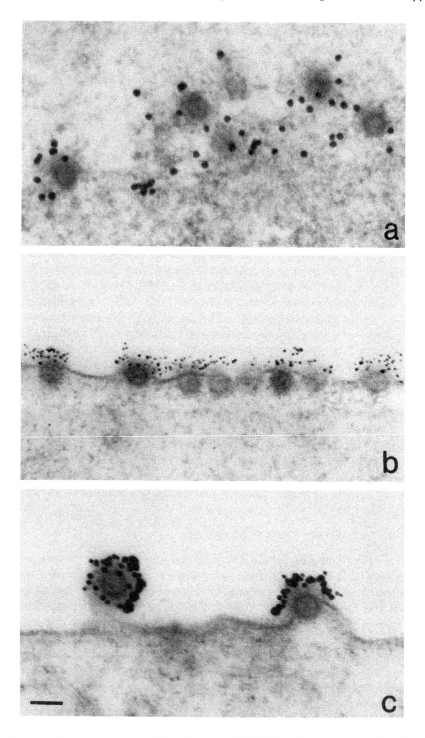

Figure 1 Immunoelectron microscopic localization of MMTV antigens expressed at the surface of mouse mammary tumor cells. Virus producing cells were treated prior to embedment with a specific anti-MMTV antiserum followed by 20 nm (a) or 5 nm (b) protein A-gold complexes. Note that the labeling density (number of gold grains per sectioned virion) was much higher using small gold particles. In (c) the immunoreaction was conducted as in (b), but gold particles were subsequently silver enhanced (7 min) on epon-embedded ultrathin sections. This latter approach resulted in the heaviest specific labeling in that it combined the high labeling density of small gold probes with the excellent visibility obtainable with silver amplification. Bar = 0.1 μm.

Table 1 IGSS Strategies in EM According to the Time At Which the
Immunoreaction and the Silver Enhancement are Performed

Immunoreaction	Silver enhancement	Authors
pre-embedding	pre-embedding	Van Den Pol et al. (1985)
pre-embedding	post-embedding	Marchetti et al. (1987)
post-embedding	post-embedding	Lackie et al. (1985); Danscher et al. (1985)

specimens, which is useful for the detection of intracellular antigens in pre-embedding studies (van den Pol, 1985) or working on ultrathin cryosections (Slot and Geuze, 1983). The use of ultrasmall gold probes (1 nm or less), recently introduced in immunocytochemistry (Leunissen et al., 1989; van Bergen en Henegouwen and van Lookeren Campagne, 1989), results in labeling density and diffusion rates even higher than those obtainable with probes built around 5 nm (Nielsen and Bastholm, 1990; Leunissen and van de Plas, 1993). The main limitation of small and ultrasmall gold probes is their poor visibility: gold particles of 3 to 5 nm are difficult to see at low primary magnification, whereas ultrasmall gold probes are not resolvable by conventional transmission electron microscopy even at the highest magnifications useful for biological purposes. This problem has been overcome by silver-enhancement which allows the enlargement of colloidal gold particles. The silver amplification procedure can be performed either on 30 to 50 μm vibratome sections prior to embedding (van den Pol, 1985) or directly on the surface of ultrathin sections embedded in araldyte (Lackie et al., 1985), epon (Danscher and Rytter Norgaard, 1985; Marchetti et al., 1987), lowcryl (Bienz et al., 1986) resins or obtained by cryoultramicrotomy (Bastholm et al., 1986). Since the immunoreaction can be also performed either before or after embedding, different IGSS strategies are possible as shown in Table 1. In the study reported below, a pre-embedding immunoreaction was followed by a post-embedding silver amplification step.

Silver-enhancement of protein A-gold complexes on epon-embedded ultrathin sections has been employed for the electron microscopic localization of MMTV antigens (Marchetti et al., 1987). Two cell lines, derived from GR and Balb/cfRIII mouse mammary tumors, were grown at confluence and stimulated with dexamethasone in order to increase the MMTV production at the cell surface (Gonda et al., 1976). Paraformaldehyde-prefixed cells were first treated with a rabbit anti-MMTV antiserum followed by protein A-gold (5 nm), each for 1 h at room temperature. Cells were then scraped, fixed with 2.5% buffered glutaraldehyde solution, postfixed with OsO_4, and embedded in Epon. A physical developer, prepared essentially according to the silver nitrate/hydroquinone method (Dansher and Schroder, 1979), containing 30% (final concentration) gum arabic was used. (The amount of the protective colloid may be reduced to 15% to prevent its adhesion to the surfaces of the specimens [Danscher and Rytter Norgaard, 1985]; in this case the use of silver lactate, a salt with a lower dissociation coefficient, may be more convenient since it allows an easier control of the reaction [Scopsi, 1989; also refer to Chapter 2]). The intensifying solution was prepared in a dark room immediately before use and filtered through Whatman paper no. 4. Ultrathin sections mounted on nickel grids were floated on drops of the developer for 5 to 60 min at 20°C in the dark. After fixation with 5% sodium thiosulfate for 1 min, the sections were washed in bidistilled water by a mild spray from a plastic spray bottle and stained with uranyl acetate and lead citrate. A three- to sixfold increment in particle size, useful for rapid low-magnification screening of the specimens, was obtained with developing times of 5 to 10 min. Longer periods of development resulted in large precipitates which could obscure ultrastructural details. The method allowed to obtain a heavy specific labeling of MMTV antigens, with a negligible background staining (Figure 1c). Using the developer for 15 to 30 min on semithin sections of the same embedded material, the reaction became visible at the light microscopic level.

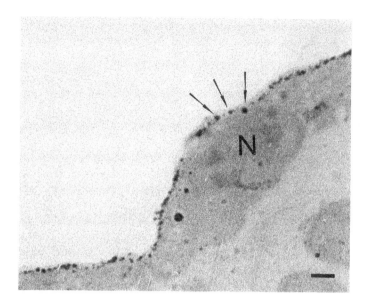

Figure 2 Light microscopic localization of MMTV antigens in infected mouse mammary tumor cells. The immunoreaction was carried out with a specific anti-MMTV antiserum followed by 5-nm protein A-gold complexes. Silver enhancement was performed on semithin sections for 30 min. Single black spots (arrows) with a diameter between 0.3 and 0.5 µm were visible on the cell surface exposed to the immunoreagents. Considering that the main diameter of MMTV particles budding from the cell membrane is 0.1 µm, each isolated black spot may correspond to a single virus particle covered by gold and silver. N = nucleus. Bar = 2 µm.

Single black spots, possibly corresponding to individual virus particles covered by gold and silver, were visible on the cell surface exposed to the immunoreagents (Figure 2). This high degree of resolution was not obtainable by immunoenzymatic methods (PAP) because of diffusion artefacts (Marchetti et al., unpublished results).

The silver amplification procedure has been successfully applied to enhance the visibility of ultrasmall gold particles in transmission electron microscopy (van Bergen en Henegouwen and van Lookeren Campagne, 1989; also refer to Chapters 5 and 11). Although ultrasmall gold probes in conjunction with silver-enhancement have not been used so far for detecting viral antigens, the results reported here as well as in other parts of this book indicate that they hold great promise for future application in virology.

18.2.2.1.2 Double Immunostaining of Viral Antigens by IGSS. Silver-enhancement of colloidal gold particles has proved useful to simultaneously localize two polioviral antigens in the same ultrathin section of Lowicryl-embedded material (Bienz et al., 1986). To this aim, they used a two-layer indirect technique, employing primary monoclonal antibodies raised in the same species and identical 10-nm gold labeled antimouse IgG as secondary reagents. The first antigenic site was labeled with the specific monoclonal antibody followed by the 10-nm gold marker which was subsequently silver-enlarged (two- to fourfold). The sections were then restained with a monoclonal antibody specific for the second antigenic site and nonenhanced 10-nm gold probes. Enlarged (i.e., silver-enhanced) and nonenlarged gold grains could be easily distinguished; moreover, large grains were visible at low magnification. Figure 3 illustrates the results of a parallel experiment conducted with 5-nm gold probes as secondary reagents.

When primary antisera raised in the same species are used in postembedding double-labeling experiments performed on the same side of the section, the first-step secondary antibody conjugated with colloidal gold has to be destroyed or removed to prevent cross-

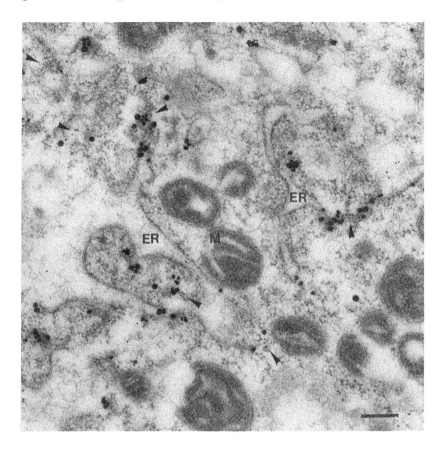

Figure 3 Two intracellular proteins (protein 2B and protein 2C) of poliovirus type 1 (Mahoney) were simultaneously immunolocalized in paraformaldehyde-fixed, Lowicryl-embedded Hep-2 infected cells. Ultrathin sections were incubated with mouse monoclonal antibodies recognizing the viral protein 2C followed by goat-antimouse immunoglobulins conjugated with 5-nm colloidal gold particles (GAM-5). After silver-enhancement, a mouse monoclonal antibody against protein 2 B was applied followed by GAM-G5. The large grains thus represent protein 2C, whereas the small ones (arrows) correspond to the 2B protein. Both antigens are associated with the viral replication complex adjacent to the dilated rough endoplasmic reticulum (ER). M: mitochondria. Figure kindly supplied by Dr. K. Bienz. Bar = 0.2 μm.

contamination (van den Pol, 1985; Wang and Larsson, 1985). Using the above-mentioned method this step does not seem to be necessary, since the deposition of silver ions around colloidal gold particles during silver-enhancement inactivates the gold-labeled secondary antibody (van den Pol, 1985; Bienz et al., 1986). The mechanism of this inactivation is unknown. It has been suggested that the silver amplification procedure forms a sphere of heavy metal around the colloidal gold particle that completely covers the immunoreagents (van den Pol, 1985). Alternatively, the secondary antibody may simply be detached from the gold (Bienz et al., 1986). On the other hand, since the deposition of metallic silver only occurs around the colloidal gold particles, the antigenicity of the specimen is unaffected by the silver-enhancement step (Bienz et al., 1986, Arias et al., 1989).

The main advantages of this method are as follows: (1) two or more antigenic sites can be detected using small gold probes, which are the most efficient labels. As we discussed before, this may be particularly useful when working with viral antigens; (2) monoclonal or polyclonal antibodies raised in the same species can be used without any treatment of the specimens; and (3) multiple immunoreactions can be performed using the same gold conjugate.

A particular physical development procedure was introduced by Bienz et al. (1986), in order to obtain uniform-sized silver grains. They used a slow working, fine grain photographic developer (Metol; Merk, Germany) containing the Ilford L4 emulsion as silver donor. This protocol yields silver grains that are more uniform in size than those obtainable by physical development with silver lactate, hydroquinone and gum arabic. However, it is both more expensive and more light sensitive (Scopsi, 1989).

18.2.2.2 Scanning Electron Microscopy

Due to the strong emission of secondary electrons, colloidal gold particles represent a good marker for scanning electron microscopy (SEM). However, only large-size gold probes (15 to 40 nm) are resolvable in conventional instruments by both secondary (SEI) and backscattered electron imaging (BEI). As we have already discussed, the labeling efficiency of such large probes is rather low. The more efficient small gold markers can be easily resolved by SEM after silver amplification (see Chapter 6). Only a few studies have been published concerning the application of IGSS in SEM for detecting viral antigens. The method has been used to study the release of bluetongue virus (BTV) from infected cells (Hyatt et al., 1989), the localization of the BTV nonstructural protein NS3 in the host cell membrane (Hyatt et al., 1992), BTV absorption, and antibody mediated-neutralization (Hyatt and Brookes, 1993). In all these studies the immunolabeling has been performed according to a standard protocol. Cells grown on glass coverslips were prefixed in 0.1% glutaraldehyde for 10 min, washed in buffer, blocked in order to reduce the background staining, and exposed to the primary antibody for 1 h at room temperature or 37°C. The cells were then rinsed in buffer and incubated with protein A-gold (10 nm) for 1 h at room temperature. After rinsing, cells were postfixed in 2.5% buffered glutaraldehyde for 20 min and exposed to 1% buffered OsO_4 for 1 h. By using the IGSS method in SEM, samples can be osmicated because OsO_4 obliterates BEI of unenhanced colloidal gold particles (de Harven and Soligo, 1989), but not BEI of silver-enhanced gold probes. The silver amplification step was performed using a commercial kit in order to get gold-silver grains of about 100 nm. The samples were then dehydrated, mounted in aluminum stubs, and coated with conductive materials. The observation was performed in SEI, BEI, and mixed (BEI+SEI) modes. The mixed image combines the topographical information provided by SEI with the atomic contrast information of BEI, which enhances the presence of the gold-silver marker. By this method the authors obtained a highly specific labeling, much more intense than that observed using 40-nm unenhanced gold probes, with extremely low background staining. It has been observed that the signal-to-noise ratio is strongly dependent on the fixatives, buffers, and blockers used (Hyatt and Brookes, 1993). Immunogold-silver staining of uninfected cells, performed in the absence of antibodies, produced a high level of background when the cells were prefixed with 2.5% glutaraldehyde for 10 min. On the contrary, a prefixation step with 0.1% glutaraldehyde for 10 min resulted in negligible background signal. As regards the buffer systems, the better results in terms of background were obtained when the glutaraldehyde was diluted in PBS rather than in cacodilate buffer (Herrera et al., 1990). Among the different blockers used to reduce the unspecific labeling, the cold water fish skin gelatine was found to be the most efficient one (Hyatt and Brookes, 1993).

Several fields of virology may take advantage of IGSS-SEM, especially as related to the localization of viral antigens expressed at the cell surface, viral release, adsorption, and neutralization studies.

18.3 Concluding Remarks

In this chapter we have discussed the application of IGSS for detecting viral antigens in infected cells. There is no doubt that the gold-silver marker system has considerable potential in virology.

In light microscopy the higher sensitivity of IGSS facilitates the detection of viral antigens expressed at very low levels in infected cells, when other immunocytochemical techniques fail. Moreover, the resolution of the method is so high, especially on resin-embedded sections, that a fine localization of viral antigens can be performed. As reported above, single viral particles labeled with gold and silver can be resolved at the light microscopic level.

In both transmission and scanning electron microscopy the silver amplification procedure applied to small gold probes yields denser, more efficient labeling of viral antigens, than that obtainable by larger sized, unenhanced gold markers. The use of ultrasmall gold particles in conjunction with silver amplification will probably further improve the efficiency of immunolabeling, especially in pre-embedding experiments on vibratome sections or in immunocryoultramicrotomy due to the higher penetration power of such probes.

Acknowledgments

We thank Mrs. Rosetta Biondi for her excellent technical help. This work was partially supported by A.I.R.C. (Italian association for cancer research).

References

Arias, J., Scopsi, L., Fisher, J. A., and Larsson, L.-I. 1989. Light- and electron-microscopical localization of calcitonin, calcitonin gene-related peptide, somatostatin and C-terminal gastrin/cholecystokinin immunoreactivities in rat thyroid. *Histochemistry* 91: 265.

Bastholm, L., Scopsi, L., and Nielsen, M. H. 1986. Silver-enhanced immunogold staining of semithin and ultrathin cryosections. *J. Electron Microsc. Tech.* 4: 175.

Bienz, K., Egger, D., and Pasamontes, L. 1986. Electron microscopic immunocytochemistry. Silver enhancement of colloidal gold marker allows double labeling with the same primary antibody. *J. Histochem. Cytochem.* 34: 1337.

Chamlian, A., Benkoel, L., Ikoli, J. F., Brisse, J., and Jacob, T. 1993. Nuclear immunostaining of hepatitis C infected hepatocytes with monoclonal antibodies to C100-3 nonstructural protein. Comparison of immunogold silver staining with other immunocytochemical methods. *Cell Mol. Biol.* 39: 243.

Chen, W.-J., Chen, S.-L., Fang, A.-H., and Wang, M.-T. 1993. Detection of dengue virus antigens in cultured cells by using protein A-gold-silver staining (pAgs) method. *Microbiol. Immunol.* 37: 359.

Cleator, G. M., Klapper, P. E., Thornton, J., Cropper, L., and Reid, H. 1986. Immunogold-silver staining in the diagnosis of herpes encephalitis. *J. Neurol. Neurosurg. Psychiat.* 49: 1209.

Cleator, G. M., Klapper, P. E., Sharma, H., and Longson, M. 1987. A rat model of herpes encephalitis with special reference to its potential for the development of diagnostic brain imaging. *J. Neurol. Sci.* 79: 55.

Cohn, R. G. 1987. Silver enhancement of small diameter gold particles conjugated for immunogold stain signal amplification in scanning electron microscopy. *J. Electr. Microsc. Tech.* 7: 132.

Dankner, W. M. and Spector, S. A. 1989. Applications of immunogold-silver enhancement: testing of monoclonal antibodies and detection of human cytomegalovirus in histologic specimens. *Am. J. Anat.* 185: 310.

Danscher, G. and Schroder, H. D. 1979. Histochemical demonstration of mercury induced changes in rat neurons. *Histochemistry* 60: 1.

Danscher, G. and Rytter Norgaard, J. O. 1985. Ultrastructural autometallography. A method for silver amplification of catalytic metals. *J. Histochem. Cytochem.* 33: 706.

de Harven, E. and Soligo, D. 1989. Backscattered electron imaging of the colloidal gold marker on cell surface. In: *Colloidal Gold: Principles, Methods and Applications.* Vol. 1. p. 229. Hayat, M. A. (Ed.), Academic Press, San Diego.

De Mey, J., Hacker, G. W., De Weale, M., and Springall, D. 1986. Gold probes in light microscopy. In: *Immunocytochemistry-Modern Methods and Applications.* p. 71. Polak, J. M. and Van Noorden, S. (Eds.), Wright, Bristol.

De Waele, M., De Mey, J., Reynaert, Ph., Dehou, M.-F., Gepts, W., and van Camp, B. 1986. Detection of cell surface antigens in cryostat sections with immunogold-silver staining. *Am. J. Clin. Pathol.* 85: 573.

Ghitescu, L., Galis, Z., and Bendayan, M. 1991. Protein AG-gold complex: an alternative probe in immunocytochemistry. *J. Histochem. Cytochem.* 39: 1057.

Gonda, M. A., Arthur, L. O., Zeve, V. H., Fine, D. L., and Nagashima, K. 1976. Surface localization of virus production on a glucocorticoid-stimulated oncornavirus-producing mouse mammary tumor cell line by scanning electron microscopy. *Cancer Res.* 36: 1084.

Hacker, G. W., Springall, D. R., van Noorden, S., Bishop, A. E., Grimelius, L., and Polak, J. M. 1985a. The immunogold-silver staining method. *Virchows Arch. Pathol. Anat.* 406: 449.

Hacker, G. W., Polak, J. M., Springall, D. R., Tang, S.-K., van Noorden, S., Lackie, P., Grimelius, L., and Adam, H. 1985b. Immunogold-silver staining (IGSS)-a review. *Microskopie (Vienna)* 42: 318.

Hacker, G. W. 1989. Silver-enhanced colloidal gold for light microscopy. In: *Colloidal Gold: Principles, Methods and Applications.* Vol 1. p. 297. Hayat, M. A. (Ed.), Academic Press, San Diego.

Herrera, M. I., Cuevas, L., Santa-Maria, I., and Najera, R. 1990. Levels of expression of human immunodeficiency virus antigens in lymphocytes from healthy carriers by monoclonal antibody probing and backscattered electron imaging. In: *Proc. 12th Int. Congr. Electron Microsc.* Vol. 3. p. 350. Peachey, L. D. and Williams, D. B. (Eds.), San Francisco Press.

Hodges, G. M., Southgate, J., and Toulson, E. C. 1987. Colloidal gold-a powerful tool in scanning electron microscope immunocytochemistry: an overview of bioapplications. *Scanning Microsc.* 1: 301.

Hoefsmit, E. C. M., Korn, C., Blijleven, N., and Ploem, J. S. 1986. Light microscopical detection of single 5 and 20 nm gold particles used for immunolabeling of plasma membrane antigens with silver enhancement and reflection contrast. *J. Microsc. (Oxford)* 143: 161.

Holgate, C. S., Jackson, P., Cowen, P. N., and Bird, C. C. 1983. Immunogold-silver staining: New method of immunostaining with enhanced sensitivity. *J. Histochem. Cytochem.* 31: 938.

Horisberger, M. 1981. Colloidal gold: a cytochemical marker for light and fluorescent microscopy and for transmission and scanning electron microscopy. *Scanning Electron Microsc.* 2: 9.

Horisberger, M. 1989. Colloidal gold for scanning electron microscopy. In: *Colloidal Gold: Principles, Methods and Applications.* Vol 1. p. 217. Hayat, M. A. (Ed.), Academic Press, San Diego.

Hyatt, A. D., Eaton, B. T., and Brookes, S. M. 1989. The release of bluetongue virus from infected cells and their superinfection by progeny virus. *Virology* 173: 21.

Hyatt, A. D., Gould, A. R., Coupar, B., and Eaton, B. T. 1992. Localization of the non-structural protein NS3 in bluetongue virus infected cells. *J. Gen. Virol.* 72: 1992.

Hyatt, A. D. and Brookes, S. M. 1993. Immunoscanning electron microscopy: the application of colloidal gold and silver enhancement. In: *Immuno-Gold Electron Microscopy in Virus Diagnosis and Research.* p. 411. Hyatt, A. D. and Eaton, B. T. (Eds.), CRC Press, Boca Raton, FL.

Lackie, P. M., Hennessy, R. J., Hacker, G. W., and Polak, J. M. 1985. Investigation of immunogold-silver staining by electron microscopy. *Histochemistry* 83: 545.

Larochelle, R. and Magar, R. 1993. The application of immunogold silver staining (IGSS) for the detection of transmissible gastroenteritis virus in fixed tissues. *J. Vet. Diagn. Invest.* 5: 16.

Leunissen, J. L. M., van de Plas, P. F. E. M., and Borghgraef, P. E. J. 1989. Auro Probe One: a new and universal ultrasmall gold particle based (immuno)detection system for high sensitivity and improved penetration. *AuroFile* 2 (1). Janssen Life Sciences, Beerse, Belgium.

Leunissen, J. L. M. and van de Plas, P. F. E. M. 1993. Ultrasmall gold probes and cryo-ultramicrotomy. In: *Immuno-Gold Electron Microscopy in Virus Diagnosis and Research.* p. 327. Hyatt, A. D. and Eaton, B. T. (Eds.), CRC Press, Boca Raton, FL.

Levkutova, M., Levkut, M., and Bajova, V. 1992. Immunogold silver staining method for light microscopic detection of peripheral blood lymphocyte surface antigens with monoclonal antibodies. *Folia Histochem. Cytobiol.* 30: 103.

Mabruk, M. J. E. M. F., Flack, A. M., Glasgow, G. M., Smyth, J. M. B., Folan, J. C., Bannigan, J. G., O'Sullivan, M. A., Sheahan, B. J., and Atkins, G. J. 1988. Teratogenicity of the semliki forest virus mutant ts22 for the foetal mouse: induction of skeletal and skin defects. *J. Gen. Virol.* 69: 2755.

Mabruk, M. J. E. M. F., Glasgow, G. M., Flack, A. M., Folan, J. C., Bannigan, J. G., Smyth, J. M. B., O'Sullivan, M. A., Sheahan, B. J., and Atkins, G. J. 1989. Effect of infection with the ts22 mutant of semliki forest virus on development of the central nervous system in the fetal mouse. *J. Virol.* 63: 4027.

Magar, R. and Larochelle, R. 1992. Immunohistochemical detection of porcine rotavirus using immunogold silver staining (IGSS). *J. Vet. Diagn. Invest.* 4: 3.

Marchetti, A., Bistocchi, M., and Tognetti, A. R. 1987. Silver enhancement of protein A-gold probes on resin-embedded ultrathin sections. An electron microscopic localization of mouse mammary tumor virus (MMTV) antigens. *Histochemistry* 86: 371.

Nielsen, M. H. and Bastholm, L. 1989. Improved immunolabeling of ultrathin cryosections using antibody conjugated with 1 nm gold particles. In: *Proc. 7th Int. Congr. Electron Microscopy.* p. 928. Peachey, L. D. and Williams, D. B. (Eds.), San Francisco Press.

Nielsen, M. H. and Bastholm, L. 1990. Improved immunolabeling of ultrathin cryo-sections using antibody conjugated with 1-nm gold particles. In: *Proc. 12th Int. Congr. Electron Microsc.* p. 928. Peachey, L. D. and Williams, D. B. (Eds.), San Francisco Press.

Nielsen, M. H. and Bastholm, L. 1993. Multiple labeling of thin sections. In: *Immuno-Gold Electron Microscopy in Virus Diagnosis and Research.* p. 231. Hyatt, A. D. and Eaton, B. T. (Eds.), CRC Press, Boca Raton, FL.

Oram, J. D. and Crooks, A. J. 1974. A comparison of labeled antibody methods for the detection of virus antigens in cell monolayers. *J. Immunol. Methods* 25: 297.

Petrusz, P., Ordonneau, P., and Finley, J. C. W. 1980. Criteria of reliability for light microscopic immunocytochemical staining. *Histochem. J.* 12: 333.

Romasco, F., Rosenberg, J., and Wibran, J. 1985. An immunogold silver staining method for the light microscopic analysis of blood lymphocyte subsets with monoclonal antibodies. *Am. J. Clin. Pathol.* 84: 307.

Sato, Y., Mukai, K., Watanabe, S., Goto, M., and Shimosato, Y. 1986. The AMEX method: A simplified technique of tissue processing and paraffin embedding with improved preservation of antigens for immunostaining. *Am. J. Pathol.* 125: 431.

Scopsi, L. and Larsson, L.-I. 1985. Increased sensitivity in immunocytochemistry. Effects of double application of antibodies and of silver intensification on immunogold and peroxidase-antiperoxidase staining techniques. *Histochemistry* 82: 321.

Scopsi, L., Larsson, L.-I., Bastholm, L., and Nielsen, M. H. 1986. Silver-enhanced colloidal gold probes as markers for scanning electron microscopy. *Histochemistry* 86: 35.

Scopsi, L., 1989. Silver-enhanced colloidal gold method. In: *Colloidal Gold: Principles, Methods and Applications.* Vol 1. p. 251. Hayat, M. A. (Ed.), Academic Press, San Diego.

Slot, J. W. and Geuze, H. J. 1983. The use of protein A-colloidal gold (PAG) complexes as immunolabels in ultra-thin frozen sections. In: *Immunohistochemistry.* p. 323. Cuello, A. C. (Ed.), Wiley, New York.

Slot, J. W. and Geuze, H. J. 1984. Gold markers for single and double immunolabeling of ultrathin cryosections. In: *Immunolabeling for Electron Microscopy.* p. 129. Polak, J. and Varndell, I. M. (Eds.), Elsevier, Amsterdam.

Smyth, J. M. B., Sheahan, B. J., and Atkins, G. J. 1990. Multiplication of virulent and demyelinating semliki forest virus in the mouse central nervous system: consequences in BALB/c and SJL mice. *J. Gen. Virol.* 71: 2575.

Springall, D. R., Hacker, G. W., Grimelius, L., and Polak, J. M. 1984. The potential of the immunogold-silver staining method for paraffin sections. *Histochemistry* 81: 603.

Stump, R. F., Pfeiffer, J. R., Seagrave, J., and Oliver, J. M. 1988. Mapping gold-labeled IgE receptors on mast cells by scanning electron microscopy: Receptors distributions revealed by silver enhancement, backscattered electron imaging, and digital image analysis. *J. Histochem. Cytochem.* 36: 493.

van Bergen en Henegouwen, P. M. P. and Leunissen, J. L. M. 1986. Controlled growth of colloidal gold particles for labeling efficiency. *Histochemistry* 85: 81.

van Bergen en Henegouwen, P. M. P. and van Lookeren Campagne, M. 1989. Subcellular localization of phosphoprotein B-50 in isolated presynaptic nerve terminals and in young adult rat brain using a silver-enhanced ultra-small gold probe. *AuroFile* 2 (6). Janssen Life Sciences, Beerse, Belgium.

van den Pol, A. N. 1985. Silver-intensified gold and peroxidase as dual ultrastructural immunolabels for pre- and postsynaptic neurotransmitters. *Science* 228: 332.

Vella, C., Brown, C. L., and McCarthy, D. A. 1992. Coxsackievirus B4 infection of the mouse pancreas: acute and persistent infection. *J. Gene. Virol.* 73: 1387.

Wang, B. L. and Larsson, L.-I. 1985. Simultaneous demonstration of multiple antigens by indirect immunofluorescence or immunogold staining. Novel light and electron microscopical double and triple staining method employing primary antibodies from the same species. *Histochemistry* 83: 47.

Yokota, S. 1988. Effect of particle size on labeling density for catalase in protein A-gold immunocytochemistry. *J. Histochem. Cytochem.* 36: 107.

Ziegler, B., Gay, R. E., Huang, G.-Q., Fassbender, H.-G., and Gay, S. 1989. Immunohistochemical localization of HTLV-1 p19 and p24-related antigens in synovial joints of patients with rheumatoid arthritis. *Am. J. Pathol.* 135: 1.

Chapter 19

Nonmicroscopical Colloidal Gold Autometallography (AMG$_{Au}$): Use of Immunogold-Silver Staining in Blot Staining and Immunoassay

Cornelia Hauser-Kronberger, Gerhard W. Hacker, Erich Arrer, and Gorm Danscher

Contents

19.1 Introduction

Protein blotting (Towbin et al., 1979) is a routinely used tool in research, which is also being used increasingly in clinical laboratories to assess infections, autoimmune diseases, allergy, and other abnormal conditions. There are numerous variations of the technique. "Western blotting" (Gershoni and Palade, 1983) and "dot blotting" (Towbin and Gordon, 1984) are the most widely used procedures. Electrophoretic separation of proteins in polyacrylamide gel (PAGE) is a standard procedure for characterization of proteins and protein mixtures and electrotransfer of the protein pattern generated in the gel onto a porous membrane, in turn generating new perspectives to refine the detection procedures. The possibility of binding antibodies, hormones, DNA-binding proteins and others to the proteins on the membrane is greatly facilitated. In the large majority of the applications, i.e., immunoblot or Western blot, the ligand used is an antibody. The separation by molecular mass and immunological detection provides a powerful tool to check immunologically related substances. A simple dot blot method is often used for semiquantifying unknown amounts of antigens. In this chapter we will review the use of colloidal gold combined with autometallographic silver intensification for detecting proteins on blots.

0-8493-2449-1/95/$0.00+$.50

Another nonmicroscopical field where silver amplification of colloidal gold may be applied is the immunoadsorbent assay, which is briefly described in this chapter.

The detection of protein pattern on a blot can be performed in a number of ways. The conventional procedures described earlier for staining polyacrylamide gels are unsuitable for staining blots. Both Coomassie blue and conventional silver staining methods give high backgrounds due to nonspecific dye absorption and unspecific silver precipitates. However, Ponceau S, India ink, and Amido black and even more sensitive streptavidin-biotin and colloidal gold are highly suitable stains.

Immunodetection is required for identification and visualization of a specific protein on a blot. The most commonly used labels for visualization of proteins on immunoblots are radioisotopes, fluorochromes, enzymes, or chemoluminescence. These labels require further processing by autoradiography, fluorography, or chromogenic development. However, they are considered to be hazardous to human health. Using colloidal gold, a red signal is produced without further development, thus simplifying the process. The possibility of silver intensification by autometallography (Danscher, 1981a, 1984; Danscher et al., 1993) greatly increases the sensitivity and the visibility of the staining signal. Danscher was the first to detect metallic gold particles in tissue sections by silver amplification (Danscher, 1981b). Two years later, autometallography was used to silver amplify colloidal gold particles in the immunogold-silver staining (IGSS) technique (Holgate et al., 1983). Simultaneously Danscher and Norgaard (1983) introduced colloidal gold labeled RNAse with autometallographic silver amplification. The very same method is not only useful for the *in situ* detection of antigenic substances such as peptides and proteins, lectin-binding carbohydrates, or hybridized nucleic acid sequences within tissue sections or cytological material, but also for use in conventional biochemical and molecular biological assays, such as the varied blotting techniques and immunoadsorbent assays.

19.2 History

In 1984, the use of the IGSS method, more correctly the IGS-"AMG_{Au}" method, in immunoblotting was suggested (Moeremans et al., 1984; Brada and Roth, 1984). Immune overlay assays mostly use labeled protein A or a secondary antibody for detecting antigen-bound primary antibodies. These procedures are in principle the same as those used in immunocytochemical methods. Because immunoperoxidase and (strept)avidin-biotin-peroxidase methods are simple and rapid, they are often recommended. Moeremans et al. (1984) found that immunogold staining followed by autometallography can significantly improve the detection of antigen-antibody complexes on protein blots. Roth (1982) was the first to describe a method using protein A adsorbed to colloidal gold, which presently is used for localizing antigens in tissue sections by light and electron microscopy. In a further development, Brada and Roth (1984) demonstrated rapid and highly sensitive detection of immuno complexes on nitrocellulose sheets by consecutive usage of protein A-gold complexes and autometallography.

Moeremans et al. (1985) also described a relatively simple staining method for protein blots on nitrocellulose membranes utilizing colloidal metal sols (gold or silver) stabilized with Tween-20 at a pH of 3.0. A comparison between their method and commonly used dye staining and silver staining of the polyacrylamide gel showed that colloidal gold staining is more sensitive than silver staining of a polyacrylamide gel. The method is based on the selective high affinity binding of colloidal metal particles, which produces a reddish (gold) or dark gray (silver) color. In this setup, colloidal gold or colloidal silver are used in relatively high concentrations. However, if the colloidal metal is amplified by autometallography, diluted solutions can be used, which greatly enhances the sensitivity of the system.

19.3 Colloidal Gold Autometallography (AMG$_{Au}$)

The AMG$_{Au}$ technique is described in detail in this book. One must remember to include controls whenever applying this extremely sensitive technique. Autometallographic silver amplification can be performed in several different ways; the basic chemistry being the adherence of reducing molecules (e.g., hydroquinone) and silver ions (released from a silver salt) to the peptized surface of colloidal gold particles. Every time a reducing molecule releases its electrons into the valence electron cloud of the gold particles of the nanometer size, they will end up in the valence electron domain of an adhering silver ion thereby reducing it to a metallic silver atom, and simultaneously including it in the colloidal gold crystal lattice (Danscher et al., 1993).

The use of colloidal gold labeling of macromolecules introduced by Faulk and Taylor (1971) is, of course, not limited to light and electron microscopy, and AMG$_{Au}$ is restricted only by the presence of colloidal gold particles. The application of AMG$_{Au}$ for nonmicroscopical purposes has therefore been quite successful. The different possibilities of autometallographic development that can be applied are described in Chapters 1 and 2 in this book. For silver amplification, we recommend using silver lactate (Danscher, 1981a; Danscher et al., 1993; and in this book) or silver acetate autometallography (Hacker et al., 1988; and in this book); many commercial developers give a lower detection efficiency and often give rise to unspecific reactions. A working scheme for using autometallography in blotting techniques is given in *Protocol 1*.

19.3.1 Protocol 1: Silver Acetate Autometallography (Hacker et al., 1988)

1. Rinse preparations carefully with double-distilled water (5 times for 30 sec each, followed by 3 times for 3 min each).
2. Solution A: Dissolve 100 mg of silver acetate (Fluka 85140, Switzerland) in 50 ml of double-distilled water. This solution should be made fresh for every run. Silver acetate crystals can be dissolved easily by continuous stirring for ~15 min.
3. Citrate buffer (pH 3.8): Dissolve 25.5 g citric acid 1.water and 23.5 g trisodium citrate 2.water in double-distilled water to make 100 ml. Adjust to pH 3.8 with aqueous citric acid solution. This buffer can be kept at 4°C for at least 2 to 3 weeks. Check the pH before use.
4. Solution B: Dissolve 250 mg hydroquinone (Fluka 53960) in 50 ml citrate buffer.
5. Just before use, mix solution A with solution B.
6. Silver amplification: Cover the blot with this mixture and develop for 4 to 10 min. For prolonged development (longer than 15 min), it is advisable to protect the developer from bright daylight by using a paper or plastic dark box. Staining intensity can be easily checked in normal laboratory light.
7. As soon as optimum staining is reached, the autometallographic solution is removed and amplification is stopped by immersion for ~1 min in photographic fixer (e.g., Agefix, Agfa Gevaert, FR Germany) diluted 1:20 in distilled water, or in 2.5% aqueous solution of sodium thiosulfate.
8. Preparations are then rinsed with water.

19.4 Western Blots and Dot Blots

Dot blot procedures are simple and do not need any electrophoretic separation step. This technique is very convenient for large scale screening assays. Immunoblotting is applicable in a variety of experiments.

1. Identification of antigens by monoclonal and polyclonal antibodies
2. Identification of antibodies by specific antigens

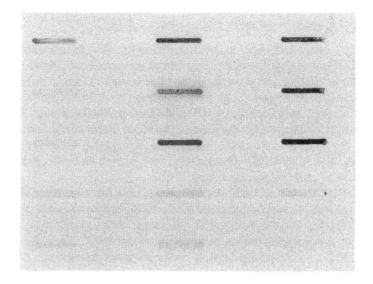

Figure 1 Slot blot stained by IGS-AMG using silver acetate autometallography. Reactivity of different dilutions of polyclonal and monoclonal antibodies with immunologically related and unrelated peptides at concentrations of 0.2 µMol is shown, using secondary antirabbit and antimouse IgG antibodies labeled with 15-nm gold particles. Negative control is located at the right bottom corner.

3. Identifying sugar residues of glycoproteins by lectins
4. Studying protein-protein interactions
5. Studying protein-DNA/RNA interactions
6. Studying enzyme-substrate interactions (Bjerrum and Heegaard, 1988)

Immunoblotting may be divided into five basic steps:

1. Sample preparation and electrophoretic separation, usually on a polyacrylamide gel
2. Transfer to a membrane by capillary blotting, diffusion blotting, vacuum blotting; mostly electroblotting (semidry, or conventional tank buffer)
3. Blocking of the residual binding capacity of the membrane
4. Binding of antibody or ligand
5. Detection of bound antibody with secondary labeled reagents

A grayish to black signal is produced when using the IGS-AMG$_{Au}$ (=IGSS) method for immunodetection of blotted antigens. In contrast to autoradiography IGS-AMG$_{Au}$ is faster and less hazardous to the operator, a fact which also applies to enzyme-based methods, e.g., utilizing the potentially cancerogenous substance diaminobenzidine-tetrahydrochloride (DAB). Using different autometallography developing periods on otherwise identically treated immunoblots, the staining intensity and therefore also the sensitivity can be adjusted. Furthermore, major and minor bands can be detected at the same time in parallel blots silver-amplified for different times without the necessity of repeating the whole experiment (Figure 1).

Autometallographic development after IGS is more sensitive than peroxidase-antiperoxidase (PAP) or avidin-biotin-complex (ABC)-immunoperoxidase systems (Moeremans et al., 1984). Immunogold-staining-autometallography (IGS-AMG) produces a highly contrasted and stable signal with practically no background. In addition, pg levels of antigen can be detected due to the high sensitivity (Jones and Moeremans, 1988). Because of their reliability and convenience, IGS-AMG techniques are suitable not only for research laboratories but also for pathological laboratories. The actual staining procedure is carried out after protein transfer to nitrocellulose or other membranes. Depending on the material of the membrane, various staining procedures show different degrees of sensitivity or background.

Nitrocellulose (NC) membranes are widely used although nylon-based membranes have some advantages in handling because of their mechanical strength and higher binding capacity. However, their high background staining exclude their use in staining procedures based on Coomassie Brilliant Blue, Amido Black, and also colloidal gold. Auro dye is limited to NC and PVDF (polyvinylidene difluoride) membranes. Depending on the staining procedure, the pore size (0.45 µm; 0.2 µm) of the sheet used may also have some influence on the sensitivity.

After protein transfer, extensive washing and blocking of the free binding capacity, Western blot sheets are cut into strips and incubated with the specific antibody and then stained. Extensive washing is necessary to remove adhering polyacrylamide gel particles and residual sodium dodecylesulfate (SDS).

Dot or slot blots are frequently used to test the avidity, sensitivity, and specificity of polyclonal or monoclonal antibodies. Histochemists use these techniques to define eventual cross-reactivities with immunologically related antigens. They are performed by application of serial dilutions of known antigens on nitrocellulose membranes using a slot blotting chamber or a dot blot applicator with subsequent fixation, washing, and IGS-AMG or enzyme-based immunochemical methods. The sensitivity of IGS-AMG methods for dot blot analyses is better than peroxidase- or streptavidin-based methods (Moeremans et al., 1984; Jones and Moeremans, 1988). In *Protocol 2*, practical guidelines are given for using IGS-AMG in Western blots, slot blots, and dot blots. Using a streptavidin-biotin-IGS-AMG method (i.e., one of the methods described by Hacker et al., 1988), it is also possible to detect nucleic acid sequences in tissue sections as well as in blotting methods.

19.4.1 Protocol 2: Western Blot and Dot Blot Immunogold-Silver Staining

Nitrocellulose (e.g., Schleicher & Schüll, Amersham, Bio-Rad, etc.) or nylon-based membranes (e.g., *Nytran*, Schleicher & Schüll; *Hybond*, Amersham) can be used as an immobilizing matrix.

1. *Western blots:* Transfer the proteins from the gel to nitrocellulose or nylon-based membranes, preferably using electroblotting. Cut a narrow strip of the blotted area. Fix the strip with fixing solution (e.g., 4% neutral phosphate-buffered formaldehyde) for 30 min, and then wash in phosphate-buffered saline (PBS) (pH 7.4) three times for 3 min each. If nitrocellulose sheets are used, a strip of the transfer membrane may be stained with a silver stain to check the efficiency of the transfer.
   Dot blots and slot blots: Antigen in different concentrations diluted in PBS is applied on a nitrocellulose sheet using a slot blotting chamber (e.g., Schleicher & Schuell, Dassel, FRG, code no. SRC072/0 "Minifold II") or a dot blot applicator. Air dry for 30 min, then fix as above in fixing solution for 30 min, and wash in PBS three times for 3 min each.
   Washing steps for Western blots and dot blots are performed in Petri dishes, and slot blots are washed within the chamber.
2. Immerse in Tris-buffered saline (TBS) (pH 7.6) containing 0.1% cold water fish gelatin (TBS-gelatin) for 5 min.
3. Incubate the membrane, or strips of it, under constant agitation in washing buffer containing 1% normal serum and 0.1% bovine serum albumin (BSA) for 30 min at room temperature.

4. Drain and incubate with primary antibody, optimally diluted in TBS or PBS containing 0.1% BSA and 0.1% sodium azide (NaN$_3$). The correct dilution should be ascertained in titration series, which is carried out for each antibody used. A typical incubation time at room temperature is 2 h.
5. Wash the membrane in TBS-gelatin three times for 3 min each.
6. Dilute the immunogold staining reagent (e.g., 1:50 to 1:200) in TBS-gelatin. Gelatin inhibits nonspecific binding of gold probes to certain proteins, and also diminishes unspecific silver precipitation during autometallography, without affecting the immunoreactivity. Incubate the membrane in the diluted gold-reagent, in case of larger membrane stripes under constant agitation. Incubation may be carried out overnight or shorter. If a shorter incubation time is required (e.g., 1 h), a more concentrated immunogold probe must be used.
7. Wash the membrane three times for 3 min each in TBS-gelatin.
8. Wash thoroughly in double-distilled water and perform silver amplification, e.g., silver acetate autometallography *(Protocol 1).*

   Controls: As a negative control, the primary antibody solution should be replaced with TBS containing 1% normal serum (prepared from the same species as the secondary antibody). A positive control using an antibody of known reactivity is helpful to judge the experimental conditions.

19.5 Whole Protein Staining

Moeremans et al. (1985) have found that colloidal gold particles buffered at pH of 3.0 and stabilized with Tween 20 can be used as a highly sensitive total protein stain on nitrocellulose membranes, and a similar application has also been developed independently by Rohringer and Holden (1985) and Rohringer (1989). All components for successfully carrying out this procedure are commercially available (*AuroDye*™, Amersham, UK). The procedure has proven useful as it may detect more spots in 2D protein blots than conventional silver staining can detect on 2D polyacrylamide gel (Jones and Moeremans, 1988). The IGSS method can also be combined with the total protein staining by colloidal gold on the same membrane (Daneels et al., 1986).

19.6 Nanogold

Nanogold and *undecagold* are new approaches of using gold particles for the immunodetection of antigens in tissue sections. These approaches give superior results as compared to those obtained with traditional colloidal gold probes (Hainfeld et al., 1992; and in this volume). Nanogold with IGSS has been successfully used in Western and dot blots, and microtiter plate assays. Because of their small size, and coupling to antibody at the hinge sulfhydryl group that preserves antibody activity, more antigens are coated than with conventional colloidal gold, and higher sensitivities may be obtained. A sensitive dot blot (0.2 pg target detected = 10^{-18} mol) with Nanogold-IGSS is shown by Hainfeld et al. (1993).

Nanogold-labeled proteins may be run on gels and stained directly with autometallography. Since Nanogold is small, it has very little negative effects on the binding capacity of the antibody and bands are retarded only slightly, generally corresponding to the weights of the clusters (undecagold has a MW of 5 kDa, and Nanogold is 15 kDa); sometimes, no retardation is seen. Covalent attachment of the gold label to the macromolecule means that the gold does not dissociate. There is no need to transfer the bands to blot paper before autometallographic silver enhancement. Since the development is rapid, gold cluster labeled bands may be developed within a few minutes. Also, only gold labeled bands are developed, so a parallel gel with Coomassie staining can indicate which proteins

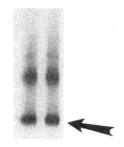

Figure 2 Nanogold-labeled Fab′ and F(ab′)2 used in native PAGE (poly-acrylamide gel electrophoresis). Gel was developed directly in *LI silver* for 8 min. Sensitivity is about 0.1 pg per band, and Nanogold-labeled proteins are selectively stained. Lower band (arrow) is Fab′-Nanogold; upper band is F(ab′)2-Nanogold. Nanogold and *LI silver* were from Nanoprobes Inc., Stony Brook, NY. Courtesy of Dr. James Hainfeld, Stony Brook, NY.

were labeled (Weinstein et al., 1989; Wilkins and Capaldi, 1992). Novel applications of Nanogold for derivatization of ribosomes for crystallographic studies describe the synthesis of a radioactively labeled undecagold cluster for application in X-ray structure analysis of ribosomes (Boeckh and Wittmann, 1991).

Since gold clusters can be covalently attached to specific chemical groups, a number of unique products can be made for nonmicroscopical applications. Nanogold may be attached to proteins, peptides, nucleic acid bases, carbohydrates, lipids, and other molecules. This situation is in contrast to the adsorption of rather large molecules usually required for colloidal gold technology. A commercial source (Nanoprobes Inc., Stony Brook, NY) is now selling "golden lipids" which are being used by some liposome technology companies. In addition, unique products are currently under development (Figure 2).

Nanovan is an organic vanadate negative stain also developed by Nanoprobes. It has a higher density than protein, can function as a good (but lower density) negative stain, and has usually been applied to isolated molecules or fibers viewed with the electron microscope. It is less dense than gold, so gold labeling (or IGS-AMG) may be distinctively recognized and is not obscured, as may be the case with higher density stains such as uranyl acetate, especially when viewing small gold-silver particles (Hainfeld et al., 1993).

19.6.1 Protocol 3: Nanogold™-Labeled Bands on Gels or Blots (J. Hainfeld, Stony Brook, NY, Personal Communication)

1. After labeling with Nanogold, remove unbound gold particles by using column chromatography, sucrose gradient, or other purification means. Leaving excess free Nanogold in the sample may interfere with the intended gel staining.
2. Run gel as usual with one precaution: Nanogold is degraded by β-mercaptoethanol (or dithiotreitole = DTT), so the sample must not be mixed with a reducing agent, i.e., a nonreducing gel must be run. Normal concentrations of other ingredients (SDS, etc.) are acceptable. Samples are boiled in SDS before loading on the gel. Nanogold is not degraded appreciably by this procedure, and heating is recommended since bands run more reliably after heating.
3. The gel may be electrotransferred to nitrocellulose if desired, although this is unnecessary.
4. Rinse gel with several changes of distilled water. Place the gel or blot in a suitable dish and apply enough freshly prepared developer. Apply silver amplification, e.g., as suggested in *Protocol 1*.
5. Watch development of bands which should appear brown-black. Usual development time is 2 to 10 min.
6. When optimal staining is reached, interrupt development in photographic fixer.
7. Rinse with distilled water.

19.7 Silver Enhanced Gold-Labeled Immunoadsorbent Assay

A novel immunoassay, the silver-enhanced gold-labeled immunoadsorbent assay (SEGLISA, or IGS-AMG-IA) has recently been described (Rocks et al., 1990a,b; Patel et al., 1991–1993). It is based on conventional microtitration technology commonly used in enzyme-linked immunoadsorbent assays (ELISA) but uses the IGS-AMG method instead of enzyme-labeled conjugates and substrate reactions. Normal immunoadsorbent assay procedures are applied, and the gold-labeled reagent becomes attached to the cell walls. The signal is amplified by autometallography, which results in the formation of a deposit of metallic silver. The plates are read either visually or with a conventional colorimetric plate reader, which measures the apparent absorbance of the thin gold-silver layer (Patel et al., 1993).

The IGS-AMG-IA shows some advantages compared to standard ELISAs: (1) None of the problems associated with variation of enzymatic activity are observed; (2) the stages involved are less time consuming, and the lengthy substrate incubation is substituted by a simpler and quicker autometallographic procedure; (3) the plates show a permanent record of tests to be stored for future reference; and (4) possible health hazardous enzyme substrates are avoided.

References

Bjerrum, O. J. and Heegard, N. H. H. 1988. *Handbook of Immunoblotting of Proteins.* Vol. 1. Technical descriptions. CRC Press, Boca Raton, FL.

Boeckh, T. and Wittmann, H.-G. 1991. Synthesis of a radioactive labeled undecagold cluster for application in X-ray structure analysis of ribosomes. *Biochem. Biophys. Acta* 1075: 50.

Brada, D. and Roth, J. 1984. "Golden blot"-detection of polyclonal and monoclonal antibodies bound to antigens on nitrocellulose by protein A-gold complexes. *Anal. Biochem.* 142: 79.

Daneels, G., Moeremans, M., De Raeymaeker, M., and De Mey, J. 1986. Sequential immunostaining (gold/silver) and complete protein staining on Western blots. *J. Immunol. Meth.* 89: 89.

Danscher, G. 1981a. Histochemical demonstration of heavy metals. A revised version of the sulphide silver method suitable for both light and electron microscopy. *Histochemistry* 71: 1.

Danscher, G. 1981b. Localization of gold in biological tissue. A photochemical method for light and electron microscopy. *Histochemistry* 71: 81.

Danscher, G. and Norgaard, J. O. 1983. Light microscopic visualization of colloidal gold on resin-embedded tissue. *J. Histochem. Cytochem.* 31: 1394.

Danscher, G. 1984. Autometallography. A new technique for light and electron microscopical visualization of metals in biological tissue (gold, silver, metal sulphides and metal selenides). *Histochemistry* 81: 331.

Danscher, G., Hacker, G. W., Norgaard, J. O., and Grimelius, L. 1993. Autometallography silver amplification of colloidal gold. *J. Histotechnol.* 16: 201.

Faulk, W. P. and Taylor, G. M. 1971. An immunocolloid method for the electron microscope. *Immunochemistry* 8: 1081.

Gershoni, J. M. and Palade, G. E. 1983. Protein blotting: principles and applications. *Anal. Biochem.* 131: 1.

Hacker, G. W., Grimelius, L., Danscher, G., Bernatzky, G., Muss, W., Adam, H., and Thurner, J. 1988. Silver acetate autometallography: an alternative enhancement technique for immunogold-silver staining (IGSS) and silver amplification of gold, silver, mercury and zinc in tissues. *J. Histotechnol.* 11: 213.

Hainfeld, J. F. and Furuya, F. R. 1992. A 1.4-nm gold cluster covalently attached to antibodies improves immunolabeling. *J. Histochem. Cytochem.* 40: 177.

Hainfeld, J. F., Furuja, F. R., Carbone, K., Simon, M., Lin, B., Braig, K., Horwich, A. L., Safer, D., Blechschmidt, B., Sprinzl, M., Ofengand, J., and Boublik, M. 1993. High resolution goldlabeling. In: *Proc. 51st Ann. Meet. Micros. Soc. Amer.* Bailey, G. W. and Rieder, C. L. (Eds.), San Francisco Press.

Holgate, C. S., Jackson, P., Cowen, P. N., and Bird, C. C. 1983. Immunogold-silver staining: A new method of immunostaining with enhanced sensitivity. *J. Histochem. Cytochem.* 131: 938.

Jones, A. and Moeremans, M. 1988. Colloidal gold for the detection of proteins on blots and immunoblots. In: *Methods in Molecular Biology.* Vol. 3. p. 441–479. Walker, J. (Ed.), Humana Press, Clifton, NJ.

Moeremans, M., Daneels, G., Van Dijck, A., Langanger, G., and De Mey, J. 1984. Sensitive visualization of antigen-antibody reactions in dot and blot immune overlay assays with immunogold and immunogold-silver staining. *J. Immunol. Meth.* 74: 353.

Moeremans, M., Daneels, G., and De Mey, J. 1985. Sensitive colloidal metal (gold or silver) staining of protein blots on nitrocellulose membranes. *Anal. Biochem.* 145: 315.

Patel, N., Rocks, B. F., and Bailey, M. P. 1991. A silver enhanced, gold labeled, immunosorbent assay for detecting antibodies to rubella virus. *J. Clin. Pathol.* 44: 334.

Patel, N., Rocks, B. F., and Iversen, S. A. 1992. The direct measurement of low density lipoprotein in whole blood by a silver-enhanced gold-labeled immunoadsorbent assay. *Ann. Clin. Biochem.* 29: 283.

Patel, N., Rocks, B. F., and Bailey, M. P. 1993. Sandwich silver enhanced gold labeled immunoadsorbent assay for determination of human growth hormone. *J. Histotechnol.* 16: 259.

Rocks, B. F., Bertram, V. M. R., and Bailey, M. P. 1990. Detection of antibodies to the human immunodeficiency virus by a silver enhanced gold-labelled immunosorbent assay. *Ann. Clin. Biochem.* 27: 114.

Rohringer, R. and Holden, D. W. 1985. Protein blotting: Detection of proteins with colloidal gold, and of glycoproteins and lectins with biotin-conjugated and enzyme probes. *Anal. Biochem.* 144: 118.

Rohringer, R. 1989. Detection of proteins with colloidal gold. In: *Colloidal Gold: Principles, Methods, and Applications.* Vol. 2. p. 398–429. Hayat, M. A. (Ed.), Academic Press, San Diego.

Roth, J. 1982. Applications of immunocolloids in lightmicroscopy. Preparation of protein A-silver and protein A-gold complexes and their application for the localization of single and multiple antigens in paraffin sections. *J. Histochem. Cytochem.* 30: 691.

Towbin, H., Staehelin, T., and Gordon, J. 1979. Electrophoretic transfer of proteins from polyacrylamide gels to nitrocellulose sheets: Procedure and applications. *Proc. Natl. Acad. Sci. U.S.A.* 76: 4350.

Towbin, H. and Gordon, J. 1984. Immunoblotting and dot immunoblotting. Current status and outlook. *J. Immunol. Meth.* 72: 313.

Weinstein, S., Jahn, W., Hansen, H., Wittman, H. G., and Yonath, A. 1989. Novel procedures for derivatization of ribosomes for crystallographic studies. *J. Biol. Chem.* 264: 191381.

Wilkens, S. and Capaldi, R. A. 1992. Monomaleimidogold labeling of the g subunit of the Escherichia coli F1 ATPase examined by cryoelectron microscopy. *Arch Biochem Biophys.* 299: 105.

Chapter 20

Application of an Image Analyzer to Immunogold Labeling

Hiroshi Shimizu and Takeji Nishikawa

Contents

20.1 Introduction

Immunogold is one of the most valuable probes used as a marker for immunocytochemistry in electron microscopy, because of its advantages over immunoperoxidase (Eady et
al., 1992; Horisberger et al., 1977; Shimizu et al., 1989, 1993b). Commercially available gold
probes range from 1 to 40 nm in diameter. Although visibility is improved with larger gold
probes, small gold probes provide better penetration into the tissue in pre-embedding
labeling (Sakai et al., 1986; Shimizu et al., 1990; Akiyama et al., 1991; Kellenberger and
Hayat, 1991; McGrath et al., 1993). Also, the smaller the probe, the greater the number that
can conjugate with a given number of immunoglobulin molecules (Yokota, 1988). Consequently, denser labeling can be obtained with smaller gold probes.

The 5-nm particle, the smallest gold probe commercially available that is detectable
under transmission electron microscopy, is widely used for immunoelectron microscopic
studies. However, such probes are too small to be identified in electron dense structures
or in low-magnification electron micrographs. The evaluation of the level of background
labeling requires the objective observation of small particles over an entire electron micrograph and is difficult to accomplish. To overcome these problems while retaining the
advantages of 5-nm gold probes, an image analyzer to enhance the color and size of such
small gold particles was successfully used (Shimizu et al., 1993a). The visualization of
particles was greatly improved with this application.

20.2 Materials and Methods

20.2.1 Immunoelectron Microscopy

An image analyzer can be applied to immunogold electron micrographs during both pre-
and postembedding procedures. Postembedding immunoelectron microscopy was performed

0-8493-2449-1/95/$0.00+$.50

as described previously (Shimizu et al., 1989, 1991; Ishida-Yamamoto et al., 1991; Ishiko et al., 1993). Briefly, normal human skin samples were cut into small pieces (<1 mm³), cryofixed in liquid propane cooled to –190°C, using a KF-80 cryofixation apparatus (Reichert-Jung, Vienna, Austria). The frozen skin was then subjected to freeze-substitution in 100% methanol for 48 h at –80°C using a CS-auto freeze-substitution apparatus (Reichert Jung). The skin sample was embedded in Lowicryl K11M (Chemische Werke Lowi, Waldkraiburg, Germany) at –60°C. Polymerization was initiated under ultraviolet radiation at –60°C and continued at room temperature. Thin sections were cut vertically to the skin surface and collected on nickel grids. Sections were incubated in primary antibody, followed by incubation with 5-nm gold-conjugated secondary antibody.

Primary antibody for postembedding immunoelectron microscopy was obtained from the serum of patients with bullous pemphigoid that contained circulating IgG autoanti-body against the epidermal basement membrane (Beutner et al., 1968; Stanley et al., 1981; Labib et al., 1986; Shimizu et al., 1988). Bullous pemphigoid is a blistering disease in which patients possess IgG class circulating autoantibodies against the 230-kDa and/or the 180-kDa autoantigens in the epidermal basement membrane zone.

Pre-embedding en bloc immunoelectron microscopy was performed as previously described (Shimizu et al., 1990). Freshly obtained normal human skin was immediately cut into small pieces (<1 mm³), and incubated in primary antibody. The skin was then incubated in 5-nm gold-conjugated secondary antibody. After fixation and embedding in resin, thin sections were cut from the surface of the tissue sample. After staining with uranyl acetate and lead citrate, sections were examined and electron micrographs were obtained. As primary antibody for the pre-embedding method, we used epidermolysis bullosa acquisita serum containing circulating autoantibody against the N-terminus of type VII collagen (Shimizu et al., 1988) and rabbit polyclonal antibody against the collagenous domain of type VII collagen (Bruckner-Tuderman et al., 1987, 1989) (a kind gift from Dr. Bruckner-Tuderman). Epidermolysis bullosa acquisita is an autoimmune blistering skin disease similar to bullous pemphigoid, in that patients have circulating IgG class autoan-tibodies against the anchoring fibrils at the epidermal basement membrane zone. Type VII collagen is a major component of the anchoring fibrils (Woodley et al., 1984, 1988).

20.2.2 Procedures for the Image Analyzer

We used the Olympus CIA system, SP 500 (Tokyo, Japan), for image analysis (Tatsuya et al., 1991; Shimizu et al., 1993). Using a TV camera (Ikegami, ITC-370M, Japan), an image of the original electron micrograph was projected onto the TV screen of the color image analyzer. The image analyzer used has 512 × 480, or a total of 245,760 pixels. The analyzer recognizes images of each pixel by grading brightness from 0 (darkest) to 255 (brightest). The image of the original black and white electron micrograph was projected as an assembly of a total of 245,760 pixels possessing the original grading brightness. This projected image was recorded as record 1 in the analyzer (Figure 1A).

We used the floating threshold method to enhance the density of the small electron dense gold particles on the electron micrograph. We first used a smoothing filter proce-dure. In this procedure, a grading brightness score for each pixel was calculated as the mean grading brightness of 15 × 15 (total of 225) pixels that surrounded the pixel measured in the center. This procedure was performed by the computer for each of 245,760 pixels as shown below:

The image according to the final grading brightness score of each pixel after the floating threshold method was projected on the video monitor. Each gold particle became easier to detect after this procedure. Finally, a suitable binarization procedure was used to detect the probable distribution of the gold particles. The distribution of all gold particles detected on the video monitor was recorded in the analyzer. The image analyzer was then used to enhance the particles to a more clearly visible color and size that corresponded to the original micrograph, called record 2 (Figure 1B). Records 1 and 2 were then overlapped to build a double image that was recorded as record 3 (Figure 1C). At this stage, any color and size could be chosen, although only one characteristic in each class could be used at a given time.

The image analyzer allows one to count the number of gold particles located in a particular area, or to measure the distance of each particle to the manually delineated plasma membrane. We previously used the image analyzer to count the number of intra- and extracellular gold particles distributed on the bullous pemphigoid antigen. The distance of each particle to the nearest plasma membrane was also measured. The result displayed as a diagram in the image analyzer (Shimizu et al., 1993).

$$\frac{\text{original grading brightness}}{\substack{\text{grading brightness after} \\ \text{smoothing filter procedure}}} \times \substack{\text{revised coefficient} \\ \text{(128 in this case)}} = \substack{\text{final grading brightness} \\ \text{score of each pixel after} \\ \text{floating threshold method}}$$

20.3 Results

Examples of image analysis of immunoelectron micrographs are shown in Figures 1A through 1D. Postembedding immunogold electron microscopy revealed that the bullous pemphigoid antigen occurs in both intra- and extracellular sites in the hemidesmosomes (there are two major bullous pemphigail antigens; see Ishiko et al., 1993).

The resolution of the image of an original immunogold electron micrograph was projected on the TV screen by a total of 245,760 pixels (Figure 1A). Each gold particle, which was detected and recorded as one dot, was processed and then enhanced by the image analyzer to any size and color (Figure 1B). After overlapping records 1 (original electron micrograph; Figure 1A) and 2 (distribution of gold particles; Figure 1B), the size and color of the gold particles could be further enhanced as appropriate for the individual electron micrograph. With this procedure, we could achieve a clearer visualization of both intra- and extracellular antigen distribution on and around hemidesmosomes (Figure 1C and 1D) as compared with the original electron micrograph (Figure 1A). Background labeling was objectively detected as well. Improvement was particularly noted for low-magnification electron micrographs, compared to the original 5-nm gold-labeled micrographs.

In Figure 2, the antigen distribution on the dermal and the basal lamina ends of the anchoring fibrils was more clearly visualized after the application of a color image analyzer (Figure 2B) as compared with the original image (Figure 2A). In immunoelectron micrographs of skin labeled with antibody to the collagenous part of type VII collagen (Figure 3A), better visualization of antigen in the central banding part of the anchoring fibrils could also be achieved with the image analyzer (Figure 3B).

For quantitative analysis of antigen distribution, the image analyzer can automatically count the number of gold particles in a certain area. For example, when the plasma membrane was manually delineated on the original image of Figure 1A, the image analyzer counted the total number of detected gold particles in the intra- or extracellular area and measured the distance between each particle and the nearest plasma membrane. The distribution of bullous pemphigoid antigen in the micrograph, including the number and the distance of gold particles to the nearest plasma membrane, could then be automatically displayed as a diagram (Shimizu et al., 1993).

20.4 Discussion

The 5-nm gold particle is the smallest commercially available probe detectable with the conventional transmission electron microscope. While these particles have been previously demonstrated mainly in high-magnification electron micrographs (Shimizu et al., 1989, 1990; Bruckner-Tuderman et al., 1989; Hieda et al., 1989; De Panfilis et al., 1991), they are too small to be observed in electron dense structures or at a low magnification. Once the image analyzer detects the small gold probe, the computer can enhance the spot to any suitable size and color while observing the image on the TV screen. Once suitably enhanced, it is much easier to evaluate particle distribution, even at a low magnification.

Although, 15- or 20-nm gold probes are easier to see in postembedding labeling immunoelectron microscopy, the larger the size employed, the smaller the number of antigens that can be labeled on the surface of the thin section. In en bloc pre-embedding immunoelectron microscopy, larger gold probes penetrate the sample less effectively than small 5-nm gold probes. The image analyzer can overcome the inevitable disadvantages of using the small gold probe.

Silver staining of gold particles is another useful method (Holgate et al., 1983; Shimizu et al., 1992; McGrath et al., 1993). In some cases, silver-enhancement provides a better demonstration of various antigens as compared to the immunolabeling with gold probe alone (Tanaka et al., 1990; Shimizu et al., 1992; this volume). However, the size of the enhanced gold particles varies thereby making it unsuitable for quantitative analysis. Since an image analyzer can be used on immunoelectron micrographs previously taken (Shimizu et al., 1993), the possibility of quantitative analysis is one of the advantages. The application of image analysis to immunogold electron microscopy offers advantages for future study, not only in skin research but also in other fields as well.

20.5 Summary

We described the use of an image analyzer to improve the visualization of the immunogold labeling of small (5-nm gold) probes on pre- and postembedding immunoelectron micrographs. Using a TV camera connected to a color image analyzer, we projected an image of an original immunoelectron micrograph onto a TV screen. The image was recorded in the analyzer as record 1. After using a floating threshold method to reduce the contrast of the substructure, we could easily detect the electron dense 5-nm gold particles. The size and color of the particles were then enhanced by the analyzer and their image was tagged as record 2. Records 1 and 2 were then overlapped on the TV screen to produce a double image. Compared to results obtained with the small, electron dense 5-nm gold particles in the original electron micrograph, the ultrastructural localization of various antigens was seen more clearly and easily, even on low-magnification electron micrographs. The level of background labeling could also be evaluated more accurately and objectively. Quantitative analysis was possible by counting the number of gold particles that labeled a certain area and using the analyzer to present the results as a diagram. We also discussed the advantages and wide applications of this technique.

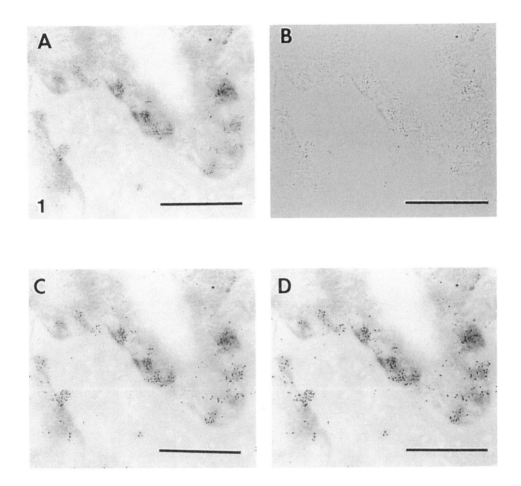

Figure 1 Image processing using image analysis of gold particles in postembedding immunoelectron micrographs labeled with bullous pemphigoid serum. An example of image processing of gold particles in immunoelectron micrographs using the image analyzer. (A) The image of an original electron micrograph projected on the TV screen. (B) Using the floating threshold method with the smoothing filter procedure, small electron dense gold particles were more easily detectable with the image analyzer. (C) After overlapping (A) and (B), the size of the detected gold particles could be varied as needed. Red was the chosen color in this figure. (D) Any suitable color could be selected at the final stage of the enhancement. Blue was used in this figure. Bars = 1 μm.

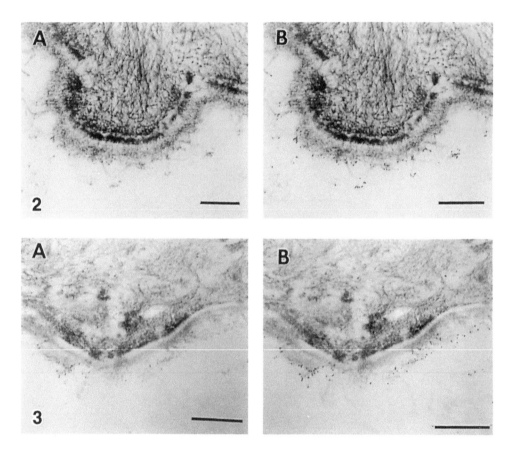

Figure 2. Pre-embedding immunoelectron micrographs labeled with epidermolysis bullosa acquisita serum that possesses autoantibodies against type VII collagen of anchoring fibrils. (A) Original image (upper left). (B) After application of image analysis, labeling on both dermal and lamina densa ends of anchoring fibrils are easily detectable (upper right). Bars = 1μm. **Figure 3.** Pre-embedding immunoelectron micrographs labeled with polyclonal antibody against the collagenous part of type VII collagen. (A) Original image (lower left). (B) After application of image analysis (lower right). In (B), clearer visualization of the antigen distribution along the banding portion of anchoring fibrils was achieved. Bars = 1 μm.

References

Akiyama, M., Hashimoto, T., Sugiura, M., and Nishilkawa, T. 1991. Ultrastructural localization of pemphigus vulgaris and pemphigus foliaceus antigens in cultured human squamous carcinoma cells. *Br. J. Dermatol.* 125: 233.

Beutner, E. H., Jordon, R. E., and Chorzelski, T. P. 1968. The immunopathology of pemphigus and bullous pemphigoid. *J. Invest. Dermatol.* 51: 63.

Bruckner-Tuderman, L., Schnyder, U. W., Winterhalter, K. H., and Bruckner, P. 1987. Tissue form of type VII collagen from human skin and dermal fibroblasts in culture. *Eur. J. Biochem.* 165: 607.

Bruckner-Tuderman, L., Mitsuhashi, Y., Schnyder, U. W., and Bruckner, P. 1989. Anchoring fibrils and type VII collagen are absent from skin in severe recessive dystrophic epidermolysis bullosa. *J. Invest. Dermatol.* 93: 3.

De Panfilis, G., Manara, G. C., Ferrari, C., and Torresane, C. 1991. Adhesion molecules on the plasma membrane of epidermal cells. III. Keratinocytes and Langerhans cells constitutively express the lymphocyte function-associated antigens. *J. Invest. Dermatol.* 96: 512.

Eady, R. A. J. and Shimizu, H. 1992. Electron microscopic immunocytochemistry. In: *Principles and Practice.* pp. 207–222. Oxford University Press, Oxford, U.K.

Hieda, Y., Tsukita, S., and Tsukita, S. 1989. A new high molecular mass protein showing unique localization in desmosomal plaque. *J. Cell. Biol.* 109: 1511.

Holgate, C. S., Jackson, P., Cowen, P. N., and Bird, C. C. 1983. Immunogold-silver staining- a new method of immunostaining with enhanced sensitivity. *J. Histochem. Cytochem.* 31: 938.

Horisberger, M., and Rosset, J. 1977. Colloidal gold, a useful marker for transmission and scanning electron microscopy. *J. Histochem. Cytochem.* 25: 295.

Ishida-Yamamoto, A., McGrath, J. A., Chapman, S. J., Leigh, I. M., Lane, E. B., and Eady, R. A. J. 1991. Epidermolysis bullosa simplex (Dowling-Meara type) is a genetic disease characterized by an abnormal keratin-filament network involving keratins K5 and k14. *J. Invest. Dermatol.* 97: 959–968.

Ishiko, A., Shimizu, H., Kikuchi, A., Ebihara, T., Hashimoto, T., and Nishikawa, T. 1993. Human autoantibodies against the 230-kDa bullous pemphigoid antigen (BPAG1) bind only to the intra-cellular domain of hemidesmosome, whereas those against the 180-kDa bullous pemphigoid antigen (BPAG2) bind along the plasma membrane of hemidesmosome in normal human and swine skin. *J. Clin. Invest.* 91: 1608.

Kellenberger, E. and Hayat, M. A. 1991. Some basic concepts for the choice of methods. In: *Colloidal Gold: Principles, Methods, and Applications.* Vol. 3. pp. 1–30. Hayat, M. A. (Ed.), Academic Press, New York.

Labib, R. S., Anhalt, G. J., Patel, H. P., Mutasim, D. F., and Diaz, L. A. 1986. Molecular heterogeneity of the bullous pemphigoid antigens as detected by immunoblotting. *J. Immunol.* 136: 1231.

McGrath, J. A., Ishida-Yamamoto, A., O'Grady, A., Leigh, I. M., and Eady, R. A. J. 1993. Structural variations in anchoring fibrils in dystrophic epidermolysis bullosa: correlation with type VII collagen expression. *J. Invest. Dermatol.* 100: 366–372.

Sakai, L. Y., Keene, D. R., Morris, N. P., and Burgeson, R. E. 1986. Type VII collagen is a major structural component of anchoring fibrils. *J. Cell Biol.* 103: 1577.

Shimizu, H., Hayakawa, K., and Nishikawa, T. 1988. A comparative immunoelectron microscopic study of typical and atypical cases of pemphigoid. *Br. J. Dermatol.* 119: 717.

Shimizu, H., McDonald, J. N., Kennedy, A. R., and Eady, R. A. J. 1989. Demonstration of intra- and extracellular localization of bullous pemphigoid antigen using cryofixation and freeze substitution for postembedding immunoelectron microscopy. *Arch. Dermatol. Res.* 281: 443.

Shimizu, H., McDonald, J. N., Bunner, D. B., Black, M. M., Bhogal, B., Leigh, I. M., Whitehead, P. C., and Eady, R. A. J. 1990. Epidermolysis bullosa acquisita antigen and the carboxy terminus of type VII collagen have a common immunolocalization to anchoring fibrils and lamina densa of basement merarane. *Br. J. Dermatol.* 122: 577.

Shimizu, H., Hashimoto, T., Nishikawa, T., and Eady, R. A. J. 1991. Human monoclonal anti-basement membrane zone antibodies derived from virally transformed lymphocytes of a patient with bullous pemphigoid recognize epitopes associated with hemidesmosomes. *Br. J. Dermatol.* 124: 217.

Shimizu, H., Ishida-Yamamoto, A., and Eady, R. A. J. 1992. The use of silver-enhanced 1-nm gold probes for light and electron microscopic localization of intra- and extracellular antigens in skin. *J. Histochem. Cytochem.* 40: 883.

Shimizu, H. and Nishikawa, T. 1993a. Application of an image analyzer to gold labeling in immunoelectron microscopy to achieve better demonstration and quantitative analysis. *J. Histochem. Cytochem.* 41: 123–128.

Shimizu, H. and Nishikawa, T. 1993b. Recent advances in electron microscopic immunocytochemistry in dermatology. *Eur. J. Dermatol.* 3: 635.

Stanley, J. R., Hawley-Nelson, P., Yuspa, S. H., Shevach, E. M., and Katz, S. I. 1981. Characterization of bullous pemphigoid antigen, a unique basement membrane protein of stratified squamous epithelia. *Cell* 24: 897.

Tanaka, T., Korman, N. J., Shimizu, H., Eady, R. A. J., Klaus-Kovtun, V., Cehr, K., and Stanley, J. R. 1990. Production of rabbit antibodies against carboxy terminal epitopes encoded by bullous pemphigoid cDNA. *J. Invest. Dermatol* 94: 617.

Tatsuya, I. and Nakamura, K., 1991. Objective quantification of grade of atypia in epithelial tumors of the stomach by image processing. *Jpn. J. Cancer Res.* 82: 199.

Woodley, D. T., Briggaman, R. A., O'Keefe, E. J., Inman, A. O., Queen, L. L., and Gammon, W. R. 1984. Identification of the skin basement membrane autoantigen in epidermolysis bullosa acquisita. *N. Engl. J. Med.* 310: 1007.

Woodley, D.T., Burgeson, R.E., Lunstrum, G., Bruckner-Tuderman, L., Reese, M.J., and Briggaman, R.A. 1988. Epidermolysis bullosa acquisita antigen is the globular carboxyl terminus of type VII procollagen. *J. Clin. Invest.* 81: 683.

Yokota, S. 1988. Effect of particle size on labeling density for catalase in protein A-gold immunocytochemistry. *J. Histochem. Cytochem.* 36: 107.

Index

Intensity of staining, see Staining intensity
Ion source, see specific silver salts
Isoamylacetate, 234

J

Jet cooling, 140

K

Kinetochores, 77
Kinetodesmal fibers, 243

L

Labeling density, see Density, labeling
Lactate salts, 13, 17, see also Silver lactate
Laminin, 153
Lead citrate, 111, 113
Lectin gold-silver staining, 259–260, 266
Lectin histochemistry, 30
Lege artis, 15
Liesegang's developer, 13
Lighting conditions, silver-enhancement
 procedure, 113
Light microscopy
 applications, 12–15
 autometallography, 22
 epipolarization, increased detection
 efficiency with, 197–204
 freeze–substituted materials, 145–146
 fungal microorganisms, 267–268
 gold toning, 211–212, 214–215
 microwave fixed specimen processing,
 171–172
 nanogold and undecagold cluster
 applications, 86–87, 89
 one-nanometer colloidal gold-labeled
 sections, 110–111
 quantitation of atrial natriuretic peptide
 staining in cardiac myocytes, 121
 viral antigens in infected cells, 276–279
Liquid crystal calibration slides, 176–179
Liquid nitrogen and liquid helium, 139–140
Loop technique, 15
Lowicryl HM20, 104
Lowicryl K4M, 62
Lowicryls, 59, 61
 background staining with, 145
 cryosection embedding, 150
 embedding for epipolarization
 microscopy, 200
 silver loss in, 112
 tissue dehydration for embedding in, 64

Low-temperature dehydration, 139–144
LR-Gold, 59, 61–63, 200
LR-White, 58, 61–63
 particle size heterogeneity, 65–66
 tissue dehydration for embedding in, 64
Lymphocyte plasma membrane-associated
 antigens, 248, 250–252

M

Magnification
 atrial natriuretic peptide in cardiac
 myocytes, 125
 epipolarization microscopy, 199–200
 104
 recognition of silver-enhanced particles,
 156
Mast cells, 172–173
β-Mercaptoethanol, 94–95
Mercury, autometallographic
 demonstration, 22, 23
Mesangial antigen, 200
MES buffer, 220, 221, 226, 228
Metals, 17
 artifacts caused by, 16
 microwave-fixed materials, 178
 tissue, autometallography, 22–23
Methanol, 141, 142
Metol/Ilford L4 solution, 111
Microbiology, see Fungal microorganisms;
 Viral antigens in infected cells
Microorganisms, see Fungal
 microorganisms; Viral antigens in
 infected cells
Microtubules
 backscattered SEM, 231–243
 protozoan, 233
 spindle, 77, 86–87
Microwave-aided binding, colloidal gold-
 protein-substance P, 183–194
 binding site visualization, 185–190
 biological activity, 190–193
 preparation of GPSP, 184–185
Microwave calibration, 176–179
Microwave cooking, 28, 60–62, 65
 antigen retrieval in aldehyde fixed
 specimens, 175
 enhancement of tissue adherence to glass
 slides, 173–176
 stimulation of immunogold and affinity
 gold silver staining, 175
Microwave fixation
 with freeze fixation, light miscoscopy
 study of mast cells, 172–173
 methods, 168–172

For Product Safety Concerns and Information please contact our EU representative GPSR@taylorandfrancis.com Taylor & Francis Verlag GmbH, Kaufingerstraße 24, 80331 München, Germany

T - #0063 - 160425 - C0 - 254/178/18 [20] - CB - 9780849324499 - Gloss Lamination